Information Technology Evaluation Methods and Management

Wim Van Grembergen

University of Antwerp (UFSIA) and
University of Leuven (KUL),
Belgium

IDEA GROUP PUBLISHING
Hershey • London • Melbourne • Singapore

Acquisitions Editor:	Mehdi Khosrowpour
Managing Editor:	Jan Travers
Development Editor:	Michele Rossi
Copy Editor:	Amy Poole
Typesetter:	Tamara Gillis
Cover Design:	Deb Andree
Printed at:	Sheridan Books

Published in the United States of America by
    Idea Group Publishing
    1331 E. Chocolate Avenue
    Hershey PA 17033-1117
    Tel: 717-533-8845
    Fax: 717-533-8661
    E-mail: jtravers@idea-group.com
    Web site: http://www.idea-group.com

and in the United Kingdom by
    Idea Group Publishing
    3 Henrietta Street
    Covent Garden
    London WC2E 8LU
    Tel: 44 20 7240 0856
    Fax: 44 20 7379 3313
    Web site: http://www.eurospan.co.uk

Copyright © 2001 by Idea Group Publishing. All rights reserved. No part of this book may be reproduced in any form or by any means, electronic or mechanical, including photocopying, without written permission from the publisher.

Library of Congress Cataloging-in-Publication Data

Van Grembergen, Wim, 1947-
    Information technology evaluation methods and management / Wim Van Grembergen.
      p. cm.
    Includes bibliographical references and index.
    ISBN 1-878289-90-X (paper)
    1. Information technology--Management. 2. Information technology--Evaluation. I. Title.

HD 30-2 . V364 2000
658'.05--dc21   00-053958

British Cataloguing in Publication Data
A Cataloguing in Publication record for this book is available from the British Library.

 # NEW from Idea Group Publishing

- **Developing Quality Complex Database Systems: Practices, Techniques and Technologies/** Shirley Becker, Florida Institute of Technology/ 1-878289-88-8
- **Human Computer Interaction: Issues and Challenges/** Qiyang Chen, Montclair State University/ 1-878289-91-8
- **Our Virtual World: The Transformation of Work, Play and Life via Technology/** Laku Chidambaram, University of Oklahoma and Ilze Zigurs/1-878289-92-6
- **Text Databases and Document Management in Modern Organizations/** Amita Goyal Chin, Virginia Commonwealth University/1-878289-93-4
- **Computer-Aided Method Engineering: Designing CASE Repositories for the 21st Century/** Ajantha Dahanayake, Delft University/ 1-878289-94-2
- **Managing Internet and Intranet Technologies in Organizations: Challenges and Opportunities/** Subhasish Dasgupta, George Washington University/1-878289-95-0
- **Information Security Management: Global Challenges in the New Millennium/** Gurpreet Dhillon, University of Nevada Las Vegas/1-878289-78-0
- **Telecommuting and Virtual Offices: Issues & Opportunities/** Nancy J. Johnson, Capella University/1-878289-79-9
- **Managing Telecommunications and Networking Technologies in the 21st Century: Issues and Trends/** Gerald Grant, Carleton University/-878289-96-9
- **Pitfalls and Triumphs of Information Technology Management/** Mehdi Khosrowpour/1-878289-61-6
- **Data Mining and Business Intelligence: A Guide to Productivity/** Stephan Kudyba and Richard Hoptroff/1-930708-03-3
- **Internet Marketing Research: Theory and Practice/** Ook Lee, University of Nevada Las Vegas/1-878289-97-7
- **Knowledge Management & Business Model Innovation/** Yogesh Malhotra/1-878289-98-5
- **Strategic Information Technology: Opportunities for Competitive Advantage/** Raymond Papp, Quinnipiac University/1-878289-87-X
- **Design and Management of Multimedia Information Systems: Opportunities and Challenges/** Syed Mahbubur Rahman, Minnesota State University/1-930708-00-9
- **Internet Commerce and Software Agents: Cases, Technologies and Opportunities/** Syed Mahbubur Rahman, Minnesota State University, & Robert J. Bignall, Monash University/ 1-930708-01-7
- **Environmental Information Systems in Industry and Public Administration/** Claus Rautenstrauch and Susanne Patig, Otto-von-Guericke University Magdeburg/ 1-930708-02-5
- **Strategies for Managing Computer Software Upgrades/** Neal G. Shaw, University of Texas Arlington/1-930708-04-1
- **Unified Modeling Language: Systems Analysis, Design and Development Issues/** Keng Siau, University of Nebraska-Lincoln and Terry Halpin, Microsoft Corporation/ 1-930708-05-X
- **Information Modeling in the New Millennium/** Keng Siau, University of Nebraska-Lincoln and Matti Rossi, Helsinki School of Economics/ 1-878289-77-2
- **Strategies for Healthcare Information Systems/** Robert Stegwee and Ton Spil, University of Twente/ 1-878289-89-6
- **Qualitative Research in IS: Issues and Trends/** Eileen M. Trauth, Northeastern University/ 1-930708-06-8
- **Information Technology Evaluation Methods and Management/** Wim Van Grembergen, University of Antwerp/1-878289-90-X
- **Managing Information Technology in a Global Economy** (2001 Proceedings)/Mehdi Khosrowpour/1-930708-07-6

Excellent additions to your library!

**Receive the Idea Group Publishing catalog with descriptions of these books by calling, toll free 1/800-345-4332
or visit the IGP Web site at: http://www.idea-group.com!**

Information Technology Evaluation Methods and Management

edited by
Wim Van Grembergen

PART I: BENEFITS REALIZATION

1. A review of IS/IT investment evaluation and benefits management issues, problems, and processes 2
   Chad Lin & Graham Pervan, Curtin University of Technology, Australia

2. The benefits realization approach to IT investments 25
   John Thorp, DMR Consulting, Canada

3. The IT evaluation and benefits management life cycle 44
   Judy McKay & Peter Marshall, Edith Cowan University, Australia

PART II: IT EVALUATION RESEARCH AND METHODS

4. A review of research issues in evaluation of information systems 58
   Vassilis Serafeimidis, University of Surrey, UK

5. Methodologies for investment evaluation: A review and assessment 78
   Egon Berghout, Delft University of Technology, & Theo-Jan Renkema, Eindhoven University of Technology, The Netherlands

PART III: ALTERNATIVE WAYS OF TRADITIONAL IT EVALUATION

6. The institutional dimensions of information systems evaluation 99
   Vassilis Serafeimidis, University of Surrey, UK

7. Evaluating IS quality: Exploration of the role of expectations on stakeholders' evaluation 111
   Carla Wilkin, Rodney Carr & Bill Hewett, Deakin University, Australia

8. Evaluating evolutionary information systems: A post-modernist perspective 130
   Nandish Patel, Brunel University, UK

9. A framework to evaluate the informatization level 144
   Soo Kyoung Lim, University of Wisconsin-Madison, USA

PART IV: EVALUATION OF NEW TECHNOLOGIES

10. Using cost benefit analysis for enterprise resource planning project evaluation: A case for including intangibles 154
    Kenneth Murphy & Steven Simon, Florida International University, USA

11. Evaluating the management of enterprise systems with the balanced scorecard 171
    Michael Rosemann, Queensland University of Technology, Australia

12. A balanced analytic approach to strategic electronic commerce decisions: A framework of the evaluation method 185
    Mike Raisinghani, University of Dallas, USA

PART V: IT EVALUATION THROUGH THE BALANCED SCORECARD

13. Information technology governance through the balanced scorecard 199
    Wim Van Grembergen & Ronald Saull, University of Antwerp (UFSIA), Belgium

14. Using a balanced scorecard framework to leverage the value delivered by IS 212
    Bram Meyerson, QuantiMetrics, South Africa

15. Management of large balanced scorecard implementations: The case of a major insurance company 231
    Peter Verleun, Egon Berghout & Maarten Looijen, Delft University of Technology, & Roel van Rijnbach, Nationale-Nederlanden, The Netherlands

16. Integrating the balanced scorecard and software measurement frameworks 240
    Nancy Eickelmann, Motorola Labs, USA

17. A comparative analysis of the balanced scorecard as applied in government and industry organizations 253
    Nancy Eickelmann, Motorola Labs, USA

Preface

The evaluation of IT and its business value are recently the subject of many academic and business discussions. Investments in IT are growing extensively, and business managers worry about the fact that the benefits of IT investments might not be as high as expected. This phenomenon is often called the Investment Paradox or the IT Black Hole: large sums are invested in IT that seem to be swallowed by a large black hole without rendering many returns. Getting value from IT and the evaluation of IT will be the concern of this book.

This book, *Information Technology Evaluation Methods and Management,* brings together seventeen papers on IT evaluation written by academics and practitioners from different countries including Australia, Belgium, Canada, the Netherlands, South Africa, the United Kingdom, and the United States.

Potential contributors were reached through a Call for Chapters issued on the Web and distributed at the 1999 International Resources Management Association (IRMA) Conference in Hershey (US) and at the 1999 International Symposium on the IT Balanced Scorecard in Antwerp (Belgium). Presenters with interesting papers on IT evaluation at the 2000 edition of both conferences were invited to expand their paper into a chapter and submit it for this book.

The authors of the different chapters have been included in the review process and have reviewed and critiqued the manuscripts of their colleague-authors. I wish to thank the contributors to this book for submitting their chapter(s) and for assisting me in the review process as well.

The different contributions in this book discuss, besides the more traditional methods that focus on financial measures such as the return on investment, a number of alternative evaluation methods and the recent introduced measurement and management system, and the IT balanced scorecard.

The book is divided into five parts and seventeen chapters:

Part I: Benefits Management introduces new approaches to benefits realization through information technology and consists of three chapters:

Chapter 1: A Review of IS/IT Investment Evaluation and Benefits Management Issues, Problems, and Processes (Graham Pervan, Curtin University of Technology, Australia). Problems in managing and realizing IT investments include measurement problems, lack of pressure to measure, cost of post-implementation reviews, poor IS/IT adoption practices, and organizational culture. Unfortunately, some managers see IT as a technical issue, seek financial bottom-line justifications, and see functionality as a benefit in itself. It is important to recognize that financially-orientated measures such as NPV and ROI are useful but largely ignore intangible benefits. In this chapter, different approaches to benefits management are discussed including Cranfield's Process Model of Benefits Management, the Active Benefit Realization approach, and DMR's Benefit Realization Model. The application of structure and discipline to the process, through models such as these, will improve the measurement of IS/IT benefits in organizations and lead to more effective investment of organizations' scarce resources in key elements of information systems and technology.

Chapter 2: The Benefits Realization Approach – Leveraging the Potential Value of Information Technology (John Thorp, DMR Consulting, Canada). As we enter the so-called "New Economy", the question that is increasingly being asked is whether the investments in information technology are providing the greatest possible value. This is not a technology issue, it is a business issue. If organizations are to realize the full potential of the latest wave of e-Business applications, they must approach the implementation of such applications differently than they have in the past. In this chapter, a new approach to the management of IT investments, a Benefits Realization Approach is introduced. Benefits realization requires a new mind-set, one that is focused on business results but that recognizes that IT alone cannot deliver these results. The elements of the proposed DMR Benefits Realization approach include: (1) moving beyond stand-alone IT project management to business program management; (2) proactively managing change as an integral part of business programs; (3) shifting from free-for-all competition for resources to a disciplined portfolio management approach; and (4) adopting a full cycle governance view of managing projects, programs and portfolios, with clear accountability and effective measurement.

Chapter 3: The IT Evaluation and Benefits Management Life Cycle (Judy McKay & Peter Marshall). This chapter describes and analyses a framework to achieve adequate linkage between IS/IT planning, evaluation of investments on an ongoing basis, and also active realization of benefits to the organization over time. This framework is called the IT Evaluation and Benefits Life Cycle, and shows how to integrate planning, evaluation and benefits activities. It is argued that this mix of planning, evaluation and benefits management is vital, as each of these components adopts a somewhat different (albeit important) focus on the other. The position adopted in this chapter reflects the authors' belief in a need to meld or simultaneously juggle these three perspectives if more effective utilization of the IT resource is to occur.

Part II: IT Evaluation Research and Methods reviews the current research in the evaluation of information systems and the methodologies for IT investment evaluation. This part consists of two chapters:

Chapter 4: A Review of Research Issues in Evaluation of Information Systems (Vassilis Serafeimidis, University of Surrey, UK). Information systems evaluation is an important element of information systems governance and management. Therefore, the academic literature is not short of publications in the area. This chapter critically discusses a variety of views of information systems evaluation that have appeared in the literature during the last ten years. The chapter aims at providing a comprehensive classification and discussion of the theoretical and practical developments in the field. The analysis is based on a conceptual framework deriving from a definitional approach to evaluation. Each area of evaluation will be investigated across: purpose/reasons (Why?), the subject (What?), the criteria/measurement (Which Aspects?), the time frame (When?), the people (Who?), and the methodologies/tools (How?). The framework considers the socio-technical nature of information systems and the multidimensional character of evaluation. Three streams of evaluation research are discussed: the traditional technical/functional, economic/financial and a number of interpretive alternatives (e.g. flexible non-bureaucratic, contemporary meta-methodologies) which attempt to propose answers to the various criticisms of the rational/positivistic streams.

Chapter 5: Methodologies for IT Investment Evaluation: A Review and Assessment (Egon Berghout & Theo-Jan Renkema, Delft University of Technology). The contribution of this chapter is twofold. First, the different concepts that are used in IT evaluation are discussed and more narrowly defined. When speaking about IT investments, concepts are used that originate from different disciplines. In many cases there is not much agreement on the precise meaning of the different concepts used. However, a common language is a prerequisite for the successful communication between the different organizational stakeholders in evaluation. In addition to this, the chapter reviews the current methods and puts them into a frame of reference. All too often new methods and guidelines for investment are introduced, without building on the extensive body of knowledge that is already incorporated in the available methods. Four basic approaches are discerned: the financial approach, the multi criteria approach, the ratio approach and the portfolio approach. These approaches are subsequently compared and the chapter concludes with suggestions on how to improve evaluation practice and recommendations for future research.

Part III: Alternative Ways of Traditional IT Evaluation focuses on nontraditional IT evaluation approaches such as evaluation procedures taking into account organizational properties, the evaluation of IS quality, the evaluation of evolutionary systems, and the evaluation by comparing with a framework of informatization levels. This part consists of four chapters:

Chapter 6: The Institutional Dimensions of Information Systems Evaluation (Vassilis Serafeimidis, University of Surrey, UK). Information systems evaluation is a highly complex, in conceptual and operational terms, social process. The academic research has focused, primarily, on methodological guidelines and formal evaluation procedures ignoring its organisational nature. This chapter adopts a highly interpretative approach and explores a number of organisational properties which play a key role in IS evaluation. The aim of the chapter is to increase the management awareness in terms of the organisational properties affected by IS evaluation and lead to its successful integration with other IT and business management processes. The chapter discusses the evaluation stakeholders analysis, their evaluation (cognitive) schemas and the most common (evaluation related) organisational roles (the strategist for evaluation and the evaluator). Those entities are used as foundation stones to identify one of the four possible evaluation orientations (control, social learning, sense making and exploratory) and lead to practical actions which will facilitate the successful management integration.

Chapter 7: Evaluating IS Quality: Exploration of the Role of Expectations on Stakeholder's Evaluation (Carla Wilkin, Rodney Carr & Bill Hewett, Deakin University, Australia). With organizations so reliant on information systems to perform day-to-day activities, the quality of such systems can impact significant on organizational performance. One way to determine the quality of performance of such systems is to evaluate stakeholder end user opinions. This chapter used a new configuration of the IS Success Model as the theoretical framework for appraisal of IS Quality, through evaluation by stakeholder and users. The related empirical study tested the relevance of a marketing premise to information systems, namely that evaluation of quality is best determined by understanding stakeholders' perceptions and expectations for such performance. This trial sought to determine whether quality of information systems is better measured by the disconfirmation between stakeholders' perceptions and expectations of performance, or by the evaluation only of such perceptions. The results confirm that there are differences using the two approaches. In particular, the results of this study suggest the relevance of measuring expectations. Some implications of these results are discussed.

Chapter 8: Evaluating Evolutionary Information Systems: A Post-modernist Perspective (Nandish Patel, Brunel University, UK). Information technology and information systems that are intertwined with business processes and that are subjected to continuous business change are characterized as evolutionary information systems. Such systems pose a new challenge in the field of systems evaluation. The development and use of evolutionary information systems in business organizations is traditionally subjected to objective measures of evaluation. In this chapter, issues in evaluating information technology and information systems based on business processes in modern organizations are examined using a post-modernist perspective that takes into consideration subjective factors of users and their process-based organizations. Subjective interpretation, situation, and the context of information system use are key subjective factors that are proposed in a post-modernist framework for evaluating evolutionary information systems. Heideggar's ontological consideration of human Dasein (being) form the philosophical basis of the proposed framework. Practical issues in information technology and information systems evaluation such as systems requirements, systems functionality, and system adaptability are discussed in the context of the proposed post-modernist approach.

Chapter 9: A Framework to Evaluate the Informatization Level (Soo Kyoung Lim, University of Wisconsin-Madison, USA). A rapid development of information and communication technologies followed by economical change, cultural innovation and organizational reformation; this phenomenon has been referred as Informatization. Informatization is considered as one of the important success factors for economical growth. Since the middle of 1990, performance-based management has put emphasis on the IT investment. In this respect, evaluation of an organization's informatization level is an important managerial concern. In this chapter, as reviewing various evaluation models and frameworks, meaningful indexes that can represent informatization are provided, and an evaluation model with a new and different approach is introduced.

Part IV: Evaluation of New Technologies covers the evaluation of Enterprise Resource Planning projects, and an evaluation approach to strategic electronic commerce decisions. This Part IV consists of three chapters:

Chapter 10: Using Cost-Benefit Analysis for Enterprise Resource Planning Project Evaluation: A Case for Including Intangibles (Kenneth Murphy & Steven Simon) This chapter demonstrates how cost-benefit analysis can be applied to large-scale Enterprise Resource planning (ERP) projects, and that these project justification techniques can incorporate intangible benefits with ERP systems. A brief review of the standard cost benefit analysis techniques for ex-ante project evaluation is followed by an in-depth discussion of intangibles that focuses on those factors that may be involved in justifying technology investments. Detailed information on the business case utilized by a large computer manufacturer in their decision to implement SAP system R/3 is presented. The organization under study utilized standard cost-benefit techniques including the tangible factors of productivity increases, and decreases in inventory and IT operations expense to build their case. This company significantly strengthened their position by including intangibles, e.g. user satisfaction in the cost-benefit analysis framework.

Chapter 11: Evaluating the Management of Enterprise Systems with the Balanced

Scorecard (Michael Rosemann, Queensland University of Technology, Australia). In this chapter, the balanced scorecard, a framework originally developed in order to structure the performance measurement for an enterprise or a department, is used for the evaluation of the implementation of Enterprise Resource Planning (ERP) systems. By adapting the balanced scorecard and adding a new fifth project perspective, a comprehensive evaluation of ERP software is achieved and an alternative IT evaluation approach is introduced. This approach supports the time consuming implementation of ERP systems and their benefits realization stage.

Chapter 12: A balanced Analytic Approach to Strategic Electronic Commerce Decisions: A Framework of the Evaluation Method (Mahesh Raisinghani, University of Dallas, USA). This chapter presents a comprehensive model for optimal electronic commerce strategy and extends the relatively novel Analytic Network Process (ANP) approach to solving quantitative and qualitative complex decisions in electronic strategy. A systematic framework for the identification, classification and evaluation of electronic commerce strategy using the Internet as an information, communication, distribution, or transaction channel that is interdependent with generic business strategies is proposed. The proposed methodology could help researches and practitioners understand the relation between the benefits organizations seek from an information technology and the strategies they attempt to accomplish with the technology.

Part V: IT Evaluation Through the Balanced Scorecard, rounds out this book on IT evaluation by covering the fairly new IT evaluation approach of the IT balanced scorecard and presents case studies on this issue. This part consists of five chapters:

Chapter 13: Information Technology Governance through the Balanced Scorecard (Wim Van Grembergen and Ronald Saull, University of Antwerp, Belgium). In this chapter the balanced scorecard (BSC), initially developed by Kaplan and Norton, is applied to information technology. A generic IT BSC with a user orientation perspective, an operational excellence perspective, a business contribution perspective, and a future orientation perspective is proposed. Its relationship with the business balanced scorecard is established and it is shown how a cascade or waterfall of balanced scorecards can support the IT governance process and its related business/IT alignment process. Further, the development and implementation of an IT BSC is discussed and an IT BSC Maturity Model is introduced. The chapter concludes with the findings of a real-life case. The main conclusions of the case are that the development of a balanced scorecard for information technology is an evolutionary project and that to be successful one needs a formal project organization.

Chapter 14: Using a Balanced Scorecard Framework to Leverage the Value Delivered by IS (Bram Meyerson, Quantimetrics, South Africa). In this chapter a commercial methodology is described that can be used to assess the value that a typical IS group delivers to its business partners. The proposed approaches are based on research conducted by Computer Sciences Corporation and QuantiMetrics and are also based on practical application by QuantiMetrics. The chapter focuses on a balanced scorecard framework for an IT organization and describes a number of assessment techniques in each domain of the scorecard. These assessments include IS expenditure alignment, IS and business alignment, IS process assessment, IS capability assessment, IS and business partnership and satisfaction assessments. The assessments have enabled both IS executives and senior management to measure where their organizations stand in relation to the growing convergence between business strategy and IT.

Chapter 15: Management of Large Balanced Scorecard Implementations: The Case of a Major Insurance Company (Peter Verleun, Egon Berghout, Roel van Rijnbach, & Maarten Looijen, Delft University of Technology). In this chapter, established information resource management theory is applied to improve the development and maintenance of large balanced scorecard implementations. The balanced scorecard has proved to be an effective tool for measuring business and IT performance. Maintaining a business-wide balanced scorecard measurement system over a long period implies, however, many risks. An example of such a risk is the excessive growth of scorecards as well as scorecard metrics, resulting in massive data warehouses and difficulties with the interpretation of data. This is particularly the case in large organizations. The proposed balanced scorecard management framework is in this chapter illustrated with the experience gathered from the company-wide balanced scorecard implementation within a large insurance company in the Netherlands.

Chapter 16: Integrating the Balanced Scorecard and Software Measurement Frameworks (Nancy Eickelmann, NASA, USA). Process improvement approaches such as Total Quality Management, the Capability Maturity Model and ISO-9000 share a customer focus towards measurable business process improvements that promise cost reductions and cycle time improvements. Unfortunately, these approaches are frequently not linked to the organization's high-level strategic goals. In this chapter, the balanced scorecard is introduced as a provider of a necessary structure to evaluate quantitative and qualitative information with respect to the organization's strategic vision and goals. There are two categories of measures in the balanced scorecard: the leading indicators or performance drivers and the lagging indicators or outcome measures. The performance drivers enable the organization to achieve short-term operational improvements while the outcome measures provide objective evidence of whether strategic objectives are achieved. The two measures must be used in conjunction with one another to link measurement throughout the organization thus giving visibility into the organizations' progress in achieving strategic goals through process improvement.

Chapter 17: A Comparative Analysis of the Balanced Scorecard as Applied in Government and Industry Organizations (Nancy Eickelmann, NASA, USA). This chapter provides a comparison of two case studies regarding the use of the balanced scorecard framework. The application of the balanced scorecard is evaluated for a Fortune 500 IT organization and a government organization. Both organizations have a business focus of software development. The balanced scorecard framework is applied and reviewed in both contexts to provide insight into unique organizational characteristics for government and contract software environments. How the balanced scorecard is applied in an industry context and a government context is described and contrasted. An analysis of key differences among financial perspectives, customer perspectives, internal business process perspectives, and learning and growth perspectives for both areas is conducted. A unifying thread of the study is to evaluate the use of measurement for the operational, managerial, and strategic purposes of an organization. The case studies provide additional insight to applying the balanced scorecard in a software development intensive environment.

Wim Van Grembergen, PhD
Antwerp, Belgium
August, 2000

Acknowledgments

The editor would like to acknowledge the help of all involved in the collection and review process of this book on IT evaluation, without whose support the project could not have been satisfactorily completed. A further special note of thanks goes also to all staff at Idea Group Publishing, whose contributions throughout the whole process from inception of the initial idea to final publication have been invaluable. I especially thank my colleague Mehdi Khosrowpour for his encouragement to start working on this book. I also appreciate Jan Travers' and Michele Rossi's efforts in managing the project and making sure that it is printed on time. They both were great in prodding via e-mail for keeping the project on schedule.

I wish to thank all the authors for their insights and excellent contributions to this book. I also want to thank the authors for assisting me in the review process.

I appreciated support provided for this project by the Business Faculty of the University of Antwerp (UFSIA). I also would like to thank my undergraduate students at the University of Antwerp and the University of Leuven; my executive students at the University of Antwerp Management School, the Graduate School of Business of the University of Stellenbosch, and the Institute of Business Studies in Moscow, who provided me with many ideas on the subject of IT evaluation and the balanced scorecard concepts.

In closing, I want to thank my wife Hilde and my two daughters, Astrid and Helen, for their support throughout this project. I apologize for the repeatedly used excuse during the summer of 2000: "I am sorry, but I have to work on the IT evaluation book."

Wim Van Grembergen, PhD
Antwerp, Belgium
August, 2000

Part I:

Benefits Realization

Chapter I

A Review of IS/IT Investment Evaluation and Benefits Management Issues, Problems, and Processes

Chad Lin and Graham P. Pervan
Curtin University of Technology, Australia

INTRODUCTION

Information systems / information technology (hereafter referred to as IS/IT) now represents substantial financial investment for many organisations. Information systems and technology managers have found it increasingly difficult to justify rising IS/IT expenditures. They are often under immense pressure to find a way to measure the contribution of their organisations' IS/IT investments to business performance, as well as to find reliable ways to ensure that the business benefits from IS/IT investments are actually realised. This problem has become more complex as the nature of IS/IT investments and the benefits they can deliver has evolved over time and has changed rapidly. Furthermore, the evaluation of these IS/IT investments is a complex tangle of financial, organisational, social, procedural and technical threads, many of which are currently either avoided or dealt with ineffectively by organizations (Pervan, 1998).

In this chapter we will learn what is IS/IT investment evaluation and benefits management, discuss some of the problems and challenges in this area, review some of the better known approaches to this problem, acknowledge some of the leading authors in the area, and conclude with a summary of the current status of the field and some possible directions for future research and practice.

IS/IT INVESTMENT EVALUATION

What is IS/IT Investment Evaluation?

With increasing levels of IS/IT investments and the growing significance of information systems within organisations, IS/IT investments evaluation is becoming widely recognised as a very important activity. According to Keen (1995), information technology

Copyright © 2001, Idea Group Publishing.

(IT) has "become the generally accepted umbrella term for a rapidly expanding range of equipment, applications, services, and basic technologies." Katz (1993) has suggested that information technology (IT) is an "umbrella term that includes the integrated user-machine systems for providing information to support the operation, management, analysis and decision-making functions in an organisation." Weill and Olson (1989) define IT as a collection of "all computers, communications, software, networks and all the associated expenses, including people dedicated to the management or operation of the IT."

On the other hand, investments are commitments of resources, made with the purpose of realising benefits that will occur over a reasonably long time in the future. Therefore, an investment in information technology (IT) may be referred to as any acquisition of software or hardware which is expected to expand or increase the business benefits of an organisation's information systems (IS) and render long-term benefits (Apostolopoulos and Pramataris, 1997).

Evaluation is often considered as a process to diagnose malfunctions and to suggest appropriate planning and treatment by providing feedback information and contributing to organisational planning. It is generally aimed at the identification and quantification of costs and benefits (Symons, 1994). Taking a management perspective, evaluation is about establishing by quantitative and/or qualitative means the worth of IS to the organisation (Willcocks and Lester, 1996). Symons and Walsham (1988) pointed out that the primary function of evaluation is to contribute to the rationalisation of decision making. Therefore, by combining the definitions of investment in IT and evaluation mentioned above one can define IT investment evaluation as the weighing up process to rationally assess the value of any acquisition of software or hardware which is expected to improve the business value of an organisation's information systems.

The Productivity Paradox

The measurement of the business value of IS/IT investment has been the subject of considerable debate by many academics and practitioners (van Grembergen and van Bruggen, 1998). The difficulties in measuring benefits and costs are often the cause for the uncertainty about the expected benefits of IS/IT investments and hence are the major constraints to IS/IT investments. Organisations seeking value for money in IS/IT investments have spent a lot of energy, time and money that has largely gone to waste. Therefore, evaluation is often ignored or carried out inefficiently or ineffectively because of its elusive and complex nature (Serafeimidis and Smithson, 1996).

As mentioned earlier, information systems and information technology are often costly to purchase, set up and maintain. Therefore, it is natural to suppose that these investments offer economic value and that this value overcomes the costs. However, according to Hochstrasser and Griffiths (1991), organisations often report that large-scale IS/IT deployment has resulted in replacing old problems with new, and that, overall, introducing IS/IT can be a huge disappointment since unexpected difficulties and failures are regularly encountered and expected business benefits are frequently not realised. To add to this difficulty, the determination of IS/IT investment and returns is also problematic because of the lack of consensus in defining and measuring such investment (Mahmood and Mann, 1993). While organisations continue to invest heavily in IS/IT, research studies and practitioner surveys report contradictory findings on the effect of the expenditures on organisational productivity (Grover et al., 1998b). Therefore, it is not surprising to see that the term "productivity paradox" is gaining increasing notoriety as several studies point toward fairly static productivity and rising IS/IT expenditure, e.g., Hochstrasser (1993).

Despite large investments in IS/IT over many years, it has been difficult to determine where the IS/IT benefits have actually occurred, if indeed there have been any (Willcocks and Lester, 1997). On one hand, some studies, e.g., Strassmann (1997), have suggested that IS/IT investment produces negligible benefits, while others, e.g., Brynjolfsson and Hitt (1998), have disagreed, reporting that there appears to be some sort of positive relationship between organisations' performance and IS/IT spending. In summary, it is possible that the results of these studies indicate that the relationship between IS/IT investment spending and benefits is unclear and confounded by methodological problems as well as intervening variables (Grover et al., 1998b).

It can be argued that confusion about the IS/IT benefits is due to mismeasurements of outputs and inputs (inappropriate unit of analysis), the difficulty of establishing the overall value of IS/IT, the choice of inappropriate methods of evaluation, lags in learning and adjustments, redistribution (IS/IT may be beneficial to individual firms but unproductive from the standpoint of the industry), confusions about the terms such as expenses and revenue, and dissipation of profits, mismanagement by developers and users of IS/IT, and lack of effective IS/IT evaluation and benefits realisation management practices.

Hayashi (1997) suggests that, at least on a macroeconomic level, productivity paradox does not really exist. According to Hayashi (1997), the present state of the US economy which has been enjoying an almost unprecedented period of low unemployment, manageable inflation, strong growth in corporate profits and real wage increases has presented the missing link between IS/IT investment and efficiency. Dewan and Kraemer (1998) seem to agree with this point of view in their study of 17 developed countries over the period of 1985-1992. According to their analysis, these developed countries are receiving a positive and significant return on their IS/IT investments. According to Dewan and Kraemer (1998), a potential explanation is that the estimated returns from IS/IT investments reflect other changes in the economies of developed countries that are complementary to IS/IT investments, such as infrastructure and the application of IS/IT to business processes. In other words, the positive returns are not only due to increases in IS/IT capital per worker, but also reflect simultaneous changes in education, infrastructure and other factors that complement labour and make it more productive (Dewan and Kraemer, 1998).

This debate is still ongoing. Given the financial stakes involved, determining the impact of IS/IT investments on performance and organisational processes has been and will continue to be an important research concern for both practitioners and academics (Sriram et al., 1997).

Objectives and Criteria

There may be a number of objectives for IS/IT investment evaluation:
1. As part of the process of justification for a project (Willcocks and Lester, 1996);
2. Enabling an organisation to make comparisons of the merit of a number of different investment projects competing for limited resources (King and McAulay, 1997);
3. Providing a set of measures which enable the organisation to exert control over the investment (Farbey et al., 1992);
4. As a learning device which is necessary if the organisation is to improve its system evaluation and system building capability (Willcocks and Lester, 1996);
5. Ensuring that systems will continue to perform well by selecting the best alternative in the beginning of the project (Ballantine et al., 1996);
6. Supporting the IS broader business objectives and providing for future business expansions (Ballantine and Stray, 1998); and

7. Enabling organisations to gain competitive advantage, to develop new business, to improve productivity and performance, as well as to provide new ways of managing and organising (Earl, 1988).

Similarly, an organisation may want to evaluate an IS/IT project at any stage in its development and implementation process when (Farbey et al., 1992):
1. Strategy is being developed;
2. A specific project has been defined;
3. The project is in the development stage;
4. The project has reached the point of sign off, i.e., when the responsibility is being transferred from the IT department to the user department;
5. The project has just been implemented;
6. The project has been in operation for some time; or
7. The project is nearing the end of its life and the feasibility of replacement options is being investigated.

According to Bacon (1996), the criteria used in making the decision on IS/IT investments are vital and are significant for a number of reasons:
1. The criteria used or not used, and the way in which they are applied or not applied, significantly impacts the effectiveness with which IS/IT investment decisions are made. They determine whether the optimal projects are selected and the suboptimal rejected;
2. The criteria used by an organisation in deciding upon IS/IT investments tend to reflect the effectiveness with which IS/IT resources are being used, the degree to which senior management are involved, and the level of integration between corporate/business-unit strategy and systems strategy;
3. The criteria are significant for the organisation's finance and management accounting function, in terms of its role in optimising return on investment, and its involvement in the cost-benefit analysis that may precede an IS/IT capital investment decisions; and
4. The criteria have significance in the balance that an organisation achieves in their use, particularly between financial and management criteria.

Emerging Problems/Challenges

These evaluation and management efforts regularly run into difficulties of three generic types (Willcocks and Lester, 1997):
1. Many organisations find themselves in a catch-22 situation. For competitive reasons they cannot afford not to invest in IS/IT, but economically they cannot find sufficient justification, and evaluation practice cannot provide enough underpinning, for making the investment;
2. As IS/IT infrastructure becomes an inextricable part of the organisation's processes and structures, it becomes an increasingly difficult to separate out the impact of IS/IT from that of other assets and activities; and
3. There is widespread lack of understanding of information requirements as well as IS/IT as a major capital asset, despite the high levels of expenditure.

Ballantine et al. (1996) identified a number of problems that are frequently encountered during evaluation practice. These include difficulty in identifying and subsequently quantifying relevant benefits and costs, and neglecting intangible benefits and costs. This seems to confirm the results by the study carried out by Willcocks (1992a; 1992b). These problems in IS/IT evaluation are usually complex and therefore, can affect the determination of the expected IS/IT benefits. These problems are mentioned in the subsequent paragraphs.

First, the budgeting practice of many organisations often conceals full costs (Willcocks, 1992a). IS/IT costs are no longer equitable with the budget of the IS/IT department since there are many significant hidden costs such as maintenance and training costs. For example, the amount spent on training alone in UK increased from £311 million in 1995 to £442 million in 1996 (Kelly, 1997). The biggest single cost of training is in the staff time that needs to be released for training (which are rarely fully budgeted for in IS/IT investment proposals), and may partially explain the phenomenon of the cost-creep (cost blowout) that occurs over the course of most IS/IT projects (Launders, 1997). Most organisations simply accept project overruns as the inevitable norm. In addition, there may also be political reasons for understating costs, the main one being to gain support for, and acceptance of the project from senior managers (Willcocks, 1992a).

Second, the traditional financially oriented evaluation techniques such as return on investment (ROI), discounted cash flow/internal rate of return (DCF/IRR), net present value (NPV), profitability index (PI), cost/benefit, payback period, and present worth can be problematic in measuring IS/IT investments (Kumar, 1998). The problems with these methods are that they largely exclude the significant problem of risk as well as costs and benefits that may be difficult to quantify (Brown, 1994). Those benefits which are intangible or soft appear to be written off as unquantifiable and thus beyond any effective measurement technique (Sutherland, 1994). Therefore, many people now regard the financially oriented appraisal techniques as an inappropriate tool for justifying investments in IS/IT (Remenyi et al., 1997) since these methods are unable to capture many of the qualitative benefits that IS/IT brings. There are some, possibly the major, potential benefits such as greater job satisfaction, improved customer service, improved communication, and increased competitive advantage and responsiveness from IS/IT that are not measurable using traditional financially oriented evaluation techniques. According to Whiting et al. (1996), overreliance on these traditional financially oriented evaluation methods may lead to an excessively conservative IS/IT portfolio and loss of competitiveness, and failure to perform rigorous investment appraisal may result in a highly ineffective use of resources.

These financially oriented techniques also do not assist the process of establishing how IS/IT adds net value to an organisation (Irani et al., 1997). Another major limitation is that these financially oriented techniques do not capture management's ability to alter the pace of investment, or to stop investment at some point if conditions are unfavourable (Kumar, 1996). However, there is no widely accepted methodology that is relevant in all cases.

Third, many project managers overstate costs at the feasibility stage, with the express purpose of making sure that they could deliver within time and budget (Willcocks, 1992a). This can result in wasting precious organisational resources.

Fourth, working with new technology always introduces higher levels of risk, which in turn affect the timing, costs and delivery deadlines (Griffiths and Willcocks, 1994). There are two major areas of risk that are frequently downplayed in evaluating IS/IT proposals. The first risk is the additional costs where implementation may be less than smooth. The second risk is concerned with security exposure and systems breakdown for the organisation. This risk refers to exposure to such consequences as: failure to obtain some or all of anticipated benefits due to implementation costs much higher than expected; technical systems performance significantly below the estimate; incompatibility of the system with selected hardware and software. In addition, it is important to take into account the risk of computer systems security breach and costs of computer systems breakdown for the organisation when evaluating IS/IT proposals. The risk is quite significant since it can contribute to the rising IS/IT expenditure for the whole organisation. Surveys regularly report IS/IT introduction and usage as a high risk, hidden cost process (Griffiths and Willcocks, 1994). For example, a survey conducted in Europe found that computer systems downtime costs each user, on average, 3 weeks' work per year (Leung, 1997). However, the biggest risk of all, according to Ward and Murray (1997), is that the system will not deliver the desired benefits.

Fifth, many organisations have failed to devote sufficient or appropriate evaluation time and effort to IS/IT given that it represents a major capital asset in many organisations (Irani et al., 1997). According to Willcocks (1992a), senior management rarely know how much capital is tied up in IS/IT resources. Failure to appreciate the size and the presence of the time lag of the investment can readily lead to IS/IT investments being under-managed. Another major problem is that many organisations fail to appreciate the timing or timescale of the likely business benefits from IS/IT investments. Time horizons or time spans used for cash flow analysis by financially oriented techniques are typically set at three to five years whereas the IS/IT investment projects take several years to fully implement, so benefits do not become financially apparent for two to three years, but may continue for up to ten years. Keen (1991) suggests that infrastructure investments such as networks and telecommunications may need to be evaluated separately and funded by top management as a long-range capital investment in line with corporate policy requirements.

Finally, the lack of IS/IT planning and hence the failure to create a strategic climate in which IS/IT investment can be related to organisational direction can also lead to measurement problems during the IS/IT investments evaluation process (Willcocks, 1992a). Organisations must have IS/IT planning and strategies to facilitate the management and control of their resources and investments (Gottschalk, 1999). According to Mirani and Lederer (1993), the organisational investment climate and the alignment with stated organisational goals also have a key bearing on how investment is organised and conducted, and what priorities are assigned to different IS/IT investment proposals. Organisations should only invest in those IS/IT projects that can be shown to have clear links with the overall business strategy (Fitzgerald, 1998). According to a survey by Hinton and Kaye (1996), only one in four respondents attempt to establish whether an investment is in line with overall IT strategy. Another survey conducted by Hochstrasser and Griffiths (1991) have found that 66% of organisations do not formulate an IS/IT strategy. A direct consequence of the lack of IS/IT strategies is that organisations are neither satisfied with the current procedures for implementing priorities nor with the management of IT benefits. The same survey also found that whereas the presence of an IT strategy does not automatically guarantee a problem-free process, the organisations with an IT strategy in place suffer considerably fewer setbacks when implementing new IT process.

A System's Life-Cycle Approach

There are several reasons why no single measure of the value of IS/IT has appeared. When IS/IT operations are measured as a profit centre or as a cost centre, significant differences arise and each has to show numbers tied to management's control (Carlson and McNurlin, 1992). In addition to many unquantifiable measures, the increasing desire of top executives to explain their IS/IT investments in terms of business measures and productivity gains has broadened to encompass where value has arisen in many segments of the firm. In these circumstances, where management wants broad measures, many IS/IT managers consider the high cost of measurement not worth the effort, and therefore progress is slow.

8 Lin and Pervan

Therefore, according to Willcocks and Lester (1997), there is a need for a family of measures that cover technical and business performance of IT in an integrated manner. Measures are needed that point to cost effectiveness and containment, as well as embrace additional key IT/business performance criteria. A diagrammatic representation of their integrated evaluation lifecycle approach is shown below (Figure 1).

This evaluation lifecycle approach attempts to bring together the rich and diverse set of ideas, methods, and practices that are to be found in the evaluation literature to date, and point them in the direction of an integrated approach across systems' lifetime. The approach would consist of several interrelated activities (Willcocks and Lester, 1997):

- Identification of the costs and benefits of IS/IT through prioritising projects and aligning them with strategic directions of the organisation;
- Identification of types of generic benefit, and matching these to appropriate techniques for assessing these benefits;
- Development of integrated measures based on key criteria (financial, service, delivery, learning and technical);
- Linking of integrated measures to the needs of the development, implementation and post-implementation phases;
- Operationalisation of each set of measures to range from the strategic to the operational level;
- Determining who is responsible for tracking these measures, and regularly reviewing the results; and
- Regular review of the existing IS/IT investment, and linking this to the strategic business direction and associated performance objectives of the organisation.

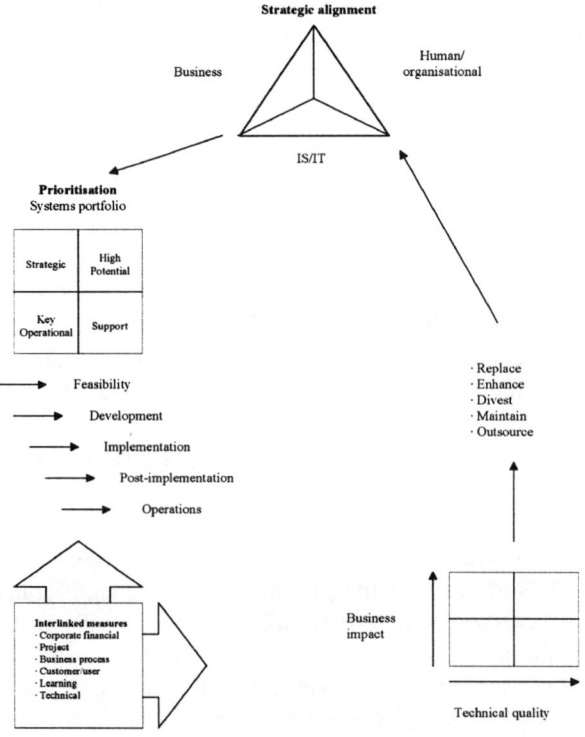

Figure 1: IS/IT investment evaluation life-cycle (Adapted from: Willcocks and Lester, 1997)

Lyon and Mooney (1994) seem to have agreed, at least in part, with this type of lifecycle IS/IT investment evaluation strategy. They argue that organisations can make significantly worthwhile efforts at evaluating IS/IT investments when such efforts are carried out as part of the formulating IS/IT planning process. Their "back to basics" approach presents a triangulation approach to IS/IT planning which is based around business need and business benefit, and is comprised of:

- A review of current business procedures and the definition of desired business procedures and associated system requirements;
- Post-design rationalisation of the detailed system requirements; and

- Management of the implementation of the new system and realisation of business benefits.

In addition, in order to make this integrated evaluation lifecycle work, it is important to involve the stakeholders in processes that operationalise the evaluation criteria and techniques. The managers can often stay in touch with relevant changes or pressures on their businesses by thinking about stakeholders, those individuals and groups whose success is bound up with the performance of the business. That is, to involve the stakeholders in processes that "breathe life into, adapt over time, and act upon" the evaluation criteria and techniques (Willcocks and Lester, 1997). The benefits that accrue to various stakeholders of a firm such as customers, suppliers, and employees have become more significant in recent years as IS/IT is no longer confined to an isolated area, but is permeating the whole value chain in modern business. The problem becomes how to tie stakeholders into supporting the implementation and subsequent operation of specific systems while achieving managerial objectives. This is an important issue because (1) any IS/IT value analysis without an assessment of all relevant stakeholder benefits is incomplete and is likely to understate the full extent of the benefits, and (2) business managers need information not only for measuring and evaluating IS/IT benefits, but more importantly, they need guidance on how to manage the investments and capture the benefits in the bottom line. Therefore, it is important to treat IS/IT payoff as a portfolio of benefits that are distributed across several stakeholder groups.

The first step in the Willcocks and Lester (1997) integrated evaluation lifecycle approach is to establish strategic alignment and linking business/organisational strategy with assessing the feasibility of any IS/IT investment. This should enable the organisation to plan for effective assessment and management of IT benefits.

There are several methodologies that can help to link strategy and feasibility of the IS/IT investments. Some of these methodologies are briefly described below:

- *SESAME* (Willcocks et al., 1992b): IBM developed this in order to provide a more flexible approach to cost/benefit analysis. Here the costs and benefits of an IT-based system are compared against an equivalent manual system. This method bases much of the assessment on user opinion, which may involve users more in the process of assessment. However, user evaluation may not, in itself, be a sufficient benchmark of IT effectiveness (Willcocks and Lester, 1994).
- *Matching Objectives, Projects and Techniques* (Figure 2) (Butler Cox Foundation 1990): This method basically attempts to match the projects with the appropriate evaluation techniques.
- *Return on Management (ROM)* (Strassman, 1990): This is a measure of performance based on the added value to an organisation provided by management. The assumption is that in

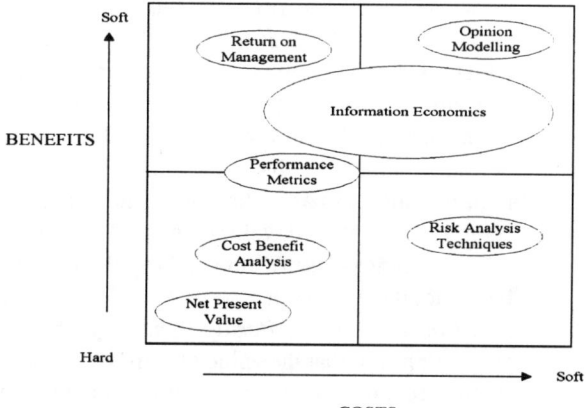

Figure 2: Matching projects to techniques (Source: Butler Cox Foundation, 1990)

the modern organisation information costs are the costs of managing the enterprise. If ROM is calculated before and after IT is applied to an organisation, then the IT contribution to the business, so difficult to isolate using more traditional measures, can be assessed. ROM is calculated in several stages. First, using the organisation's financial results, the total value added is established. This is the difference between net revenues and payments to external suppliers. The contribution of capital is then separated from that of labour. Operating costs are then deducted from labour value added, to leave management value added. ROM is management value added divided by the costs of management. However, there are some problems with the method of obtaining this figure, and whether it really represents what IT has contributed to business performance (Willcocks and Lester, 1994).

- *Information Economics Approach* (Parker et al., 1988): Here value is seen as a broader concept based on the effect IS/IT investment has on the business performance of the enterprise. It seeks to identify and measure or rank the economic impact of the changes brought about by the introduction of the new system on an organisation's performance. The first stage is building on traditional cost-benefit analysis with four additional techniques (value linking, value acceleration, value restructuring and innovation valuation) for establishing an enhanced return on investment calculation (EROI). It then enhances the cost-benefit analysis still further through business domain assessment (BDA) and technology domain assessment (TDA). BDA considers factors such as strategic match, competitive advantage, management information, competitive response, and project and organisational risk. TDA considers factors such as strategic IS architecture, definitional uncertainty, technical uncertainty, and IS infrastructure risk. The value or "economic impact of IT" is then determined by combining the EROI, BDA and TDA. This method is intended to cope with systems which provide benefits by improving the linkage and communication between departments or even as with EDI between organisations. Although this approach implies a mechanistic appraisal, and contains a subjective basis for many of the scores, it does provide a useful checklist for assessing the wider impact of introducing systems, rather than focusing only on limited financial data.

- *Kobler Unit Framework* (Hochstrasser, 1994): The proposed framework consists of four modules. Each module corresponds to a stage in the evaluation process. Activity within the first module consists in evaluating a proposed project against a checklist of previously identified critical success factors. The purpose of the activities in module two is to ensure that the appraisers have a clear grasp of the true costs of the proposed IT system prior to evaluation of the investment. Activity in module three is concerned with the identification and specification of business performance indicators which can be used to evaluate the performance and benefits of the proposed IT system. Activities in module four enable the comparison of the relative merits of alternative IT systems. According to Whiting et al. (1996), the Kobler Unit framework is practical, can be implemented readily, and it is easy to see how it can be adapted to the specific requirements of a particular organisation. However, it does not take into account of the stage in the system development cycle at which an appraisal is performance and its overly complex classification of IT systems into nine potentially overlapping areas may be difficult to carry out.

- *Multi-object, Multi-criteria Methods* (Farbey et al., 1992): This method starts from the assumption that the value of a project can be measured in terms other than money. It allows decision makers to appraise the relative value of different outcomes in terms

of their own preferences.
- *Value Analysis* (Farbey et al., 1992): This method attempts to evaluate a wide range of benefits including intangible benefits. The method is based on the notion that it is more important to concentrate on value (added) than on cost saved.
- *Options Theory* (Dos Santos, 1994): This method attempts to give senior management a way to better estimate the value of infrastructure investments - investments required before applications can be built - and then track that value. The greatest value of the approach is not necessarily a project's value but the discussions about the project's investment. In addition, since time is an important component of options theory, the further off the expiration of an option the more valuable it is.

Having completed the selection of the IS/IT investments that will support business goals and which are aligned with the business objectives, these investments should then be prioritised. The notion of a systems portfolio implies that IS/IT investment can have a variety of objectives. The practical problem becomes one of prioritisation - of resource allocation among the many objectives and projects that are put forward (Willcocks and Lester, 1997). It is crucial here for senior management to target resources to the best and most productive IS/IT projects that will achieve the most benefits for the organisation (Grover et al., 1998a). The McFarlan (1984) strategic matrix (Figure 3) is a much-used and useful framework for focusing management attention on the IS/IT evaluation question: "where does and will IS/IT give added value?".

The strategic matrix is useful for classifying systems that then demonstrate, through discussion, where IS/IT investment has been made and where it should be applied (Willcocks, 1992b). It reduces the potentially infinite options available to a reasonable, relevant number of alternatives. The objective of such a classification is also to determine the criticality of the relationship between the investment and business success and hence determine how the application should be managed, including how the investments will be appraised (Ward, 1994). Different types of IS/IT systems contribute more or less directly to an organisation's core business, and techniques for investment appraisal need to vary in accordance with that directness (Whiting et al., 1996). According to Ward (1994), one should consider how the benefits arise in the different segments of the application portfolio. All classifications express a similar notion of the degree of distance from a direct contribution to current core business. The different types of systems as categorised by the application portfolio or strategic nature are as follows:

- Strategic systems - the benefits are the result of innovation and change in the conduct of business to gain a competitive edge;
- Key operational systems - the benefits result from carrying out business processes more effectively overall, and normally result from rationalisation, integration or reorganisation of existing processes;
- Support systems - the benefits mainly come from carrying out business tasks more efficiently by removing them, or by automation to reduce the cost of carrying them out; and

Figure 3: McFarlan's strategic matrix (Adapted from: McFarlan, 1984)

Strategic	High Potential
applications critical to sustaining future business strategy	applicaitons that may be important in achieving future success
Key Operational	**Support**
applications on which the organisation currently depends for success	applications that are valuable but not critical to success

- High potential systems - these systems do not actually deliver finished, operational systems and hence real benefits, and so these systems are dealt with as high risk IS/IT investment by treating them as R&D projects.

One of the several methods which can be used to prioritise the IS/IT investments is IT investment mapping (Figure 4). According to Peters (1996), one dimension of the map is benefits ranging from the more tangible arising from productivity enhancing applications, to the less tangible from business expansion applications while another dimension is the orientation of the investment toward the business.

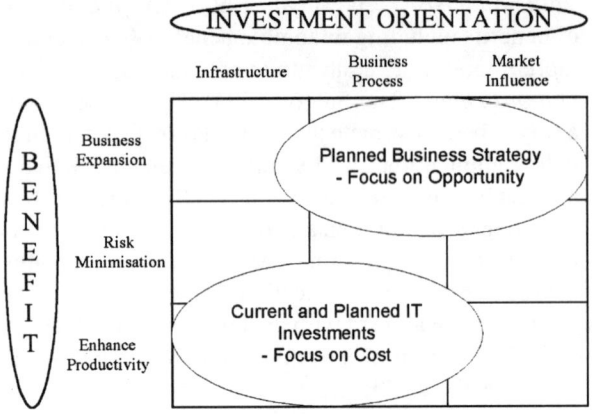

Figure 4: IT investment mapping (Adapted from: Peters, 1996)

Strassmann (1997) suggests that the competency in managing fundamentally important administrative processes is the key to the management of IS/IT investments. Willcocks (1992a) states that one useful way forward on IS/IT benefits management is to match techniques to objectives and types of projects (Figure 5). A starting point is to allow business strategy and purpose to define the category of IS/IT investment. There are five main purposes for IS/IT benefits management: (1) surviving and functioning as a business; (2) improving business performance by cost reduction/increasing sales; (3) achieving a competitive leap; (4) enabling the benefits of other IS/IT investments to be realised; and (5) being prepared to compete effectively in the future.

It is important, at this stage, to distinguish between the different types of IS/IT investments if appropriate evaluation criteria are to be applied when justifying projects. According to Willcocks (1992a), the matching IS/IT investments can then be categorised into five main types: (1) mandatory investments: for example, accounting systems to permit reporting within the organisation; (2) investments to improve performance: for example, laptop computers for sales people, partly with the aim of increasing sales; (3) competitive edge investment: for example, the American Air-

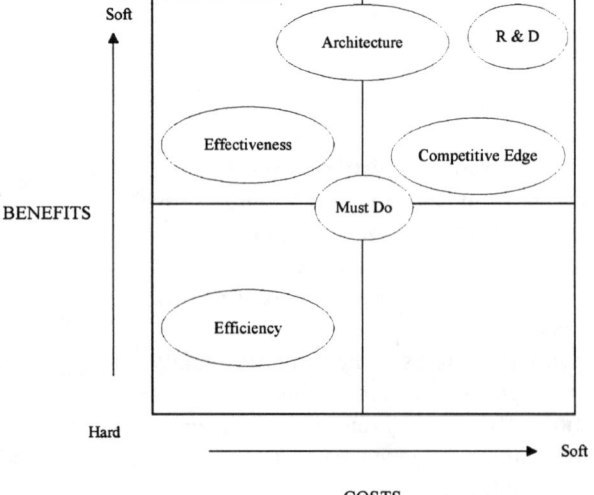

Figure 5: Classifying IS/IT projects (Source: Willcocks, 1992b)

lines' airline reservation system; (4) infrastructure investment: this would give organisations several more degrees of freedom for manoeuvre in the future; and (5) research investments: for example, CASE tools. Hochstrasser (1990), on the other hand, provides different categories: (1) infrastructure; (2) cost replacement: for example, automating manual activities; (3) economy of scope: for example, a relational database performing an extended range of tasks; (4) customer support; (5) quality support; (6) information sharing and manipulation; and (7) new technology: for example, smart cards and home banking. However, Fitzgerald (1998) has suggested that there are really only two types IS/IT projects - efficiency and effectiveness projects.

There are others who argue that the evaluation method should vary according to circumstances. Farbey et al. (1992) suggest that a good way of evaluating a project is to match a project with an evaluation method. The process has three stages:

1. Represent the circumstances of the project which is to be evaluated: this has five broad dimensions: (a) the role of the evaluation - the choice of a suitable technique will depend upon whether the evaluation is taking place early in the project's planning or late, which level at which the evaluation is being carried out, and whether tactical or strategic; (b) environment - this refers to the decision-making/cultural environment in which the project has to be evaluated; (c) the nature of the system; (d) organisation characteristics - industry situation and leadership role of the organisation; and (e) cause and effect relationships - this defines the extent to which the benefits are directly related to the system being evaluated and the degree of uncertainty with which the impact of the new system can be predicted.
2. Locating the techniques: this is to determine the circumstances in which evaluation is to be carried out against several possible evaluation techniques; and
3. Matching: this stage involves finding a preferred evaluation technique.

Hochstrasser (1993) has developed a framework for justifying and prioritising IS/IT investments. It is called *Quality Engineering* (QE). This framework is designed to assess rigorously new investment ideas concerned with improving the quality of business processes. The aim is to provide a basis for deciding trade-offs between varying levels of quality to be achieved and limited resources to be employed. There are four main modules for this framework: (1) quality standards - this addresses critical success factors in the form of a wide range of corporate quality standards that must be adhered to when proposing new IS/IT initiatives; (2) quality awareness - it is designed to raise the awareness of the wider implications of IS/IT projects on a number of issues i.e. the true costs of the project; (3) quality performance indicators - this identifies a set of measurable performance indicators for different classes of IS/IT project i.e. how these indicators are to be measures; and (4) quality value - this builds on the previous three modules and calculates an explicit value for new IS/IT initiatives by taking into account i.e. the value of the primary objectives to be achieved by the proposed system.

According to Silk (1990), there are a total of seven types of justification that might be used for IS/IT projects. The merit of the seven types of justification is that they encourage managers to sharpen up the business case to a degree to which they still feel confident with the number. Often this will mean stopping short of financial figures and admitting that a value judgement is then necessary. These seven types of justification are as follows:

1. *Must-do*: it relates to an investment which is unavoidable - one required by legislative change or one that is essential to remain as a player in the chosen sector of business;
2. *Faith*: the investment is justified as an act of faith based on the judgement or vision of senior management;

3. *Logic*: the causal logic by which a business improvement will arise from the proposed IS is identified as the basis for the business case;
4. *Direction*: an appropriate observable quantity is identified which is then measured to check whether the business has indeed moved in the intended beneficial direction;
5. *Size*: the size of the change in observable quantity is estimated and this is checked quantitatively when the system is in operation;
6. *Value*: the quantified changes are given some considered weighting, so that disparate benefits can be compared with each other; and
7. *Money*: in this final stage, each of the benefits is given a financial value. Not only can they be compared with each other, but also the impact on overall business financial statements and performance measures can be calculated.

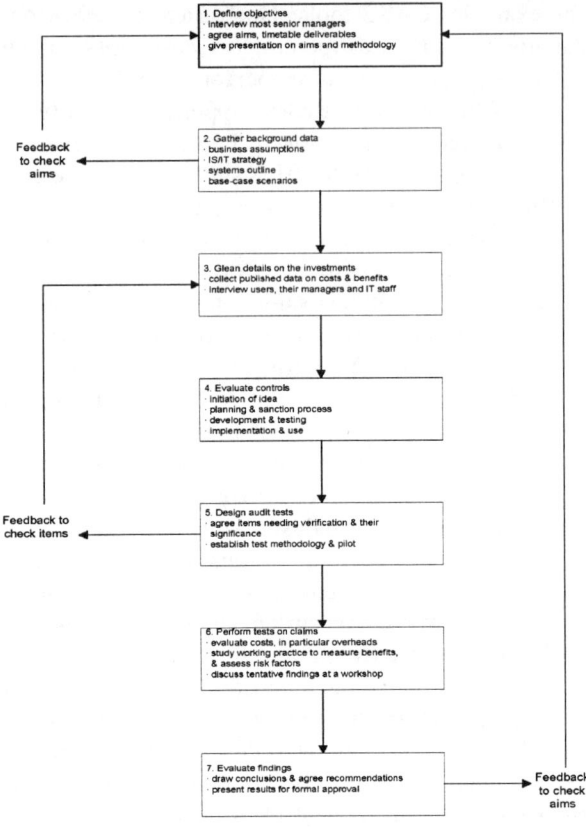

Figure 6: Basic approach to post-implementation appraisal (Source: Norris, 1996)

After alignment and prioritisation assessment, the feasibility of each IS/IT investment then needs to be examined. Many research studies show that the main weakness here has been the overreliance on and misuse of traditional financed based cost-benefit analysis (Willcocks and Lester, 1997). At this stage, active involvement of a group of stakeholders is essential in judging and identifying methods in evaluating IS/IT.

Following this, the evaluation needs to be conducted in a linked manner across systems development lifecycle and into systems implementation and operational use (Willcocks and Lester, 1997). This suggests the development of a series of interlinked measures that reflect various aspects of IS/IT performance, and that are applied across systems lifetime. These are tied to processes and people responsible for monitoring performance, improving the evaluation systems and also helping to achieve and manage the benefits from the investment. A good measure to use here is the *balanced scorecard* (BSC) approach by Kaplan and Norton (1996).

In order to produce a new "dash board" of IT performance indicators throughout the systems development lifecycle, Kaplan and Norton (1996) proposed this method to evaluate an organisation's progress from four different perspectives: financial, internal processes, customer, and innovation and learning. All these measurements (evaluations) are framed in a strategic management system that drives improvement and that allows the management of an organisation to prepare for the future (van Grembergen and van Bruggen, 1998). It may be applied not only to assess the contribution of a specific IS/IT investment project, but also to evaluate the performance and guide the activities of an IS/IT department (Martinsons et al., 1999). To do this, the BSC approach uses a three-layered structure: (1) the mission of the organisation; (2) the mission is then translated into objectives; and (3) the objectives can be measured through well-chosen indicators.

This approach brings together, in a single management report, many of the seemingly disparate elements of a company's competitive agenda. This would also force senior management to consider all the important operational measures together as well as letting them see whether improvements in one area may have been achieved at the expense of another (Kaplan and Norton, 1997).

Therefore, the BSC approach, according to Willcocks and Lester (1997), would be an ideal tool for the integrated evaluation lifecycle approach to measure the performance of IS/IT investments. The strength of the balanced scorecard is that it responds to the need for a number of high-level measures to be developed, and reflects different viewpoints on the organisation and performance. It can also be made to (1) respond adequately to the frequently voiced need for quantified measures; (2) provide measures that can be carefully tracked beyond the investment appraisal stage and into the system's lifecycle; and (3) provide a framework of goals and high-level measures on which a hierarchy of more detailed measures can be erected (Willcocks and Lester, 1994). However, the BSC approach also faces several obstacles. It requires a substantial commitment from key stakeholders and few modifications of its four perspectives in order to achieve business success (Martinsons et al., 1999).

Post-implementation arises out of implementation assessment on an ongoing basis, with an already existing set of evaluators in place (Willcocks and Lester, 1997). This is an assessment of the IS/IT project's success or failure. It can provide valuable opportunities for much-needed organisational learning on IS/IT within the organisation (Willcocks and Mark, 1989). Using post-implementation reviews, data are collected, recorded and analysed to compare expected results against actual benefits and returns. They provide valuable feedback on the value being achieved by expenditure on information systems (Norris, 1996). Existing IS/IT-related activity can also devour the majority of the financial resources available for IS/IT investment. Very often such failures derive from not having in place, or not operationalising, a robust assessment approach that enables timely decision on systems and service divestment, outsourcing, replacement, enhancement, and/or maintenance (Willcocks and Lester, 1997). Such decisions need to be based on at least two criteria - the technical quality of the system/service, and its business contribution - as well as being related back to the overall strategic direction and objectives of the organisation. Norris (1996) offered a basic seven-step approach (Figure 6) to conducting a post-implementation appraisal: (1) define objectives: to gain a clear statement of the specific objectives of the review; (2) gather background data: to obtain a general understanding of the business situation, the aims and history of the investment, and the logical description and physical components of the system; (3) glean details on the investment: to develop a more detailed understanding of the system; (4) evaluate controls: to identify and evaluate the controls that were, and are being, exercised; (5) design audit tests: to design its auditing procedures by using the most appropriate techniques in order to verify the statements on the costs, benefits and controls; (6) perform audit tests on claims; and (7) evaluate findings: to agree the conclusions that can be drawn from the detailed findings.

BENEFITS MANAGEMENT AND BENEFITS REALISATION

Problems and Challenges

While pre-investment appraisal and post-implementation review are important for evaluation purposes, they are insufficient in terms of ensuring that the benefits required are realised and delivered to the organisation (Ward and Griffiths, 1996). Assessing the effective delivery of useful benefits from these services to the business is very difficult (Remenyi and Whittaker, 1996). A survey conducted by Wilson (1991) put measuring benefits as one of the most important barriers to setting up and implementing IS strategy. Some of the reasons put forward for the failure to monitor whether the projected benefits of IS/IT were being realised by the organisations are:

(1) It is too difficult to assess benefits after a project has been implemented (Norris, 1996);
(2) It is not necessary as the project was implemented according to plan (Norris, 1996);
(3) It is too costly to undertake the proper post-implementation reviews on benefits (Norris, 1996);
(4) Many organisations tend to give very little attention to the intangible benefits when investment decisions are made (Beaumont, 1998);
(5) Many organisations have poor IS/IT adoption practices (Fink, 1998); and
(6) It is against many organisations' culture to act as both the watchdog and implementer for benefits delivery.

IS/IT is just one of the enablers of process change (Grover et al., 1998b) and it only enables or creates a capability to derive benefits. Increases in benefits can only be obtained if the process is changed. According to Ward et al. (1996:215), the essence of benefits realisation is "not to make good forecasts but to make them come true ... and IS/IT on its own does not deliver benefits." Benefits may be considered as the effect of the changes, i.e. management of changes - the difference between the current and proposed way that work is done (Ward and Griffiths, 1996). Earl (1992) has also taken the view that benefits are associated with business change and not the technology itself. Things only get better when people start doing things differently (Ward and Murray, 1997).

As benefits are frequently long term, uncertain and intangible (Sassone, 1988), the future benefits are too wide-ranging to be estimated with any accuracy. Therefore, IS/IT projects should be evaluated in the context of accumulated costs and benefits from related initiatives, not judged on single initiatives (Galliers et al., 1996). In order to determine if the desired benefits have been achieved in practice, it is necessary to measure and evaluate post-project. If no measurable effects can be identified post-project, other than the implementation of the technology itself, then it would be safe to assume that no benefits have actually been realised (Ward et al., 1996).

Increasingly, as IS/IT expenditure has risen dramatically and as the use of IS/IT has penetrated to the core of organisations, the search has been directed towards not just improving evaluation techniques and processes, but also towards the management and realisation of benefits (Fitzgerald, 1998). Very few organisations have a benefits management approach, and much attention is paid to ways of justifying investments, with little effort being extended to ensuring that the benefits expected are realised (Lin et al., 2000). As the result, there is a massive imbalance between IS/IT investment and benefits derived from that investment (Sutherland, 1994).

Table 1: Paradigm Shift for Benefits Realisation (Source: Truax, 1997)

Traditional Benefits Realisation Principles	New Benefits Realisation Principles
Benefits are stable over time.	The potential benefits from an investment change over time.
The investment determines the nature and scope of the benefits.	The organisation and its business context determine the benefits.
Financial returns represent the most valid justification for an investment.	All the outcomes of an investment represent potential sources of value.
It is sufficient to manage the investment to generate the benefits.	The organisation must be proactive in realising benefits.

According to Truax (1997), there are a number of problems for organisations not getting the benefits they expected from their IS/IT investments:
- Immediate results of an investment are rarely the expected benefits;
- Necessary means for benefits realisation are not identified;
- Benefits do not occur where and when they are planned;
- The "right" benefits are difficult to identify up front;
- Projects are too narrowly defined for effective delivery of benefits; and
- Organisations often have a limited ability to manage change.

Ward and Murray (1997) identified three mindset constraints that seem to operate strongly when business managers approach the issue of managing IS/IT. These can often lead to not getting the expected benefits from IS/IT investment. These are as follows:
- The management of IS/IT is a technical issue;
- The cost of IS/IT should be justified by financial bottom-line; and
- The functionality from IS/IT is a benefit in itself.

Too often these problems are compounded with the fact that organisations operate based on traditional benefits realisation principles, as outlined in Table 1.

According to Lederer and Mirani (1995), an understanding of benefits is very important for several reasons:
- It can give researchers an opportunity to characterise IS/IT projects thematically;
- It can create top management's expectations for the outcomes of IS/IT projects as it offers an opportunity to evaluate the projects, IS/IT management's ability to meet its commitments and thus its credibility;
- It may be able to help predicting the achievable IS/IT projects outcomes better and thus realise them more often; and
- It can give some guidance for IS/IT managers in proposing new projects and recommending their priorities.

In order to achieve and maximise the expected benefits from the IS/IT investments, some researchers have come up with ways of evaluating and realising the IS/IT benefits. This is often called benefits management. It has been defined as "the process of organising and managing such that potential benefits arising from the use of IS/IT are actually realised" (Ward and Griffiths, 1996). It aims to be a whole lifecycle approach to getting beneficial returns on IS/IT investments (Ward and Murray, 1997). Benefits management plans encourage the business users to focus on exactly how they will make the system pay off and contribute to the business objectives. The ability to achieve benefits from one investment will depend on the organisation's experience and knowledge of what benefits IS/IT can or cannot deliver and how they can be obtained (Ward and Griffiths, 1996). Coleman and Jamieson (1994) state: "An IS/IT project does not finish with the successful delivery of a working system; it continues as long as benefits are being accrued to the business." The following diagram (Figure 7) illustrates the relationship between IS/IT evaluation and IS/IT benefit management.

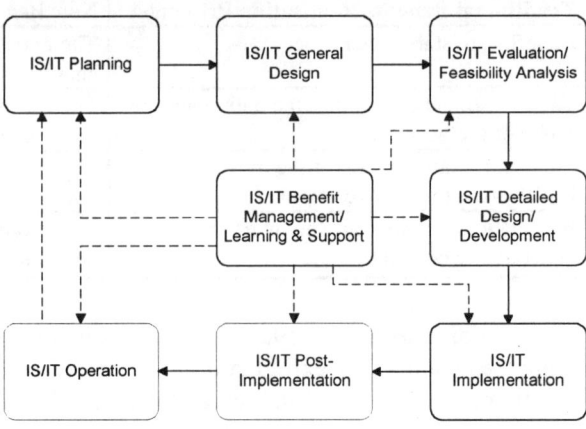

Figure 7: IS/IT evaluation and benefit management (Adapted from: Whitten et al., 1994)

According to King and McAulay (1997), for the advocates of the one best way approach in evaluating IS/IT benefits, success in IS/IT evaluation is determined by the acceptance of projects which show a positive net present value (NPV). However, as mentioned before, most researchers have argued that the financially oriented evaluation techniques such as net present value (NPV) and return on investment (ROI) have largely ignored the intangible benefits as well as potential risks (Hochstrasser, 1993). King and McAulay (1997) have further stated that, for those who suggest alternative approaches, whether quantitative or qualitative in nature, there remains an implicit assumption that selecting an appropriate evaluation technique will secure a successful choice of projects. According to the advocates of the contingency theory, success is given by applying a technique that is determined by the context within which the evaluation takes place. The process model school, on the other hand, argues that success follows from adhering to an appropriate procedure.

Process Model of Benefits Management

Without an effective benefits management process, IS/IT benefits will be within the organisation leaving the investing organisation as a whole without satisfactory payoff. According to Ward et al. (1996), the *Process Model of Benefits Management* developed by Cranfield research program can be used as the basis for guidelines on best practice in benefits management.

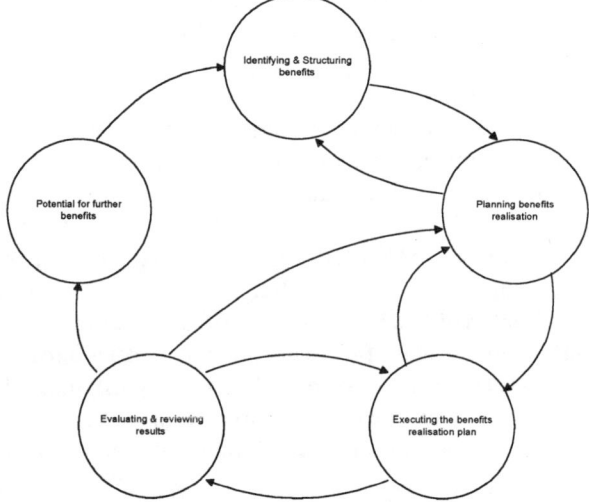

Figure 8: Process model of benefits management (Source: Ward et al., 1996)

Figure 8 shows the elements of this process model are as follows:

1. Identifying and structuring benefits: the proposed benefits and dis-benefits are identified and, for each proposed benefit, suitable business measures are devel-

oped. The list of benefits required must be agreed by the managers whose activities are affected by the system. At the same time, potential dis-benefits of the system should be considered, i.e. what adverse impacts on the business or organisation it could have. The benefits are structured in order to understand the linkages between technology effects, business changes and overall business effects. Other things need to be identified here are: (a) who should be responsible for benefits delivery; and (b) where should the benefits occur;

2. Planning benefits realisation: specific responsibility for realising the benefit is allocated within the business for each benefit. The task is to consider the stakeholders affecting delivery of each benefit, and the changes and tasks needed to ensure delivery. In order to make a fully informed decision as to the viability of the proposed project, the required business changes are planned for and assessed, and a benefits realisation plan is produced at this stage. Only when this has been completed for all of the required benefits should funding for the IS/IT investment be sought;

3. Executing the benefits realisation plan: the necessary business changes as detailed in the benefits realisation plan are carried out, together with the implementation of the proposed IS/IT application. Monitoring progress against the activities and deliveries of the benefits realisation plan is just as important as for the IS/IT development plan, and the two plans interact. It will be necessary to replan, and issues may arise that prevent the delivery of some or even all of the benefits. It is also possible that, at this stage, further benefits are identified;

4. Evaluating and reviewing results: after the full implementation of IS/IT and business changes, the previously developed business measures are used to evaluate the effects of the project. Review of 'before and after' measures provides an explicit device for evaluating whether the proposed business benefits have actually been realised. This evaluation, which should involve all key stakeholders, has several purposes: (a) to maximise the benefits of the particular project; (b) to provide experience for other future projects; (c) to identify what was achieved, what has not been achieved, and why; and (d) to identify any unexpected benefits that have actually been achieved. This post-implementation review should not become a "witch-hunt" and must be an objective process with future improvements in mind; and

5. Potential for further benefits: it may become apparent that, after the post-project review, further benefits are now achievable, which were not expected in the beginning. This stage provides the opportunity to plan for and realise these further benefits as well as to learn from the overall project process.

By using this process model, it is possible to diagnose why some projects are successful in delivering benefits and others are not. It is also possible to show how the less successful could be addressed with remedial action to obtain benefits that are being lost, and in most cases further benefits could be uncovered (Ward and Griffiths, 1996).

Active Benefit Realisation (ABR)

Remenyi et al. (1997) have advocated that their approach, known as *Active Benefit Realisation* (ABR), can be utilised to continually assess and manage potential benefits arising from the use of IS/IT. The ABR, based on contingency philosophy, can be used to maximise value from IS/IT investment by ensuring that the information systems development process, from the beginning to benefit delivery, is managed effectively and efficiently. Fundamental to this approach is that the principal stakeholders of the information system are identified at the onset and that they accept and agree upon their continuous involvement.

The ABR approach can be divided into three distinct phases. They are as follows:
1. Setting the course - this involves the development of sets of precise requirements under the headings of a business picture, a financial picture and a project picture. These pictures are statements, models in a loose sense, of the context, the required benefits and the specification of the appropriate metrics to be used to evaluate, monitor and control benefits realisation. Once the three pictures have been produced, a decision is made and an agreement reached as to whether or not to launch the project;
2. Formative evaluation - this involves assessing the progress of the project. During this phase, all the stakeholders are able to develop views as to how the project is progressing and to exchange these views in open and constructive discussion. There are three possible outcomes: (a) update the three initial pictures; (b) the project may need to be substantially reformed and there may not be sufficient funds, time or skills available. This means that a material change is required to the way the original business solution is currently perceived and defined; and (c) the project may have become for one or more reasons irrelevant to the organisation's business requirements and should result in project termination; and
3. Moving forward - this provides a feedback loop which should be available, not only during development, but also throughout the entire life of the project.

DMR's Benefit Realisation Model

According to Truax (1997), senior management needs a new set of worldviews - in the form of richer investment decision-making frameworks and a well-rounded focus on benefits. This means the full range of benefits and the actual process of benefits realisation. Such an investment model must clearly map out a complete web of benefits and the logical chain of results: from immediate, predictable outcomes to intermediate and final benefits. That map must display the paths linking an investment to the achievement of identified benefits, as well as provide a framework for supporting the management of the change process. According to DMR (1997), to implement benefits realisation in organisation, new approaches are needed in four key areas:
1. Business cases for investment programs - this means complementing traditional return on investment (ROI) and payback analysis with assessments of other sources of value, including softer benefits and the clusters of benefits flowing over time from implementation of key business strategies;
2. Methods of investment program management - individual projects must be organised into structured investment programs in order to implement true benefits management cycles. These programs need to include many types of projects, not just IS/IT projects but also training, organisational change and business process redesign;
3. Benefits realisation modelling - a robust model of the benefits realisation process for each major investment program is needed. These are developed with DMR's unique technique, Results Chain. The model maps key linkages among management and investment initiatives, numerous intermediate outcomes, expected contributions, assumptions about business and market conditions and the end-benefits. The Result Chain model helps define options, giving a big picture of risk/reward relationships and the true value of alternative programs; and
4. Measurement systems and accountabilities - to ensure that business performance improves in line with the resource commitments, it is important to make adjustments to two of the most visible change agents in any organisation: measurement systems

and accountabilities for results. These adjustments are tailored to each organisation, often using the Results Chain model as a reference point.

CONCLUSIONS

In this chapter we have discussed the basic principles of IS/IT investment evaluation and benefits management. The need for better methods of IS/IT investment evaluation has arisen from problems such as the 'productivity paradox' where existing measures fail to reveal the gains made from these investments. The reasons for IS/IT investment evaluation may be more than just solving this paradox, however, and may include project justification, project comparisons, control, learning, and competitive advantage.

Problems in this area include the budgeting practices of organizations concealing full costs, traditional financial evaluation techniques (such as NPV, IRR, ROI and others) excluding intangible benefits and risks, deliberate overstatement of costs by project managers, the uncertainty involved in new technology projects, lack of time and care in the evaluation process, and lack of IS/IT planning. Willcocks and Lester's (1997) proposed approach has been presented as a framework for a number of other methods and techniques that can be combined them into a process for success in this area.

Problems in managing and realising of IT investments include measurement problems (particularly of intangible benefits), lack of pressure to measure, cost of post-implementation reviews, poor IS/IT adoption practices, and organisational culture. Unfortunately, some managers see IT as a technical issue, seek financial bottom-line justifications, and seeing functionality as a benefit in itself. More recent benefits realisation principles include the recognition that (1) potential benefits change over time, (2) organization/business contexts determine benefits, (3) all outcomes represent potential sources of value, and (4) organizations must be proactive in realising benefits. It is important to recognise that financially-orientated measures such as NPV and ROI are useful but largely ignore intangible benefits.

Different approaches to benefits management have been discussed, including Cranfield's Process Model of Benefits Management, the Active Benefit Realisation approach, and DMR's Benefit Realisation Model. Recent research indicates that formal approaches are not often used (Pervan and Lin, 2000), and it is argued that the application of structure and discipline to the process, through models such as the three aforementioned, will improve the measurement of IS/IT benefits in organizations and lead to more effective investment of organisations' scarce resources in key elements of information systems and technology.

REFERENCES

Apostolopoulos, T.K., and Pramataris, K.C. (1997). Information technology investment evaluation: investments in telecommunication infrastructure. *International Journal of Information Management*, *17*(4), 287-296.

Bacon, C.J. (1994). Why companies invest in information technology. In L. Willcocks (Ed.), *Information management: the evaluation of information systems investments* (Ch2, pp31-47). Chapman & Hall, London.

Ballantine, J.A., Galliers, R.D., and Stray, S.J. (1996). Information systems / technology evaluation practices: evidence from UK organizations. *Journal of Information Technology, 11*, 129-141.

Ballantine, J.A., and Stray, S. (1998). Financial appraisal and the IS/IT investment decision making process. *Journal of Information Technology, 13*, 3-14.

Beaumont, N.B. (1998). Investment decisions in Australian manufacturing. *Technovation, 18*(11), 689-695.

Brown, A. (1994). Appraising intangible benefits from information technology investment. *Proceedings of the 1st European Conference on IT investment evaluation*. Henley, UK, September, 187-199.

Brynjolfsson, E., and Hitt, L.M. (1998). Beyond the productivity paradox: computers are the catalyst for bigger changes. *Communications of the ACM, 41*(8), 49-55.

Butler Cox Foundation (1990). *Getting Value from Information Technology*. Research Report 75, June. Butler Cox: London.

Carlson, W., and McNurlin, B. (1992). Basic Principles for Measuring IT value. *I/S Analyzer*, October, 1-16.

Coleman, T., and Jamieson, M. (1994). Beyond return on investment. In L. Willcocks (Ed.), *Information management: the evaluation of information systems investments* (Ch10, pp189-205). Chapman & Hall, London.

Dewan, S., and Kraemer, K.L. (1998). International dimensions of the productivity paradox. *Communication of the ACM, 41*(8), 56-63.

DMR (1997). *Driving up investment success rates*. DMR consulting, 1-4.

Dos Santos, B.L. (1994). Assessing the value of strategic information technology investments, In L. Willcocks (Ed.), *Information management: the evaluation of information systems investments*, (Ch7, pp133-148). Chapman & Hall, London.

Earl, M.J. (1988). *Management Strategies for Information Technology*. Prentice Hall: London.

Earl, M.J. (1992). Putting IT in its place: a polemic for the nineties. *Journal of Information Technology, 7*, 100-108.

Farbey, B., Land, F., and Targett, D. (1992). Evaluating investments in IT. *Journal of Information Technology, 7*, 109-122.

Fink, D. (1998). Guidelines for successful adoption of information technology in small and medium enterprises *International Journal of Information Management, 18*(4), 243-253.

Fitzgerald, G. (1998). Evaluating information systems projects: a multidimensional approach. *Journal of Information Technology, 13*, 15-27.

Galliers, B., Newell, S., and Robertson, M. (1996). The information challenge - making your IT investment pay off. *Hot Topics, 1*(2), March, Source: [On-Line] http://www.wbs.warwick.ac.uk/HotTopics.

Gottschalk, P. (1999). Implementation of formal plans: the case of information technology strategy. *Long Range Planning, 32*(3), 362-372.

Griffiths, C., and Willcocks, L. (1994). Are major information technology projects worth the risk? *Proceedings of the 1st European Conference on IT investment evaluation*. Henley, UK, September, 256-269.

Grover, V., Teng, J.T.C., and Fiedler, K.D. (1998a). IS investment priorities in contemporary organizations. *Communication of the ACM, 41*(2), 40-49.

Grover, V., Teng, J., Segar, A. H., and Fiedler, K. (1998b). The influence of information technology diffusion and business process change on perceived productivity: the IS executive's perspective. *Information and Management, 34*, 141-159.

Hayashi, A.M. (1997). Squeezing profits from IT. *Datamation*, July, 42-47.

Hinton, M., and Kaye, R. (1996). Investing in information technology: a lottery? *Management Accounting*, November, 52-54.

Hochstrasser, B. (1990). Evaluating IT investments - matching techniques to projects. *Journal of Information Technology, 5*, 215-221.

Hochstrasser, B. (1993). Quality engineering: a new framework applied to justifying and prioritising IT investment. *European Journal of Information Systems, 2*(3), 211-223.

Hochstrasser, B. (1994). Justifying IT investments. In L. Willcocks (Ed.), *Information management: the evaluation of information systems investments* (Ch8, pp151-169). Chapman & Hall, London.

Hochstrasser, B., and Griffiths, C. (1991). *Controlling IT Investment: Strategy and Management*. Chapman & Hall, London.

Irani, Z., Ezingeard, J.N., and Grieve, R.J. (1997). Integrating the costs of a manufacturing IT/IS infrastructure into the investment decision-making process. *Technovation, 17*(11/12), 695-706.

Kaplan, R. S., and Norton, D. P. (1996a). Using the balanced scorecard as a strategic management system. *Harvard Business Review*, Jan. - Feb., 75-85.

Kaplan, R. S., and Norton, D. P. (1997). The balanced scorecard: translating strategy into action. *Training and Development, 51*(1), 50-51.

Katz, A. I. (1993). Measuring technology's business value: organizations seek to prove IT benefits. *Information Systems Management*, Winter, 33-39.

Keen, P. G. W. (1991). Relevance and rigor in information systems research: Improving quality, confidence, cohesion and impact. In Nissen, Klein & Hirschheim (Eds.), *Proceedings of the IFIP TC8/ WG 8.2 Working Conference on the Information Systems Research Arena of the 90s Challenges, Perceptions, and Alternative Approaches*, Copenhagen, Denmark, 14-16 December, 1990, 27-49.

Keen, P. G. W. (1995). *Every manager's guide to information technology: a glossary of key terms and concepts for today's business leader* (2nd Ed.). Harvard Business School Press, Boston, Massachusetts.

Kelly, S. (1997). Business to blame for skills crisis. *Computing*, 4 Sept., 6.

King, M., and McAulay, L. (1997). Information technology investment evaluation: Evidence and interpretations. *Journal of Information Technology, 12*, 131-143.

Kumar, R. L. (1996). A note on project risk and option values of investments in information technologies. *Journal of Management Information Systems, 13*(1), 187-193.

Kumar, R. L. (1998). Understanding the value of information technology enabled responsiveness. *Electronic Journal of Information Systems Evaluation, 1*(1), Source: [On-Line] http://is.twi.tudelft.nl/ ejise/indpap.html.

Launders, T. (1997). Nine steps for network cost control. *Computer Weekly*, 27 March, 33.

Lederer, A. L., and Mirani, R. (1995). Anticipating the benefits of proposed information systems. *Journal of Information Technology, 10*, 159-169.

Leung, L. (1997). Downtime costs each user three weeks' work a year: PCs cost users time. *Computing*, 25 September, 10.

Lin, C., Pervan, G.P., and McDermid, D. (2000). Research on IS/IT Investment Evaluation and Benefits Realization in Australia. *Proceedings of the International Conference of the Information Resource Management Association*, Alaska, USA, May 21-24, 359-362.

Lyon, C.P., and Mooney, J.G. (1994). Information technology planning for business benefit: A case study of a 'back to basics' approach. *Proceedings of the 1st European Conference on IT investment evaluation*. Henley, September, 178-186.

Mahmood, M.A., and Mann, G.J. (1993). Measuring the organizational impact of information technology investment: an exploratory study. *Journal of Management Information Systems 10*(1), 97-122.

Martinsons, M., Davison, R., and Tse, D. (1999). The balanced scorecard: a foundation for the strategic management of information systems. *Decision Support Systems, 25*, 71-88.

McFarlan, F. W. (1984). Information technology changes the way you compete. *Harvard Business Review*, May-June, 98-103.

Mirani, R., and Lederer, A.L. (1993). Making promises: The key benefits of proposed IS systems. *Journal of Systems Management, 44*(10), 16-20.

Norris, G. D. (1996). Post-investment appraisal. In L. Willcocks (Ed.), *Investing in information systems: Evaluation and management* (Ch9, pp193-223). Chapman & Hall, UK.

Parker, M.M., Benson, R.J., and Trainor, H.E. (1988). *Information economics*. Prentice Hall: London.

Pervan, G. (1998). How chief executive officers in large organizations view the management of their information systems. *Journal of Information Technology, 13*, 95-109.

Pervan, G.P., and Lin, C. (2000) Realising the benefits of IS/IT investments in Australian Organisations. *School of Information Systems Working Paper #2002*, Curtin University of Technology, Australia.

Peters, G. (1996). From strategy to implementation: identifying and managing benefits of IT investments. In L. Willcocks (Ed.), *Investing in information systems: evaluation and management*, (Ch10, pp225-240). Chapman & Hall, UK.

Remenyi, D., Sherwood-Smith, M., and White, T. (1997). *Achieving maximum value from information systems: A process approach*. John Wiley and Sons, Chichester, England.

Sassone, P.G. (1988). Cost justification: a survey of cost-benefit methodologies for information systems. *Project Appraisal, 3*(2), 73-84.

Serafeimidis, V., and Smithson, S. (1996). The management of change for information systems evaluation practice: experience from a case study. *International Journal of Information Management, 16*(3), 205-217.

Silk, D.J. (1990). Managing IS benefits for the 1990s. *Journal of Information Technology, 5*, 185-193.

Sriram, V., Stump, R. L., and Banerjee, S. (1997). Information technology investments in purchasing: an empirical study of dimensions and antecedents. *Information and Management, 33*, 59-72.

Strassman, P.A. (1990). *The business value of computers.* The Information Economics Press: New Canaan.

Strassmann, P. A. (1997). Will big spending on computers guarantee profitability? *Datamation, 43*(2), February, 75-82.

Sutherland, F. (1994). Some current practices in the evaluation of IT benefits in South African organizations. *Proceedings of the 1st European Conference on IT Investment Evaluation.* Henley, September, 27-43.

Symons, V. (1994). Evaluation of information systems investments: Towards multiple perspectives, In L. Willcocks (Ed.), *Information management: the evaluation of information systems investments* (Ch13, pp253-268). Chapman & Hall, London.

Symons, V. and Walsham, G. (1988). The evaluation of information systems: a critique. *Journal of Applied Systems Analysis, 15*, 119-132.

Truax, J. (1997). Investing with benefits in mind: curing investment myopia. *The DMR White Paper*, 1-6.

van Grembergen, W., and van Bruggen, R. (1998). Measuring and improving corporate information technology through the balanced scorecard. *Electronic Journal of Information Systems Evaluation, 1*(1), Source: [On-Line] http://is.twi.tudelft.nl/ejise/indpap.html.

Ward, J. (1994). A portfolio approach to evaluating information systems investments and setting priorities, In L. Willcocks (Ed.), *Information management: The evaluation of information systems investments*, (Ch4, pp81-97). Chapman & Hall, London.

Ward, J., and Griffiths, P. (1996). *Strategic planning for information systems.* John Wiley & Sons Ltd, Chichester, UK.

Ward, J., and Murray, P. (1997). *Benefits management: Best practice guidelines.* ISRC-BM-97016, Information Systems Research Centre, Cranfield School of Management, Cranfield, UK.

Ward, J., Taylor, P., and Bond, P. (1996). Evaluation and realisation of IS/IT benefits: An empirical study of current practice. *European Journal of Information Systems, 4*, 214-225.

Weill, P., and Olson, M.H. (1989). Managing investment in information technology: Mini case examples and implications. *MIS Quarterly, 13*(1), 3-17.

Whiting, R., Davies, J., and Knul, M. (1996). Investment appraisal for IT systems. In L. Willcocks (Ed.), *Investing in information systems: evaluation and management* (Ch. 2, pp37-57). Chapman & Hall, London.

Whitten, J.L., Bentley, L.D., and Barlow, V.M. (1994). *Systems analysis and design methods* (3rd ed.). Richard D. Irwin, Inc., Sydney.

Willcocks, L. (1992a). Evaluating information technology investments: Research findings and reappraisal. *Journal of Information Systems, 2*, 243-268.

Willcocks, L. (1992b). IT evaluation: Managing the Catch 22. *European Management Journal, 10*(2), 220-229.

Willcocks, L., and Lester, S. (1994). Evaluating the feasibility of information systems investments: recent UK evidence and new approaches. In L. Willcocks (Ed.), *Information management: the evaluation of information systems investments* (Ch3, pp49-77). Chapman & Hall, London.

Willcocks, L., and Lester, S. (1996). The evaluation and management of information systems investments: From feasibility to routine operations. In L. Willcocks (Ed.), *Investing in information systems: evaluation and management* (Ch. 1, pp15-36). Chapman & Hall, London.

Willcocks, L., and Lester, S. (1997). Assessing IT productivity: Any way out of the labyrinth? In L. Willcocks, D.F. Feeny, & G. Islei (Eds.), *Managing IT as a strategic resource* (Ch4, pp64-93). McGraw-Hill, London.

Willcocks, L., and Mark, A. L. (1989). IT systems implementation: Research findings from the public sector. *Journal of Information Technology, 4*(2), 92-103.

Wilson, T. (1991). Overcoming the barriers to the implementation of information system strategies. *Journal of Information Technology, 6*, 39-44.

Chapter II
A Benefits Realization Approach to IT Investments

John Thorp
The Thorp Network

Information technology is today transforming all aspects of our lives — how we work, shop, play and learn. It is transforming our economic infrastructure — revolutionizing methods of supply, production, distribution, marketing, service, and management. This represents nothing less than a fundamental redesign of the entire supply chains of most industries and indeed a fundamental restructuring of many industries themselves. The potential long-term impact of information technology represents an economic and social transition as fundamental as the shift from rural agriculture to urban industry 200 years ago, during the first Industrial Revolution.

Yet today we have a problem — a big problem! Chief information officers (CIOs) are finding themselves increasingly under fire for the perceived lack of value from ever-growing investments in information technology (IT) — investments that in the U.S. now represent close to 50% of companies' new capital investment and a significant portion of their operating expense. Our investments in technology are not being consistently translated into business value. The link to business results is not clear. It is hard to demonstrate how investments in IT, or in producing information translate into economic value.

A 1996 U.S. survey by the Standish Group found 73% of IT projects were cancelled, over budget or late, with 31% being cancelled. Project failures cost an estimated $145 billion. This figure does not include the loss of anticipated business benefits, likely amounting to trillions of dollars. More recent studies confirm that project failures are continuing to occur at a similar rate, and this applies to more recent ERP, e-Commerce, Supply Chain Management and Customer Relationship Management projects as well as more traditional projects.

THE QUESTION OF VALUE

As we enter the so-called "New Economy," the question that is increasingly being asked is: "Are our investments in information technology providing the greatest possible value?" Unfortunately, the CEOs and other senior executives who are asking their CIOs to compute the value of IT investments are asking the wrong person. They should instead take a hard look in the mirror. This is not a technology issue — it is a *business* issue. Not only CEOs but all business managers should indeed be asking tough questions, but, more

Copyright © 2001, Idea Group Publishing.

importantly, they must recognize and step up to their responsibility in answering those questions — failure to do so is nothing less than an abdication of their responsibility.

At the root of this question of value is the fact that the way we apply technology has evolved. In the past, we largely automated operational tasks such as payroll, where benefits were clear and relatively easy to achieve. Applications of IT today, such as e-commerce applications, enable increasingly strategic business outcomes. Yet, while these outcomes would not be possible without the technology, the cost of the technology is only a small part of the total investment that organizations must make to achieve their desired outcome, often only 5% to 15%. The reality is that there is no such thing as an IT project any more — only business change programs of which IT is an essential, but often small, part. We are today no longer implementing technology — we are implementing change. Implementing change is a very different challenge from implementing technology.

Unfortunately, our approach to managing IT has lagged in recognizing this shift. A number of years ago I was meeting with a number of senior project managers of a very large US organization. I asked them how they defined a successful (IT) project. The answers included on time, on budget, delivering the expected functionality and "getting out with my skin intact." Wrong, I told them. A successful project is one that delivers the expected benefits to the organization. This was not a totally fair response, as they had never been asked to do this.

SILVER BULLET THINKING

We still exhibit "silver bullet thinking" when it comes to IT. We act as if, once determined, the benefits associated with an investment will automatically happen. As if addressing the WHAT is sufficient to achieve them. However, simply identifying and estimating benefits won't necessarily make them happen.

Paying attention to HOW benefits happen is as important, if not more important, than focusing on what the expected benefits are. Too often, the HOW is taken for granted. With the evolution of IT applications, a new approach to the management of IT investments has become a business imperative. Benefits realization requires a new mindset, one that is focused on business results but that recognizes that IT alone cannot deliver these results.

I am not saying here that bringing projects in as specified, on time and on budget is not important. It most certainly is. It is however necessary but not sufficient. Not only the implementation process should be managed, the benefits realization process should also be proactively managed. We must shift from a single minded focus on completing initiatives on time and on budget, to understanding the results that the business expects, HOW they can be achieved out of the initiative, and ultimately undertake a proactive management stance to ensure their realization.

As we enter the uncharted waters of the new economy, these problems will become more acute and will threaten the very survival of organizations. The need for a new approach to the management of IT investments is critical. As Peter Senge said in *The Fifth Discipline,* "Learning disabilities are tragic in children, but they are fatal in organizations. Because of them, few corporations live even half as long as a person — most die before the age of 40." The need for organizations to learn how to better manage their investments in IT-enabled change has never been greater. The life expectancy of those organizations that do not learn will be significantly reduced.

A NEW APPROACH

The answer to the question of value lies in organizations resolving two key questions:
- How do we pick the winning IT investments?
- How do we ensure that we are getting value from these investments, and know that we are doing so?

Resolving the question of value should be a business imperative for executives and for all business managers today. It is part of the continuing challenge of reinventing organizations to survive and prosper in an e-business world. Those organizations that resolve this question, including those few that are already on the road to solving it, will be the winners in the new economy. Those that do not will be history.

Any new approach must recognize that investing in IT is no longer primarily investing in a piece of hardware or software, it is investing in the *process of change itself*, a process of change in the overall business system. This is a much more complex undertaking than the relatively simple delivery of an IT system. The point at which we usually declare implementation complete is unfortunately the point at which real implementation begins.

The reality of this more complex world is that:
- **Benefits do not just happen** when a new technology is delivered.
- **Benefits rarely happen** according to plan.
- **Benefits realization is a process** that can and must be managed, just like any other business process.

The benefits mindset requires a major shift in management methods and practices. Our current project management approach focuses almost exclusively on the delivery of technology, on time and on budget. In today's world of complex change, this is not enough. We need a new approach, a Benefits Realization Approach that identifies all the projects and initiatives required to produce business results, whether they involve technology, or other elements of the business system and an approach that focuses on continuous management of the benefits realization process.

THE BENEFITS REALIZATION APPROACH

The cornerstones of a Benefits Realization Approach include:
- Moving beyond stand-alone IT project management to business **program management**, where technology initiatives contribute to business results in concert with other elements of the overall business system, including organizational, process, and people initiatives;
- **Proactively managing change** as an integral part of business programs, rather than as an afterthought in reaction to "implementation problems";
- Shifting from free-for-all competition for resources to managing a number of business programs through a disciplined **portfolio management** approach; and

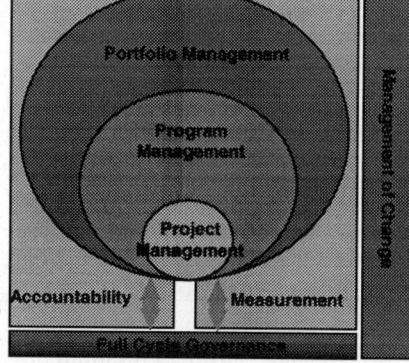

- Adopting a **full cycle governance** view of managing projects, programs, and portfolios from "concept to cash" rather than just from "design to delivery," with clear **accountability** to ensure clear business sponsorship of programs, and effective **measurement** systems.

PROGRAM MANAGEMENT

If we are to improve the odds of delivering business benefits we need more than just better project management. We must look at the full program of activities involved in changing the business system, and then manage the investment program as a whole, including all the actions required to realize benefits.

A program is a structured grouping of projects designed to produce clearly identified business results or other end benefits. The program focus is on all the activities required to deliver business benefits. A program comprises all the projects that are both necessary and sufficient to produce those benefits. It is the business program, effectively managed, that delivers the benefits to the organization.

The Keys to Program Management are:
- Identifying the desired business outcomes in terms of measurable benefits, and
- Identifying all the elements of change that are needed to deliver the benefits.

The following framework, adapted from Michael Scott Morton in his book *Corporation of the 1990s*, is helpful in understanding the business system, and defining business programs that deliver the business benefits.

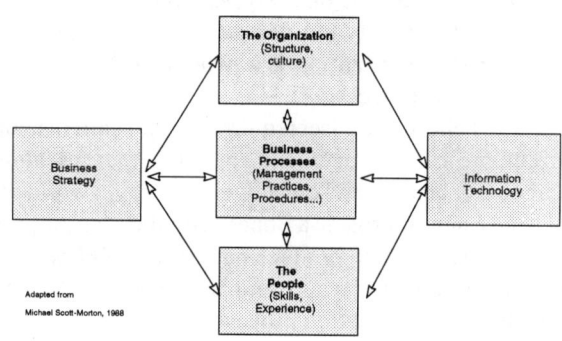

Adapted from Michael Scott-Morton, 1988

Traditionally, and all too often today, organizations focus their attention almost exclusively on the technology element of the business system. Even in the relatively new world of electronic commerce, a Gartner Group Research Note by T. Berg describing 55 electronic commerce case studies found that in only 35% of the cases were results measured in business terms (e.g. "reduced inventory costs" or "increased profitability"). The other 65% were measured in stand-alone IT project terms, addressing the questions: Did the technology work? Was it delivered on time? Was it delivered on budget? In most cases, the business element is dealt with by reference to a one-off, financially oriented business case that does not adequately value the business strategy link. That is if there is even a business case at all. The stand-alone focus on technology during the project gives little attention to the organizational, process and people issues that are so important to successful implementation.

Business programs are much more complex than technology projects. There are many linkages between the projects that comprise the program. These linkages need to be understood and managed. The extent and impact of change on the organization, its processes, and specifically on its people will also need to be understood and managed. In an e-business world, the impact of change often extends beyond the organization – to the extended enterprise. Suppliers, customers and other business partners will also be impacted. It is at this stage that the necessary actions required to manage the process of change must be identified and built into the program.

This new business program mindset requires learning and possibly most importantly — unlearning. When managers understand this view, it becomes very apparent that 85% to 95% of most activities in what had previously been thought of as IT projects are not IT related at all. They fall into the organization, process, and people categories. They realize that they are not implementing technology – they are implementing change.

DEFINING PROGRAMS – THE DMR RESULTS CHAIN™

Before organizations can implement business programs, they must first fully understand and define them. In working with DMR, one of our distinctive contributions to the Benefits Realization Approach has been to develop a technique to help prepare a clear and comprehensive model of an organization's benefits realization process. This technique, the DMR Results Chain, enables the preparation of road maps that support understanding and proactive management of business programs throughout the benefits realization process. The Results Chain technique is used to build simple yet rigorous models of the linkages among four core elements of the benefits realization process: outcomes, initiatives, contributions and assumptions.

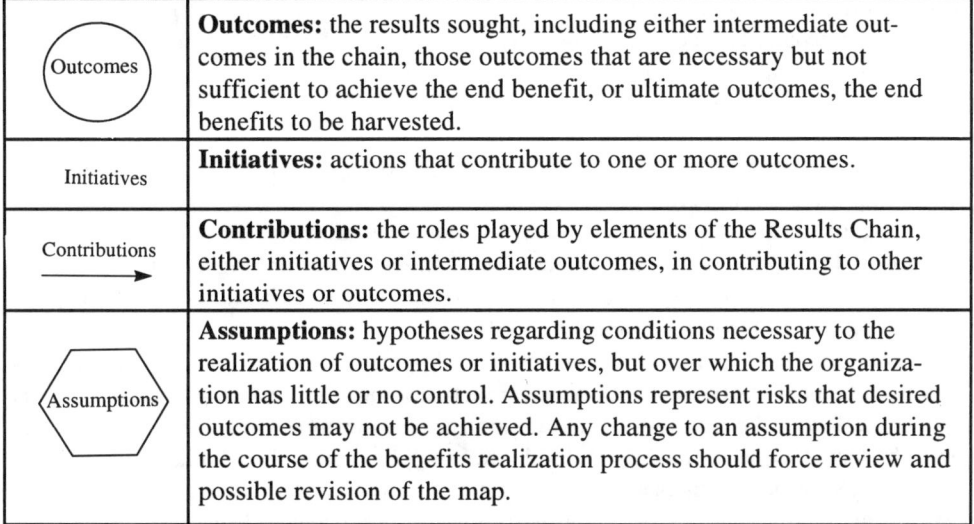

Outcomes	**Outcomes:** the results sought, including either intermediate outcomes in the chain, those outcomes that are necessary but not sufficient to achieve the end benefit, or ultimate outcomes, the end benefits to be harvested.
Initiatives	**Initiatives:** actions that contribute to one or more outcomes.
Contributions	**Contributions:** the roles played by elements of the Results Chain, either initiatives or intermediate outcomes, in contributing to other initiatives or outcomes.
Assumptions	**Assumptions:** hypotheses regarding conditions necessary to the realization of outcomes or initiatives, but over which the organization has little or no control. Assumptions represent risks that desired outcomes may not be achieved. Any change to an assumption during the course of the benefits realization process should force review and possible revision of the map.

Outcomes and contributions can be expressed in terms compatible with another widely used approach, *The Balanced Scorecard*, as described in the book of the same name by Robert Kaplan and David Norton. The Benefits Realization Approach and the Results Chain can be used to define Balanced Scorecard measurements. Unlike some implementations, where existing, and often irrelevant measures are simply regrouped into new Balanced Scorecard categories, the Benefits Realization Approach, supported by the Results Chain forces a rethinking of measurement, based on clearly defined business outcomes and contributions.

The Results Chain for a program is developed through a process of extensive interviews and workshops with business stakeholders. This process promotes discussion, consensus and commitment. It develops a shared understanding of the linkages between IT and other initiatives. Its power is in providing valuable insights through making implicit thinking explicit, and bringing hidden assumptions to the surface. This in turn facilitates communication, by getting people on the same page, and as a result, enables more informed and better decision-making.

To illustrate these points, let us review just one small fragment of a program model built using the Results Chain. It is based on a real life case of a printing firm that suffered from reduced sales. Customers were also complaining about the length of time it took to fulfill orders. The company felt that this problem was contributing to their declining sales, and that they needed to reduce their order processing cycle time. To achieve this, they decided to develop and implement a new order entry system. The initial, simple Results Chain for this case is illustrated in the figure below.

Using Results Chain terminology, the company undertook an *initiative* to develop and implement a new order entry system. The desired objective of the new order entry system was to reduce the time it took to process an order. This reduction was expected to *contribute* to reducing the order processing cycle, an *intermediate outcome*. The reduced order processing cycle was in turn expected to contribute to increased sales, the final *outcome*.

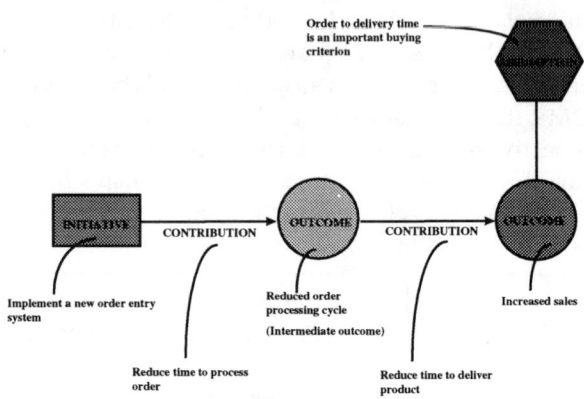

This expected contribution was also premised on the *assumption* that order to delivery time was an important buying criterion for customers.

In reality, the true Results Chain behind this case was much more complex. At the level of analysis shown, this represents a clear case of silver bullet thinking about how quickly the new order processing system can contribute to increased sales because it is unlikely that it alone will reduce the order processing cycle.

Using the Benefits Realization Approach, this model became the starting point for fleshing out other initiatives. It became apparent that some re-engineering of the order entry process itself would be required. The assumption around the impact of delivery time on sales had to be tested. Other initiatives including training, changing physical layouts, defining new roles and responsibilities and designing a new reward system were added. Along the way, further intermediate outcomes and assumptions were surfaced. These were all required for this to become a true blended investment program.

As you develop the Results Chain, different ways or paths to achieve the desired outcome may also be revealed. The Results Chain allows you to model and identify these paths. In combination with the value assessment technique described in the next section, a Results Chain lets you select the best path and to switch paths in response to changing conditions.

I recently had the opportunity to speak at a conference with Dr. Edward de Bono. Dr. de Bono's parallel thinking process, as described in his book, *Six Thinking Hats*, is another powerful tool that can be used in the development of the Results Chain. Its use can help ensure a balance between intuition and creativity, information gathering, and logical and critical assessment.

Lest the reader think that this is an easy process, it is not. Once these thoughts surface, there will be lively, often passionate debate, which will lead to tough decisions. It will be to the advantage of some, often many, to keep these thoughts hidden. One of the toughest things will be deciding what will not be done. In designing a program, and indeed in developing the strategies that programs support, it is as important, if not more important, to understand what you will not do, as what you will do. There will often be perceived winners and losers. Once surfaced and dealt with, however, the organization will be the ultimate winner.

The process will often also reveal different ways or paths to achieve the desired outcome. The Results Chain allows organizations to model and identify these paths. In combination with the four "ares" described later, a Results Chain enables selection of the best path and subsequently choosing different paths in response to new learnings and changing conditions.

Realizing all the benefits from a business program can take a long time, often a number of years. This does not mean that some benefits cannot be realized earlier. Indeed, if programs are to have any chance of success, they must be designed to achieve this. They must be broken into smaller pieces, what I call "doable chunks". They must be designed and managed to deliver measurable benefits in no more than 3 – 6 month periods.

In working with clients to develop Results Chains, I find that they have three remarkably consistent reactions:

"This is the first time I've been able to see how it all fit's together"

"I thought this was an IT project — most of the initiatives are not IT initiatives."

"I can now much more clearly assign accountability and measurements."

There is however, a major challenge at this stage. Once the Results Chain is presented, many managers think that the job is done. This is wrong — dead wrong! The Results Chain model should not be thought of as just an abstract map of business reality to be created and then filed away and forgotten. A Results Chain is not static. When completed, it becomes a living model of the benefits realization process, which should evolve over time as the context changes to reflect new benefits realization opportunities that should be tapped. It is living in that it can, and indeed should be continually revised to monitor and communicate progress and to assess the impact of changes over time. It is also living in that it can be modified to reflect changes in both investment programs and the business environment. The Results Chain accompanies an organization throughout the benefits realization process of a business program.

If organizations are to succeed, indeed survive in today's world of ongoing complex change, such a model is essential.

PORTFOLIO MANAGEMENT

The next challenge that organizations face when implementing the Benefits Realization Approach is assessing the relative value of programs, that is to say picking the winners. With the evolution towards the new economy, and the emergence of e-business, there are far more opportunities to invest in IT than there are resources to apply to those opportunities. The problem of selecting the right work to do is becoming increasingly difficult. For business decision-makers, it is a formidable challenge.

Traditional selection tools are one-off business cases, designed to support, or in many cases justify, simple go/no go decisions about major projects. These tools no longer reflect business and IT reality.

Today, we need a tool that does reflect reality. This is the portfolio view. A portfolio is a structured grouping of business investment programs selected by management to achieve defined business results, while meeting clear risk/reward standards. Portfolio management has been applied to financial investments for decades, enabling decision-makers to choose among increasingly numerous and complex options in a volatile environment. The time has come to apply this concept to manage business programs. This means looking at all potential investments in IT-enabled change, and picking and managing an optimum set of programs — the portfolio — to fit within an organization's capability to meet diverse and often conflicting demands.

As in the case of individual business programs, portfolios are not static. The term portfolio indicates the need to have a balance of opportunities to deliver the most value over time, and allowing for the vagaries of the future. The composition of the portfolio needs to be adjusted as we gain better knowledge of the investment opportunities, and to take into account changes in the business environment. With a stock and bond portfolio, financial planners look for a balance of investments to thrive in most environments. And they never stop monitoring the performance of that portfolio. The same active involvement is required for a portfolio of business programs.

The following is a typical series of steps organizations can undertake, with the aid of the Results Chain technique, to design and manage a portfolio of business programs:
- Categorize programs;
- Prepare value cases for business opportunity programs;
- Manage risk to increase value;
- Manage and leverage program interdependencies; and
- Determine, monitor and adjust portfolio composition.

One common reaction to Program Management is: "If we were to put all our programs through this process, we would never get anything done." This is absolutely correct. It would indeed take an unacceptable, probably unavailable amount of resource if we treated all programs the same way, and with the same level of rigour. Such an approach would not help in making good decisions and would prove very frustrating for all involved. Instead, we must deal with programs differently according to the degrees of freedom of management to make meaningful business decisions, and the nature of the programs in question. Effective portfolio management categorizes programs according to the types of decision that the portfolio managers can make and the nature of the investment. We apply less rigour to programs that are mandated, either by regulatory bodies or parent organizations, or programs where the linkage from the initiative to the end outcome is relatively simple. We apply much more rigour to programs that are truly discretionary, what we call business opportunity programs, or to programs where the linkage from the initiative to the end outcomes is very complex. For these programs, we need a new tool — the Value Case.

THE VALUE CASE

The traditional tool for making investment decisions is the business case. Unfortunately, the business case does not serve us well in today's complex and dynamic business environment. Traditional business cases are generally focused 90% on costs and 10% on benefits. The costs are usually fairly detailed, while the benefits are quite fuzzy. Once approved, how often are business cases ever revisited? Very rarely! Once through the hurdle-rate gate, the ROI gate, or whatever gate we had to get over for approval to proceed, we were off to the races. While there may be questions about functionality, budget and schedule, how often is the business alignment revisited? How often are assumptions revisited? Again, very rarely!

We need to replace the traditional business case with a new vehicle — the value case. The all-or-nothing bet of traditional project approval must be replaced with value cases that are operational tools, designed to allow continuous monitoring of programs through their entire life cycles. Value cases include the full life cycle costs of a program, but also put much more focus on the benefits than traditional business cases. They go beyond fuzzy statements of what is expected to clear statements of the expected outcomes. They include how the outcomes will be measured, and how we expect the various projects that make up a program to contribute to these outcomes.

When value cases replace conventional business cases to support program selection, organizations also need to embrace a new multidimensional view of business value. ROI remains important, but is not in itself more important than strategic alignment with the corporate vision, or an assessment of the risk that the expected benefits will not be realized. As we move beyond implementing technology to implementing change, risk increases greatly. Not the risk of project failure, but the risk of not being able to make or absorb all the change required to realize the benefits, or that the business environment could change. This risk must be identified, mitigated where possible, and proactively managed through the life of the program.

Again, unlike the traditional business case, value cases are not static, they are dynamic. As programs progress, we will learn more about what works and what doesn't. Some of the benefits that we had expected will turn out to be unattainable. New benefits will emerge. The business context, both internal to the organization and the external environment, will also change. Value cases are continually updated to reflect the current state of our knowledge. This topic is discussed further later in this chapter as part of the Full Cycle Governance topic.

The Results Chain is a powerful technique to understand and define business programs. It shows the desired outcomes and the possible paths that can be taken to reach those outcomes. It does not however, in and of itself, help you decide the relative value, including the opportunities and risks, of the various paths within a program, or of the potential programs within a portfolio. As such, it does not help you select programs – to pick the winners. Another technique is required to assist in gauging the odds of success for a specific investment program. This technique we refer to as the Four "Ares."

THE FOUR "ARES"

If organizations are to effectively answer the question: "Are our investments in information technology providing the greatest possible value?" they must ask four very basic questions—questions that as managers we all too often forget in today's complex and fast moving world. We must never lose sight of these four basic questions. We call these questions the Four "Ares".

There are many questions that need to be asked in developing business programs and trying to assess the relative value of paths and of programs. The four "ares" provide a structured framework to organize the questions and are supporting instruments. This provides more objectivity to the answers and facilitates comparable measurement of the answers.

The basic questions that make up the four "ares" are:

Are 1: Are we doing the right things? This question addresses the definition (or redefinition) of business, of business direction and the alignment of programs and the overall business investment portfolio with that direction.

Are 2: Are we doing them the right way? This question addresses organizational structure and process, and the integration of programs within that structure and process.

Are 3: Are we getting them done well? This question addresses organizational capability, the resources available and supporting infrastructure required to get work done efficiently.

Are 4: Are we getting the benefits? This question addresses the proactive management of the benefits realization process as a whole.

The four "ares" provide a solid structured framework for assessing value. To fully support the benefits realization process, the questions must be drilled down to a greater level of detail and incorporated into practical measurement instruments that allow consistent and therefore comparable measurement. Again, these instruments can be designed to support Balanced Scorecard measurements.

We have also found that a good practical extension of the four "ares" can be developed with more detailed questions and measurements along three dimensions: alignment, financial worth and risk. By risk, we do not mean here the traditional view of risk. The risk of a project failing. What we mean is the risk of the expected benefits of a business program not being realized. This is a very different and much more complex view of risk.

The Four "Ares"

"Alignment" — Are we doing the right things?
"Benefits" — Are we getting by?
"Integration" — Are we doing them the right way?
"Capability/Efficiency" — Are we getting them done well?

The most important thing that organizations must do if they are to master the benefits realization process is to ask the right questions, and to ask them over and over again. All too frequently, organizations rush forward blindly, basing their decisions on superficial answers to the wrong questions, and never revisit the questions except to lay blame. Taking the time to formulate and ask the right questions and continuing to ask them is critical to an effective benefits realization process

Tough questioning is critical to eliminating silver bullet thinking about IT. Asking the four "ares" helps to define the business and technical issues clearly, and thus to better define the roles and responsibilities of business executives and IT experts in the investment decision process. Are 1, Are we doing the right things? and Are 4, Are we getting the benefits? raise key business issues relating to both strategic direction and the organization's ability to produce the targeted business benefits. Are 2, Are we doing them the right way? raises a mix of business and technology integration issues that must be answered to design successful business change programs. Are 3, Are we getting them done well? directs attention to traditional IT project delivery issues, as well as to the ability of other business groups to deliver change projects.

There is a potential risk in asking all these questions. The reader may feel that there is a danger of falling into the trap of analysis paralysis. Unfortunately, some readers may even find this desirable. It is always easier to wait for the answer to just one more question before taking action. The answer to any question usually raises more questions, thus enabling action to be further deferred. Ultimately, the worst decision is no decision. Somewhere between analysis with no action, and action with no analysis we must find a balance. "Ready, aim, ready, aim" is no better than "fire, fire, fire." With a Benefits Realization Approach, "Ready, aim, fire" can and must be achieved without spending so much time on ready and aim that the target has long gone.

FULL CYCLE GOVERNANCE

If organizations are to realize the potential of program management and portfolio management concepts, and to harvest significantly increased benefits from IT-enabled business transformation initiatives, these key concepts need to be operationalized through full cycle governance.

There are a number of core components required to establish full cycle governance. These are:
- Value Cases;
- Stage gates;

Benefits Realization Approach

- Portfolio composition and program decision options; and
- Governance process and structure.

Each of these components requires significant change in how people think, manage and act.

Traditionally, investments have been managed as if their potential benefits do not change over time. With full cycle governance, organizations need to learn how to pilot business programs through a series of stage gates. These stage gates cover the life of a program, from initial design to the realization of all the potential benefits – that is from "concept to cash." At each stage gate, with an increasing level of rigour for each gate, the value case for the program is reviewed, using the four "ares" and supporting instruments. At each stage gate, there will be a number of program decision options: continue as is; continue with some changes; slow down or speed up; or stop the program. When the decision is taken to proceed, resources are progressively committed at each gate.

New governance processes and structures are required to support full cycle governance. A senior level Decision Board will be required to manage the composition of the portfolio. Business Sponsors will be required for programs. These sponsors must own the benefits of these programs. Programs also need Program Managers. They will require the skills to be able to manage complex, multi-disciplinary programs. The role of project managers will need to change to work within the program environment. In addition to all of this, organizations will need some form of what we call a Value Management Office (VMO). Benefits Realization is a new discipline. A function such as the VMO will be needed to promote the discipline, and to train, coach and counsel people in it. The VMO supports the Decision Board in making portfolio decisions, supports business sponsors in developing Value Cases, and administers and supports the stage gate process. The VMO also acts as a learning organization, refining program evaluation criteria and the overall governance process as experience is gained.

A possible structure, and the relationship between the various participants is illustrated below.

Implementation of full cycle governance requires people to think beyond the old go/no go project decisions. In today's world of continuous change, the reality is that when we get where we are going, it will rarely be where we thought we were going, and we often will not have got there the way we had initially planned. With the Benefits Realization Approach, and full cycle governance, program and project managers are encouraged to continually review and modify their initiatives in order to reduce risks and increase benefits. With full cycle governance, the decision to proceed with a program at a point in time is the right decision, based on the information available. At some future point, when more has been learned about the program, or the business environment has changed, a decision to stop the program is also the right decision. Nobody messed up. Nobody failed. The management process —full cycle governance— worked.

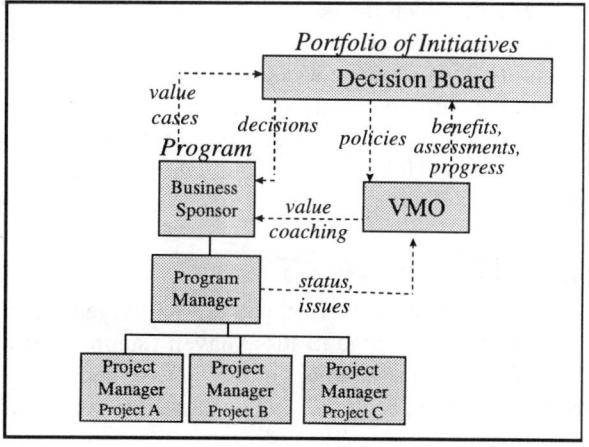

ACCOUNTABILITY

If full cycle governance is to work, and indeed if the benefits of business programs are to be realized, there must be clear and active accountability. It is also fundamental that this accountability be clearly aligned with the reward system. Without appropriate and clear accountability, full cycle governance will become no more than an experiment. It will create an illusion of progress, and take considerable time and effort, but will not result in the desired improvements to benefits realization.

There are a number of conditions that must be met if accountability is to be effective. These include:
- A clear mandate and scope;
- Clear lines of accountability;
- Relevant performance measures; and
- Alignment with the reward system.

Understanding the essence of activist accountability means changing people's industrial-age mindsets and getting them to accept the new form of outcome ownership. It means leaving behind the traditional passive approaches to business results still operating in many organizational cultures today.

A few years ago, I built a new house. Never having built a house before, my wife and I somewhat naively assumed that all we had to do was get a good architect. Work with that architect to develop the plans for the house, then hand the plans to a good builder and wait for our dream house to be built. Nothing could have been further from the truth. We did indeed find a good architect. We did develop good plans, and we did have a great builder. However, had we not been on site, almost every day for three to four months, my wife and I would not today have the house that we wanted. This was not because the architect messed up, the builder messed up, or we messed up. It was because as building progresses, as you see and learn more, ideas change, and new questions arise. Now, if we had built this house the way most organizations build technology systems what would we have done? We would indeed have handed the plans to the builder and gone away until the house was built. We would then have told the builder what was wrong with the house, fired him, and if we lived south of the 49th parallel, sued him. Would this have been fair? I don't think so. Last time I looked we were the owners of the house. Had we acted this way, we would have abdicated our responsibility as owners.

Accountability is about ownership. If we accept that we are no longer simply implementing technology, but implementing business change programs, it is clear that the business must own the benefits from such programs. The business must be involved and accountable. The IT function is still accountable for the delivery of the technology capability, but cannot be accountable for the benefits. It is the business that has to do 85 – 95% of the work required to realize the benefits, and who must therefore be accountable. This cannot however be a finger pointing type of accountability. Only when the business owners and the IT function work together as a team in true partnership will the full benefits be realized.

MEASUREMENT

"If you can't measure it, you can't manage it." Measurement is key. But measurement is not simple. Many organizations haven't stepped up to the measurement challenge because they fall back on available, but ineffective measures. The Benefits Realization Approach requires relevant, accurate, and consistent measures of the performance of each program, and of the projects within them. Without an appropriate measurement system, full cycle governance, portfolio management, program management and, as a result, benefits realization, will again be no more than a pipe dream. Clear and relevant measurement is another cornerstone for full cycle governance and the Benefits Realization Approach.

You must determine what to measure and when to measure it. The criteria for designing effective measurement systems are:
- Make sure measures exist;
- Measure the right things;
- Measure things the right way; and
- Make sure measurement systems guide decisions and action.

Recently, a manager told me that the problem with measurement was that accounting systems measure the wrong things. Well, to the extent that accounting systems measure the things that interest accountants, he may be right. But that is not the point. The real issue with measurement is that we don't take enough time at the beginning to determine the business outcomes that we expect, the intermediate outcomes that are required to achieve these business outcomes, or the contribution we expect from the various projects that make up a program. If we don't do this, how can we possibly know what to measure, and by implication, how will we ever know when to declare victory. The Benefits Realization Approach forces organizations to do this thinking up front — to focus on outcomes and how to measure them.

The fundamental concepts of benefits realization help organizations deal effectively with the issue of measuring value in four important ways:
- Identify the **outcomes** to measure, and how to measure them;
- Show the reasoning about the **linkages** relating programs and projects to outcomes, making it easier to understand what's going on;
- Make **measurement** come alive by clearly tying accountability to measured results; and
- **Take action** based upon measurements through full cycle governance.

The Results Chain helps here. A simple rule is that each outcome in a Results Chain must be described in a way that forces measurement, using a phrase containing relatively precise language. The acronym MEDIC represents the following:

M: Maintain – e.g. a level of service maintained;
E: Eliminate – e.g. a function eliminated;
D: Decrease – e.g. turnaround time decreased;
I: Increase – e.g. revenue increased;
C: Create – e.g. a certain capability created.

These terms are far preferable to the more vague terms: improved, better, and enhanced. If it is not measurable, then you can't know if it has been achieved. The essential point about measurement is that, by definition, it involves quantification in some form.

Again, it is particularly in the area of measurement that the Benefits Realization Approach can be aligned with the Balanced Scorecard. The Benefits Realization Approach, supported by the Results Chain, forces a rethinking of measurement, based on clearly defined business outcomes and contributions that can be described in Balanced Scorecard terms.

PROACTIVE MANAGEMENT OF CHANGE

I have mentioned the need to include the process of managing change in program definition. Organizations will only realize benefits through change and, equally, change will only be sustained if benefits are realized, and seen to be realized. The Benefits Realization Approach requires people to change how they think, manage and act. These will be difficult and often painful changes, and they will not happen by themselves.

We are all familiar with the expression "walking your talk." If you are to walk your talk, you must first understand the full extent of a planned change. The figure below illustrates that before you can walk your talk, you have to understand what you are talking about. It is one thing to announce that change has to happen. It is quite a different thing, and a precursor to any action, to understand the implications of what you are saying — to understand the full extent of the change you are proposing.

At the level of thinking, or the cognitive level, we become aware of a need. This often translates itself fairly rapidly into talk: "We at Thorp Inc. have to make fundamental changes to our organization." Unfortunately, the nature of those changes is rarely understood, and the definition of them is usually delegated, or more accurately abdicated. The normal reaction to this is some variation on "Here we go again," "I've only got three years to go," or "This too will pass," and, all too often, it does.

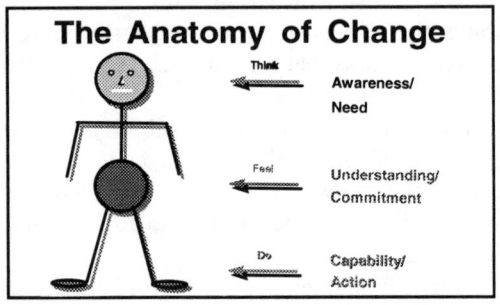

It is only when we wake up at three in the morning, reaching for the antacid, as our stomach churns with the realization of the implications of the change and the extent of what has to change, that we can begin to reach understanding. This is the precursor to commitment.

The use of the Results Chain as described previously can bring you to an earlier awakening. By earlier I don't mean two in the morning or one in the morning. I mean that you will be forced to understand the extent of the change that is planned much earlier in the life of a program. When we have the understanding necessary to build commitment, to fully understand the scope of what we are committing to, then, and only then, can we act with any reasonable chance of success. Even then, we can act only if we have the capability and the capacity to do so. Understanding the full extent and impact of a proposed change allows us to make this assessment.

Our world, our businesses, our communities, our families and indeed our own lives are becoming more complex. We are faced with complex interrelationships that are difficult, if not seemingly impossible to understand. We seek simplicity. We seek quick and painless silver bullets. In working with a client recently, I had suggested that they apply the Benefits Realization Approach to a major change program that they were embarking upon. Their reaction was that the approach was too complex. They were looking for something simple. Unfortunately, in our complex world, there are few simple solutions. If we continue to exhibit silver bullet thinking in this way, we will continue to fail. If we are to make things simpler, we must first understand the complexity that we are dealing with. Only when we understand complexity can we then simplify. The Benefits Realization Approach, and specifically the Results Chain, forces an early understanding of the complexity of what we are about to undertake.

Change management is still in many ways more an art than a science. While there are a number of different methods, techniques and tools, these must be regarded as means to an end, not ends in themselves. Organizations, groups and individuals are all different. Any approach to change management must be adapted to handle those differences. There is no one right way to manage change. Many change efforts fail because we are more interested in religiously following the method, than adapting it to organizational realities. In benefits realization terms, there is too much focus on the change management initiative, than on the outcome we desire the change to achieve.

Any approach to change management must be adapted to the needs of a specific organization. We must plan the change and always be prepared to change the plan. Most of all, we must have the vision and the commitment to stay the course.

THE EVOLUTION OF BENEFITS REALIZATION

We developed the Benefits Realization Approach in response to the problem of realizing value from IT investments. In the two years since our book, *The Information Paradox,* was published, we have found the problem to be, if anything, worse than we had thought. With the emergence of e-business applications, if we continue to do things as we have in the past, the potential for costly failure is high. While we have found a growing awareness of the problem at senior levels of organizations across the world, there is less understanding of the causes of the problem. In many cases, there is even less willingness to undertake the significant change that is required to address it. Many organizations are still looking for the "silver bullet." This could prove a costly and risky pursuit.

Organizations will not deal with the underlying issues that contribute to this problem if they continue to seek a "silver bullet," or if they continue to focus on IT alone. This is not a technology problem. It is not a problem that will be fixed by focusing on IT in isolation. It is a business problem. If they are to deal with the problem, Organizations must take a hard look at their overall governance process – from strategic planning through to program execution.

We hear all too often today that with business moving at "Internet speed", there is no time for strategic planning. I do not wholly agree with this. I do agree that we cannot do traditional strategic planning the way we have done in the past. I do not, however, believe that we should just stop doing it. To do so would result in organizations continuing to waste large amounts of money on failed e-business projects. They will miss out on significant business opportunities, and in many cases, cease to exist. What we must do is to approach strategic planning differently — very differently.

It is becoming increasingly clear to me, and to a number of organizations that I am working with, that the Benefits Realization Approach provides the missing link between business strategy and the business/IT projects that an organization undertakes. Extending the Benefits Realization thinking opens up the opportunity for a new form of business governance, one that I call *Strategic Governance*.

Many organizations today have little or no linkage between investment projects they undertake and their business strategy. We see the results of this lack of linkage in a number of ways:
- Business strategies are stated in "fuzzy" or motherhood terms;
- Desired business outcomes are neither clear, nor tracked;
- Results & measurements are unclear;
- Sponsorship & accountability are unclear;
- There is no context for overall governance;
- There are multiple steering committees, with little inter-program communication; and
- There is a lot of "finger-pointing" between various parts of the business, particularly between the IT function and other functions;
  It results in something like this at right:
- A Benefits Realization Approach, with the program and portfolio view, provides the key missing link in bridging the gap between projects and business strategy.

This is illustrated at right.

With the appropriate governance structure and processes in place, the linkage that the Benefits Realization Approach provides will enable organizations to be responsive to inevitable and constant changes in business strategy. It will enable them to move beyond traditional strategic planning and management methods, which are inadequate for the new economy, to strategic governance. A strategic governance process is dynamic, and continually manages the alignment between business strategy, the portfolio of programs, individual programs, and the projects, both business and IT projects, that make up the individual programs.

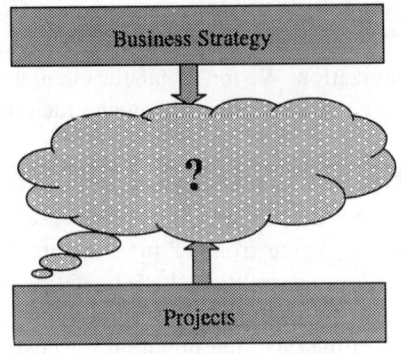

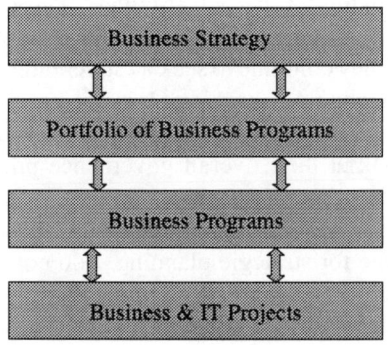

Strategic Governance

THE LEADERSHIP CHALLENGE

Earlier, I cautioned the reader not to think that adopting a Benefits Realization Approach is easy. For those who subscribe to the schools of "management by magazine," or "management by fad," this is not for you. It is easier to adopt solutions without understanding the problem. It is easier to build systems in the belief that "if we build it they will come." It is easier to assume that the system alone will deliver the desired benefits. Tragically, in summary, it is easy to fail. It is a much tougher job to succeed.

Realizing business value from investments in IT-enabled change is not a technology challenge — it is a business challenge. This is not just a challenge for the CIO and the IT department, it is a challenge for all business managers, from the CEO down. Business managers, including IT managers, must recognize that they all have a responsibility for successfully managing investments in IT-enabled change. It is only with this recognition that organizations will be able to successfully leveraging the full potential of information technology.

A Benefits Realization Approach is about making the right decisions about the right things, and then managing the decisions you make. A client of mine once said: "A lot of consultants say the same thing: start with the end in mind." A Benefits Realization Approach does not just start with the end in mind; it manages value continually with the end in mind. It's a process that goes right through to the full realization of value — that is "from concept to cash."

Adopting a Benefits Realization Approach involves a long-term, sustained change effort in how organizations think, manage and act. New processes and organizational structures will be needed to operationalize the new mindset. Major changes will be required in the areas of accountability, measurement and the process of change itself.

This change won't happen overnight, but it can and should start tomorrow. Every business has a project or program on the drawing board, underway, or in trouble that can benefit by applying such an approach. The lessons learned will serve them well in the ongoing challenge of reinventing themselves to compete for future leadership of their industries.

The need to learn how to better manage investments in IT-enabled change has never been greater. Organizations that recognize the business imperative to rethink their approach to IT investments, and that accomplish this shift, will be the winners in the new knowledge economy. Those that don't will be history!

REFERENCES

Anderson, C. (1997). Survey of Electronic Commerce: In Search of the Perfect Market. *The Economist*, May.

Arthur, B.W. (1996). Increasing Returns and the New World of Business. *Harvard Business Review*, July-August, 101-109.

A.T. Kearney Inc. (1997). The Growing Impact of Strategic Information Technology on the CEO Agenda (Chicago World Headquarters). [On-line]. Available HTTP: http//www.atkearney.com

Austin T. (1996). *Difficult Design Decisions in the Imperfectly Connected Real World*. Research Note; Client/Server (CS); Gartner Group, November 21.

Berg T. (1995). *The Business Value of Electronic Commerce, Part 1*. Research Note; InSide Gartner Group this Week (IGG); Gartner Group, October.

Brynjolfsson, E. (1993). The Productivity Paradox of Information Technology: Review and Assessment, *Communications of the ACM,* December, 36-12: 66-67. [On-line]. Available HTTP: http/ccs.mit.edu/CCSWP130/CCSWP130.html

Brynjolfsson, E. and Hitt, L. (1995). Information Technology as a Factor of Production: The Role of Differences Among Firms. *Economics of Innovation and New Technology, 3:* 183-200.

Byatt, R. (1997). IT and FM Outsourcing; A Perfect Fit. *Facilities Design and Management*, May, 16-5: 30.

Campbell-Kelly, M. and Aspray, W. (1996). *Computer: A History of the Information Machine.* New York: BasicBooks.

Compass Analysis (1996). The Compass IT Strategy Census 1996. Compass Analysis Canada Ltd. (Montreal). [On-line]. Available HTTP: http://www.compass-analysis.com

Compass Analysis (1997). Measuring Value in IT. Compass Analysis Canada Ltd. (Montreal). [On-line]. Available HTTP: http://www.compass-analysis.com

Dulaney, K. (1996). *Mobile Cost of Ownership: Higher Costs for Bigger Benefits.* Research Note; Equipment Asset Management Europe (EAME); Gartner Group, December 12.

Gibbs, W.W. (1997). Taking Computers to Task, *Scientific American,* July.

Green, K. (----). *Software Development Projects and Their Difficulties.* Cambridge, UK: Fitzwilliam College. [On-line]. Available HTTP:
http://www.fitz.cam.ac.uk/~kg201/essays/software.html

Halberstam, D. (1986). *The Reckoning.* New-York: Avon Books.

Harris, C.L. (1985). Information Power: How Companies are Using New Technologies to Gain a Competitive Edge. *Business Week,* October 14, 108.

Hubbard, D.W. (1997). Risk vs. Return. *Informationweek*, June 30, 637. [On-line]. Available HTTP: http://techweb.cmp.com/iw/637/37iursk.htm

Kaplan, R.S. and Norton, D.P. (1996). *The Balanced Scorecard: Translating Strategy into Action.* Boston: Harvard Business School Press.

Kurtzman, J. (1997). An Interview with Paul M. Romer. *Strategy & Business,* Issue 6, 91-101.

Lacity, M. and Sauter, V. (----). *Why General Managers Need to Understand Information Systems.* Saint-Louis: University of Missouri. [On-line]. Available HTTP: http://www.umsl.edu/~lacity/whymis.html

Lehr, B. and Lichtenberg, F. (1996). *Computer Use and Productivity Growth in Federal Government Agencies, 1987 to 1992.* New York: Graduate School of Business, Columbia University.

Lehr, B. and Lichtenberg, F. (1997). *Information Technology and Its Impact on Firm-Level Productivity: Evidence from Government and Private Data Sources, 1977-1993.* New York: Graduate School of Business, Columbia University.

Licht, G. and Moch, D. (1997). *Innovation and Information Technology in Services.* Mannheim, Germany: Center for European Economic Research. [On-line]. Available HTTP: http://www.csls.ca/conf-pap/licht4.pdf

Lichtenberg, F.R. (1995). The Output of Contributions of Computer Equipment and Personnel: A Firm-Level Analysis. *Economics of Innovation and New Technology,* 3: 201-217.

Lillrank, P., Lehtovaara, M., Holopainen, S., and Sippa, S. (1996). *The Impact of Information and Communication Technologies (ICT) on Business Performance.* [On-line]. Available HTTP: http://www.interactive.hut.fi/ict-bp/

Magee, F. (1996). *Failure Management: Get It Right the Third Time.* Gartner Group, July 25.

Markus, M.L. and Benjamin, R.I. (1997). The Magic Bullet Theory in IT-Enabled Transformation. *Sloan Management Review,* Winter, 55-68.

McKague, A. (1997). Only IS Gets Away with Such Poor Service. *Computing Canada,* October 10, 22-21: 15.

Newsbytes News Network. (1997). April 14. Global IT Spending Soars-European Report.

Nulden, U. (----). *Escalation in IT Projects: Can We Afford to Quit or Do We Have to Continue?* Sweden: Department of Informatics, Goteborg University. [On-line]. Available HTTP: http://www.adb.gu.se/~nulden/Research/Esca/Esca.html

Nulden, U. (----). *Failing Projects: Harder To Abandon than To Continue.* Sweden: Department of Informatics, Goteborg University. [On-line]. Available HTTP: http://www.adb.gu.se/~nulden/Research/Proj/Proj.pdf

OASIG. (1996). *The performance of Information Technology and the role of human and organizational factors.* Report to the Economic and Social Research Council of UK (Version 1.0) [On-line]. Available FTP: ftp://ftp.shef.ac.uk/pub/uni/academic/I-M/iwp/itperf.doc

Peltu, M. (1996). *Minimising the Risks of ICT Failures Having Catastrophic Consequences.* IEE Colloquium on Human, Organization and Technical Challenges in the Firm of the Future (Digest #1996/050/P 6/4). London, England: The Institution of Electrical Engineers.

Prince, C.J. (1997). IT's elusive ROI. *Chief Executive,* April, 122: 12-13.

Rai, A., Patnayakuni, R., and Patnayakuni, N. (1997). Technology Investment and Business Performance. *Communications of the ACM,* July, 40-7: 89-97.

Reimus, B. (1997). The IT System That Couldn't Deliver. *Harvard Business Review,* May-June, 22-35.

Rifkin, G. (1989). CEOs give credit for today but expect more tomorrow. *Computerworld,* October 31, 21.

Sauer, C. and Yetton, P.W. (1997). *Steps To the Future: Fresh Thinking on the Management of IT-Based Organizational Transformation.* San Francisco: Jossey-Bass Publishers.

Scott-Morton, M. (1991). *The Corporation of the 1990s: Information Technology & Organizational Transformation.* New York: Oxford University Press.

Senge, P.M. (1990). *The Fifth Discipline.* New York: Currency Doubleday Publishers.

Stewart, B. (1996). *Enterprise Performance through IT.* Gartner Group, Gartner IT Expo, Florida, Conference Paper.

Strassmann, P.A. (1997). Will Big Spending on Computers Guarantee Profitability? *Datamation,* 43-2: 75-85.

Strassmann, P.A. (----) *The Squandered Computer: Evaluating the Business Alignment of Information Technologies.* New Canaan, Connecticut: The Information Economics Press. [On-line]. Available HTTP: http://www.strassmann.com

Violino, B. (1997). Return on Investment. *Informationweek,* June 30, 637: 36-44.

Willcocks, L. and Griffiths, C. (1994). Predicting Risk of Failure in Large-Scale Information Technology Projects. *Technological Forecasting and Social Change,* 47: 205-228.

Woodall, P. (1996). Survey of the World Economy: The Hitchhiker's Guide to Cybernomics. *The Economist,* September 28.

Zuboff, S. (1988). *In the Age of the Smart Machine: The Future of Work & Power.* New York: Basic Books.

ACKNOWLEDGMENTS

This chapter is adapted from, and builds on *The Information Paradox*, published by McGraw Hill in 1999. The Benefits Realization Approach is a service offered by DMR Consulting. The DMR ResultsChain is a registered trademark of DMR Consulting.

Chapter III

The IT Evaluation and Benefits Management Life Cycle

Judy McKay and Peter Marshall
Edith Cowan University, Australia

INTRODUCTION

It appears that somewhat of a dichotomy exists in many contemporary organisations with respect to the question of investment in information and particularly in information technology (IT). On the one hand, discussions of the new information-based economy and the promise of the new e-business domain leads inevitably to enormous faith being placed in IT, or perhaps more accurately, on the critical, appropriate utilisation of IT to deliver business benefits. Such faith is illustrated by quotes such as: "Across all industries, information and the technology that delivers it have become critical, strategic assets for business firms and their managers" (Laudon and Laudon, 2000). But such enthusiasm is tempered by another view or concern that IT is not delivering on its promises, that it is *"oversold and undelivered"* (Earl, 1994), and that demonstrating the business value of IT investment is difficult in many instances. This concern that managers do not perceive that they are deriving value for money when it comes to IT investments is troubling when information and IT are often presented as the very backbone of the new economy. Such cynicism is reflected in quotes such as: *"There are many different ways to ruin a company. Speculation is the fastest, IT is the most reliable"* (Kempis et al., 1999).

Why do we experience such conflicting attitudes? Why is there so often a gap between aspirations with respect to IT, and the reality of IT implementations in many organisations? More importantly, can a sensible way forward be found, such that managers can develop greater confidence in their IT investment decisions?

Recent research suggests that an alarming proportion of companies (49%) are underperforming in both dimensions of efficiency and effectiveness of IT utilisation (Kempis et al., 1999). Yet in many organisations, investment in IT represents a large proportion of capital outlay, and indeed, IT expenditures often represent the fastest growing category of investment for the organisation (Strassmann, 1997). Thus it seems reasonable to conclude that IT assets (in terms of computer hardware, software, telecommunications facilities and human knowledge capital) are very significant, and therefore entitled to thoughtful management and careful attention to their value and contribution, and return to the organisation (Willcocks, 1994). However, concerns are all too frequently voiced by senior management about the size of their firm's investment in IT, and more specifically, about whether the firm enjoys adequate returns on this investment (Willcocks, 1996). For example, there is some evidence which suggests that large-scale IT deployment has resulted in replacing old problems with new ones, and that overall, introducing IT can be a huge disappointment since unexpected difficulties and failures are regularly encountered and expected business benefits are frequently not realised (Hochstrasser and Griffiths, 1991; Serle, 1994). Furthermore, several studies point toward fairly static productivity in business despite the rising IT expenditure (Attwell, 1996; Brynjolfsson, 1993; Cavell, 1997; Hochstrasser, 1993; King, 1996; Lillrank et al., 1998; Rai et al., 1997; Sutherland, 1994), giving rise to the notion of a 'productivity paradox' with respect to IT, and suggesting that despite large investments in IT over many years, it has been difficult to determine where the IT benefits have actually occurred, if indeed there have been any (Smyrk, 1994; Willcocks and Lester, 1997). The situation remains somewhat confusing for senior management, as there are conflicting results from research conducted in this area. While arguments are expressed suggesting that IT investment produces negligible benefits, (Weill and Olsen, 1989; Serle, 1994; Strassmann, 1997), there are also views expressed suggesting a marked positive correlation between IT investment and organisational performance (Bender, 1986; Delone and Weill, 1990).

Further research will no doubt help to clarify the situation. However until such time, management faces some real dilemmas with respect to IT. Firstly, for competitive reasons, organisations can rarely exercise a choice *not* to invest substantially in IT, even when economically they cannot find sufficient justification, and current evaluation practice cannot provide strong grounds for making the investment. Secondly, as IT infrastructure becomes an inextricable part of the organisation's processes and structures, it becomes increasingly difficult to separate out the impact of IT (both positive and negative) from that of other assets and activities. Thirdly, it would appear that comparatively few senior executives feel that they understand IT adequately, despite high levels of expenditure (Sprague and McNurlin, 1993; Willcocks and Lester, 1997). The conclusion must be drawn, therefore, that despite misgivings about return on investment and limited understanding, senior management continues to feel pressured into significant investment in IT (McKague, 1998).

A number of reasons can be posited as to why there are concerns and perceptions of an inadequate rate of return on investment in IT. Firstly, it could be that there has been an inappropriate investment in and use of information, information systems (IS) and IT in organisations, and hence concerns about the value of such investments. One often cited example of this stems from a failure to link IS/IT investments with business objectives and strategy initiatives (Edwards et al., 1995; Hochstrasser and Griffiths, 1991). Alternatively, it could be symptomatic of a lack of, or ineffective, business and/or IS/IT planning. Over time, a failure to achieve alignment of IS/IT strategies and business strategies would be argued to contribute to disappointing perceptions of IT's contribution to business performance.

Secondly, it could be that current evaluation processes are either inadequate (or nonexistent in some organisations), or that inappropriate evaluation techniques are being used (Willcocks and Lester, 1997). Perhaps a lack of confidence in the tools available leads to less than satisfactory practices. Nonetheless, if evaluation practice and procedures are inadequate, this may lead to calls for improved tools, and improved practice. Indeed, this has been the case in the IS literature (Remenyi et al., 1997), and in recent years, a proliferation in the nature and number of tools available for evaluation of IT investment has been witnessed (for example, the Balanced Scorecard (Kaplan and Norton, 1996; Olve et al., 1999), IT Investment Mapping (Peters 1994; 1996) and the Evaluation Life Cycle (Willcocks and Lester, 1997). This would be hoped to lead to improvements in practice, and for managers to be endowed with much better information as to the economic viability of an IT investment proposal. However, there could be dangers with this type of approach. It seems unlikely that the development of more appropriate methods, tools and techniques for evaluating IT investments alone will be sufficient to change practices and perceptions without being accompanied by substantial changes in managerial practices as well.

Thirdly, it may also be that an inadequate rate of return on IT investments arises because there are inadequate managerial procedures put in place to ensure the realisation of benefits from IS/IT (Ward et al., 1996; Remenyi et al., 1993). Expected benefits are nearly always identified pre-investment for new systems and technology, but rarely are proactive behaviours adopted and changes made to support the post-implementation realisation and evaluation of these anticipated benefits (Thorp, 1998).

Arguably, therefore, there are at least three key issues which will impact upon perceptions of the value of IT investments:
- that appropriate levels of business and IS/IT planning are undertaken, with the express aim of ensuring that proposals and priorities for IT investment are aligned with corporate visions, strategies, and objectives;
- that wide-ranging, qualitative and quantitative evaluation procedures and techniques to assess performance on a range of measures are adopted throughout the life cycle of IS/IT, and that the outcomes of this evaluation are actively fed into managerial decision making and action about ongoing investment in that IS/IT; and
- that organisations implement explicit procedures to ensure that adequate pre-investment consideration of benefits anticipated from IS/IT is undertaken, and more importantly, that post-implementation of that IS/IT, procedures are put in place to deliberately ensure that anticipated benefits are actively realised and managed over time.

The chapter will describe and analyse a framework to achieve adequate linkage between IS/IT planning, evaluation of investments on an ongoing basis, and also active realisation of benefits to the organisation over time. This framework is called the IT Evaluation and Benefits Management Life Cycle, and shows how to integrate planning, evaluation and benefits management activities. We would argue that this mix of planning, evaluation and benefits management is vital, as each of these components adopts a somewhat different (albeit important) focus on the other, and the position adopted in this chapter reflects our belief of a need to meld or simultaneously juggle these three perspectives if more effective utilisation of the IT resource is to occur.

BACKGROUND

In his 1994 paper, Earl outlined a progression of increasingly mature and sophisticated thinking with respect to IT utilisation in organisations. Somewhat simplistically, Earl's argument is captured below in Figure 1.

Earl (1994) seemed to be arguing for a move from the "IT is good" mindset, to one that recognised (and practiced) the need for IT investments to be derived from clearly articulated business need(s). Indeed, it could be argued that this type of thinking underpinned much of the work with respect to Information Systems Planning (ISP) that occurred during the early 1990s. Thus it was not uncommon to read that "business strategy indicates what top management are trying to accomplish...the IS/IT strategy is derived from the underlying business strategy" (Peppard, 1993). For the purposes of this chapter, the definition of ISP from Wilson (1989) will be used, when he writes that ISP "brings together the business aims of the company, an understanding of the information needed to support those aims, and the implementation of computer systems to provide that information. It is a plan for the development of systems towards some future vision of the role of IS in the organisation."

Improving ISP was thus viewed as a serious concern for non-IT and IT managers in industry (Galliers et al., 1994), and much of the focus of ISP was in successfully achieving alignment between business imperatives and IT investments. Methods, tools and techniques were articulated to support this focus (see, Ward and Griffiths, 1996; Tozer, 1996; Earl, 1996). While there was a great deal of sophistication with respect to the argumentation and approaches articulated, at times there seemed to be an assumption implicit that desirable outcomes would be achieved if only alignment could be achieved. Thus, in terms of our diagram in Figure 1, moving beyond stages 1 and 2, and embracing the thinking and actions implied by stage 3, seemed to be a way of overcoming disappointments with respect to IT investments; as a failure to achieve satisfactory linkages between business and IT initiatives has been cited as a contributing factor to a perceived lack of business benefits from IT (Edwards et al., 1995). Organisational reality, we would argue, has proved to be much more complex than this, and that other important factors serve to mitigate the delivery of benefits from IT to the organisation.

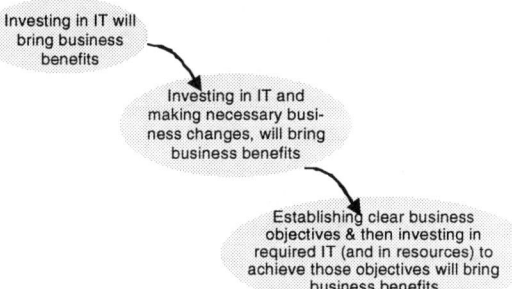

Figure 1: Increasing sophistication with respect to IT

One such factor that seems to have emerged is that of evaluation of IT investments. For the purposes of this discussion, a managerial perspective is adopted in defining IT evaluation as "about establishing by quantitative or qualitative means the worth of IT to the organization" (Willcocks, 1992). Concerns have been voiced, however which suggest that the most frequently used approaches for IT evaluation such as cost-benefit analysis, may be unsuited to application to some IT projects, and hence may fail to reveal benefits that have been derived from a particular investment (Willcocks and Lester, 1997). In addition, some research indicates that formal IT evaluation processes occur all too infrequently in many organisations (Farbey et al., 1993), and that formal evaluation is too often limited to project management-type measures of success (Willcocks and Lester, 1997), and that inadequate or no evaluation is carried out in a number of cases (Farbey et al., 1993).

Given these and other concerns about IT evaluation, and in particular that considerations about the worth of an investment needing to be more dynamic in nature and take account of changing requirements throughout the whole life cycle of an information system, rather than being based simply on pre-investment assessments and project management metrics, Willcocks and Lester (1997) proposed an evaluation life cycle. While IT evaluation on its own typically identifies costs (with a view to establishing some sort of control) and benefits as the counterbalance to costs in an attempt to justify the investment, the evaluation life cycle attempts to bring together a diverse set of methods and approaches to evaluate the entire extended systems development life cycle from a number of differing perspectives. Thus the concept of the evaluation life cycle is that evaluation should become an ongoing component of IT management, from planning, through systems development (or acquisition), operations, until finally decisions are required on when to kill off an IT investment. A range of interlinked measures are proposed which take into account the diversity of the potential benefits and costs of an IT investment (see Figure 2).

One of the strengths of this type of approach is its recognition that notions of cost and value are not static, but rather change throughout the life of a particular investment. Secondly, there is a recognition that IT may contribute to an organisation in ways other than that which is easily taken into account by traditional financially-based measures. Thus, the evaluation life cycle encourages evaluation from a customer or use user perspective, from a learning perspective, and so on. It may help in cases where financial justification is hard to make, but where other nonfinancial or intangible benefits suggest that the investment overall would be beneficial to the organisation. It also seems to promote a view of evaluation not as a static snapshot of worth, but as something which needs to permeate management thinking and reflection, and motivate decision making and action almost on a day-to-day basis. Evaluation thus viewed becomes part of a management culture, rather than a highly politicized, legitimizing activity. It is our view that the contribution of Willcocks and Lester (1997) helps to move our thinking with respect to IT management to another level of sophistication, as presented in Figure 3.

One difficulty with all evaluation is that while it may be helpful, indeed essential, to the identification of costs and expected or perceived benefits from a particular perspective, it does little to implement processes and procedures to ensure the management and realisation of those benefits over time. Hence we see the emergence of benefits management approaches which typically institute procedures to ensure the realisation and management of expected benefits throughout the

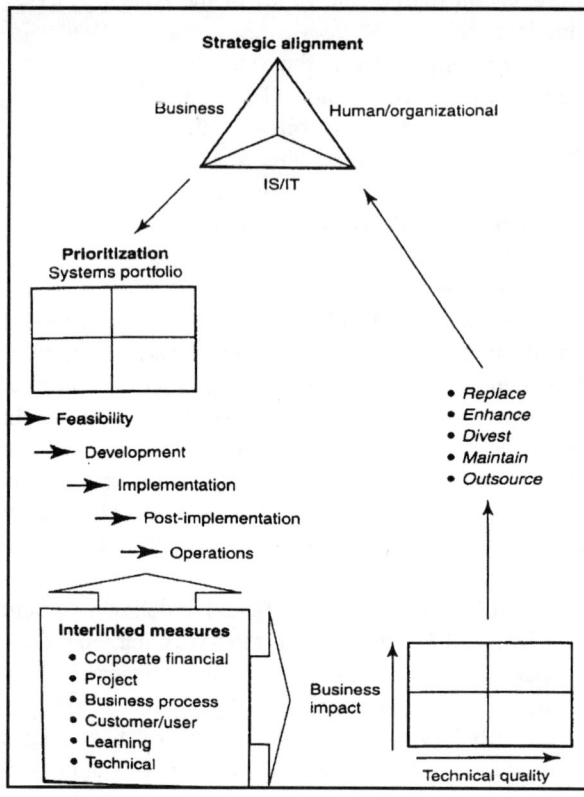

Figure 2: The IT evaluation life cycle (Willcocks and Lester 1997)

Figure 3: Linking business needs, IT investments and evaluation

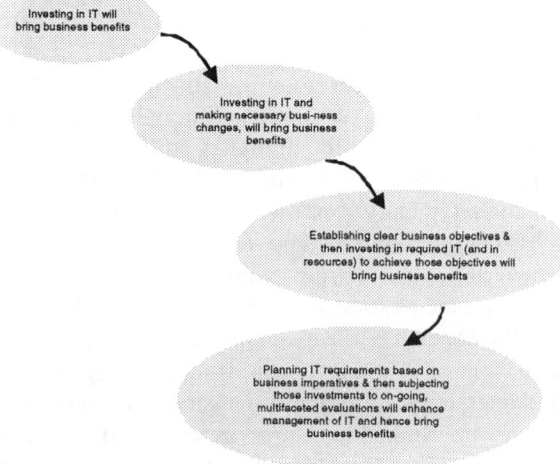

life cycle of an IT investment (Remenyi et al., 1993).

Thus procedures for the active realisation of the benefits from IS/IT investments should, together with procedures for the evaluation of such investments, be built into the routines and rituals organisations, enabling an informed adaptive response to the problems of achieving ongoing value from IS/IT investments. An ongoing programme of IT evaluation and benefits management very naturally closes the loop on the careful evaluations, reviews and adjustments that typically take place before IS/IT investments are committed to, never to be repeated or followed up as systems are developed, implemented and move into operations and maintenance phases. We believe that such a programme is a natural outcome of a fully-fledged business oriented view of the application of IS/IT.

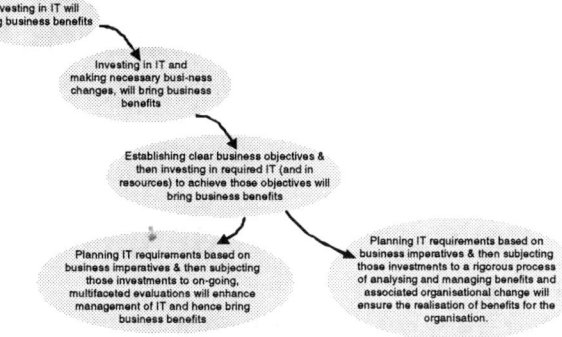

Figure 4: The need for active management of business benefits from IT

Adoption of a benefits management perspective, therefore, will move us to yet another stage of sophistication with respect to management of the IT resource in an organisation, as captured in Figure 4.

Benefits management approaches excel at identifying and managing the achievement of benefits but have few explicit means for linking these procedures to ongoing decision making about further investments needed for modifications and enhancements, or actions to terminate, divest or outsource the investment, for example. Thus, consideration of Figure 5 leads to a conclusion that there is a need to bring together evaluation and benefits management into an integrated, seamless approach to thinking and acting with respect to IT in organisations. Whereas IT evaluation is concerned with methodologies and processes used to measure the costs and the potential and/or achieved benefits from IS/IT investments, benefits management is concerned with the management and delivery of actual IS/IT benefits to the organisation. However, there is a need to merge these two approaches into a single and effective evaluation and benefits realisation approach, in order to reduce the inconvenience of going through the separate processes for evaluating, managing, and realizing the benefits from IS/IT investments.

This approach is aimed at ensuring that the organisation would properly plan and evaluate its IS/IT projects, while, at the same time, feel confident that the maximum expected benefits would also be articulated, achieved and delivered.

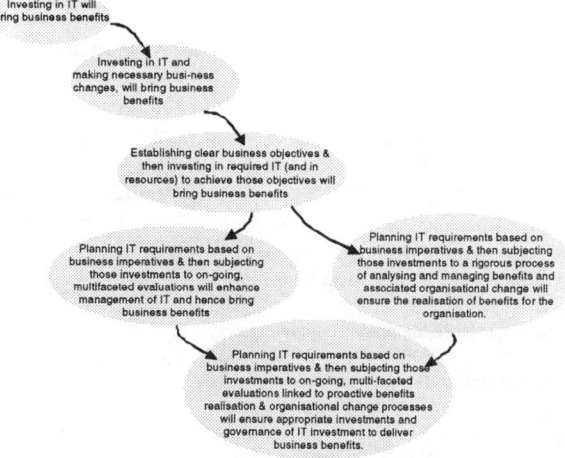

Figure 5: Integrating planning, IT evaluation and benefits management

The IT Evaluation and Benefits Management Life Cycle discussed in this chapter will justify why planning, evaluation and benefits management activities should be integrated, and how they can best be integrated.

Thus, the key to effective investment in IS/IT that is optimal in an ongoing sense is an integrated programme of IS/IT planning, evaluation and benefits management that is embedded in the day-to-day routines and rituals of the organisation. Such an integrated life cycle of activities should not only assure sensible and rational commitments to IS/IT initiatives, but also assure that such commitments remain viable, worthwhile and relevant.

IT EVALUATION AND BENEFITS MANAGEMENT LIFE CYCLE

Let us sum up the arguments presented to date by way of a simple diagram (see Figure 6).

Strategic planning and thinking about IT thus support finding answers for, or at least contemplating, key questions as business directions, objectives, considerations of how IT can either support or enable the achievement of objectives, and thus to considerations of whether a suite of coherent, strategic investments in IT is being proposed. Evaluation of IT enables greater certainty as to the value of IT investments, and by extending the evaluation process throughout the systems life cycle, the dynamic nature of the worth of IT can be established, and hence, managed. Establishing a sound business case for new and continuing investments is an important concern of evaluation. In managing benefits, our concern focuses more on harnessing potential benefits, ensuring they become realised benefits, and in so doing, recognising that the realisation of benefits needs to be considered in the context of a raft of organisational change initiatives. Our conviction is that management thinking and routine practice needs to link these sometimes disparate activities.

How does this work in practice? The first step in our integrated approach involves establishing strategic alignment between proposed IT investments and business strategy, assessing initial feasibility and identifying and structuring benefits. Arguably, bidirectional

Figure 6: Broadening considerations of the value of IT

- Where are we headed?
- What are our objectives?
- How can IS/IT support/enable achievement of objectives?
- Are we proposing coherent, strategic investments?

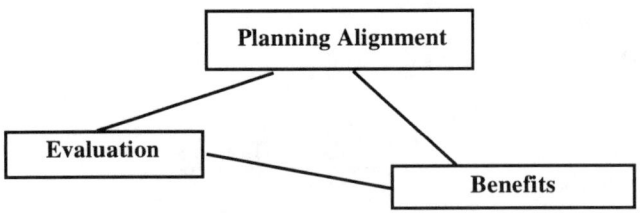

- Are we aware of the "value" of our IT investments at certain critical points in the life cycle?
- Can we establish a realistic business case?
- Are we getting things well done?

- Are we deriving the maximum benefit for the organisation?
- Are we addressing change management issues?
  -processes
  -training
  -trading partners
  -job redesign

flows and relationships exist here (see Figure 7).

Thus, demonstrably close alignment between business strategy and IT initiatives is likely to enhance perceptions of potential benefits from IT investments, and hence improve the likelihood that the project feasibility can be securely grounded in a strong business case. By contract, doubts about feasibility may encourage reconsideration of potential benefits, or indeed, as to the extent of alignment, and so on.

This process of proactively flickering between notions of achievement of objectives, possible benefits, possible costs and risk, arguably supports the prioritisation of a suite of potential investments, which must, in turn, be subjected to a more comprehensive feasibility study. Feasibility will be impacted one way or the other as understandings as to potential benefits are enhanced, with the potential existing that heightened sensitivity with respect to benefits could affect priorities for investments. Fluidity in investigating, considering and reviewing information regarding priorities, feasibility and expected benefits is expected (see Figure 8).

Assuming that a go decision is reached, then the process of systems analysis (including the establishment of requirements) and design must proceed (arguably irrespective of whether a develop or buy and tailor decision is reached). Systems development (used here to include a buy option) is itself a fluid process, and thus design decisions and changes need to be reviewed against whether or not alignment with business objectives has been undesirably affected (the problem of scope creep), whether decisions and changes impact expected benefits positively or negatively, and thus whether feasibility is in any sense compromised (see Figure 9).

The IT investment must go through a process of implementation and testing, ultimately with the aim of becoming a fully operational system. A variety of perspec-

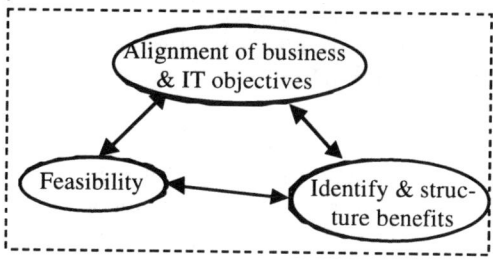

Figure 7: Starting the process of deriving value from IT investments

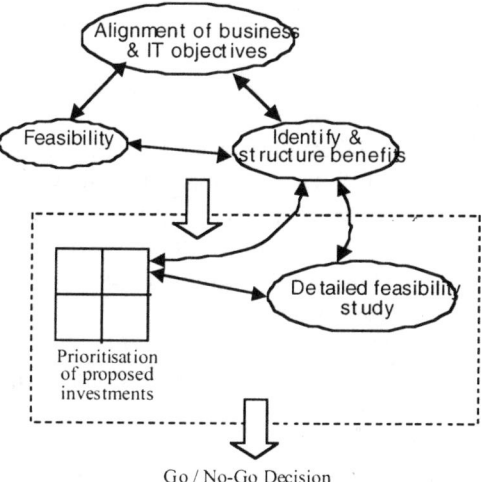

Figure 8: Prioritising IT investments according to feasibility and benefits

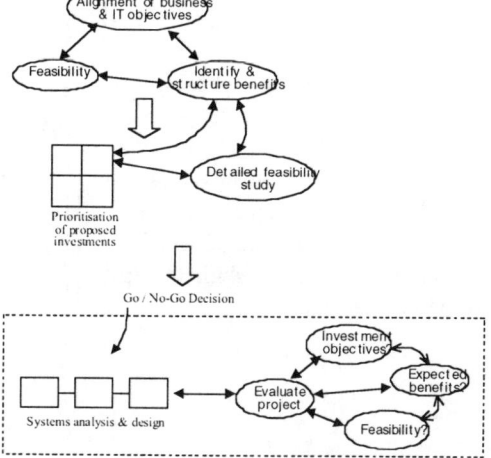

Figure 9: Considering the impacts of systems development on benefits and feasibility

tives or measures needs to be adopted to enhance management of the realisation and delivery of business benefits, to answer questions of the business impact and technical quality of an investment, ultimately leading to making decisions about the future of the investment (Should we continue to maintain the system? Does it need to be enhanced in order to continue to support the achievement of business objectives? Should it be replaced? Would it be beneficial to outsource its operations?). Any concerns about business impacts, technical quality, and/or a failure to deliver ongoing benefits may well result in outsourcing or replace decisions. This, in turn, implies for renewed planning and assessment of the business requirements and drivers to take place (see Figure 10).

Figure 10: On-going review of the value and benefits of operational systems

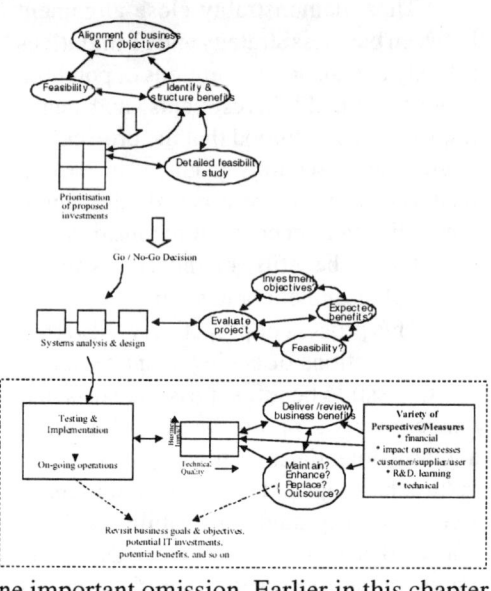

The IT Evaluation and Benefits Management Life Cycle is thus complete, with one important omission. Earlier in this chapter it was argued that benefits from IT would only be realised if appropriate organisational changes to support the technological change were planned, implemented and managed. To complete our cycle, therefore, this vital dimension of change management is added to the diagram (see Figure 11).

There is always concern with graphics such as Figure 11 that the temptation exists to view this as a structured, step-by-step approach which must be doggedly executed in order to achieve a particular desired outcome. This could hardly be further from our intentions. Rather, we advocate flexibility and fluidity, incessant critical reflection, analysis and learning, almost to the point where this whole interplay of planning and alignment, evaluation and managing benefits permeate the consciousness and actions of those associated with IT decision making and management. Regrettably, we lack the skills to capture this pictorially!

Figure 11: Identifying and managing required organisational change

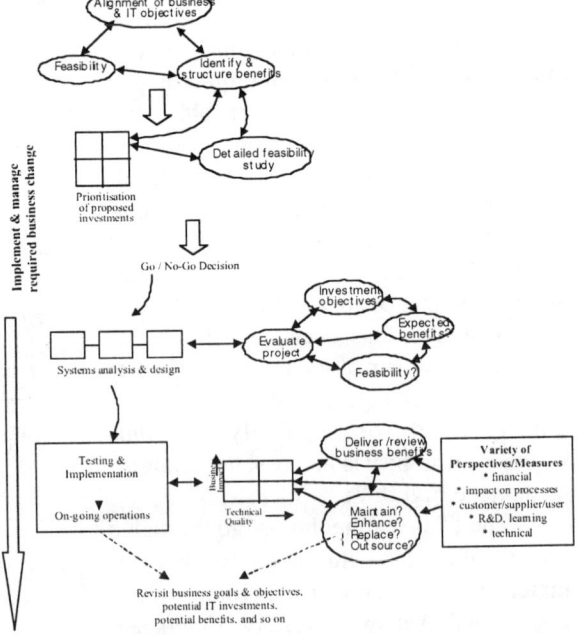

ADDRESSING ORGANISATIONAL COMPLEXITY

There are some other important issues associated with complexities and realities of modern organisations that need briefly to be mentioned. Firstly, investments in IT are rarely made completely independently of either the existing IT legacy in the organisation or of other concurrent IT investment activity. Thus, issues of alignment, evaluation and realisation of benefits need to be considered in the context of a cross-system or cross-investment basis (see Figure 12). The point we are attempting to illustrate here is that in identifying a suite of potential IT investments and in proceeding to develop or purchase a number of this suite, interrelationships and impacts need to be considered if we are to avoid the common dilemmas (Phillips, 1989). That is, if we are to maximize positive organisational impacts and returns, then making the best decision and taking the best actions overall should take precedence over deriving the best outcomes on a system-by-system basis. How often in the IT field do we hear of relatively isolated systems development processes (during which quite reasonable project management decisions are taken) becoming organisational nightmares when integrated at an operational stage with existing IT investments?

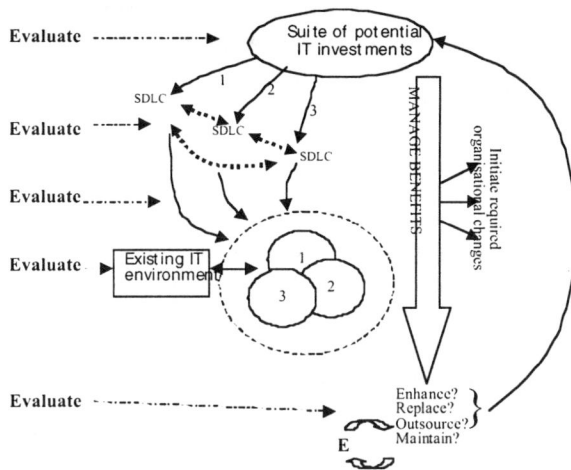

Figure 12: Managing the synergies and interconnections between a suite of IT investments

An even more potentially vexing situation occurs with interorganisational systems, virtual organisations, or in coordinated IT investment activity spanning a strategic business network (see McKay and Marshall, 2000). Not only must all the complexities and interactions of a single organisation's IT inheritance and environment be addressed, but heed must also be paid to similar issues across a range of members of the business network.

Attempting to develop a step-by-step method to address these issues seems an untenable position to adopt, but arguing that perpetual musing and decision making, cognisant of the issues, should infuse the everyday behaviour of managers does not.

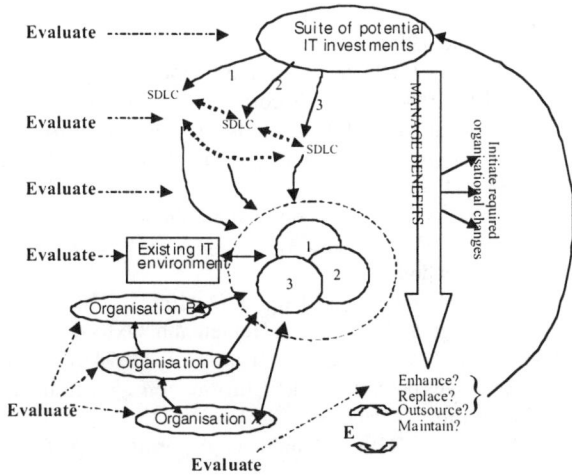

Figure 13: Considering the synergies and interconnections of IT investment spanning a strategic business network

This is our express aim in articulating the IT Evaluation and Benefits Management Life Cycle. We see its outcomes as much in terms of behavioral and cultural change, in changes to routines, as much as anything else. Neither do we envisage revolutionary upheaval in trying to make such changes. A strategically-positioned manager, quietly asking questions, gently probing the rationale of decisions and so on, can start to effect the changes we are advocating.

CONCLUSION

In an era of e-commerce, an information economy, and increasing connectivity, the pervasiveness of IT, and its strategic importance, seems to be growing at an unprecedented rate in most organisations. While this growth is clearly evidenced in increasing IOT expenditures, assessing the value of that expenditure to the organisation and the contribution and benefits that IT delivers is not such a simple task. Indeed, there are clearly concerns expressed as to whether IT does make a reasonable contribution, given its cost.

This paper has discussed the interplay between the alignment of IS/IT planning objectives and business objectives, evaluation of IT investments, and realising benefits from IT investments. Arguments have been developed that synergies could be gained from achieving a close association and interplay between these activities. We believe that the importance of this paper is that it outlines an integrated approach to IS/IT planning, evaluation and benefits management that covers all stages of the information systems life cycle, including enterprise-wide planning. Further, the paper urges practitioners to embed this approach into the routines, rituals and practices of the organisation. In this way, the many tools and techniques of planning, evaluation and benefits realisation can be brought together and implemented in a way that really makes a difference to the deployment of information systems and technology in contemporary organisations.

REFERENCES

Attwell, P. (1996). Information technology and the productivity challenge, In Kling, R. (Ed.), *Computerization and Controversy: Value conflicts and social choices,* 2nd Ed., Academic Press, San Diego, California.

Bender, D. (1986). Financial impact of information processing. *Journal of Management Information Systems,* 3(2): 232-238.

Brynjolfsson, E. (1993) The productivity paradox of information technology, *Communications of the ACM,* 36(12): 67-77.

Cavell, S. (1997). Computing intelligence, *Computing,* 25 Sept., p 38.

Earl, M.J. (1994). Putting information technology in its place: a polemic for the nineties. In R.G. Galliers and B.S.H. Baker (eds.) *Strategic Information Management: Challenges and Strategies in Managing Information Systems.* Butterworth Heinemann, Oxford, pp76-90.

Earl, M.J. (1996). An Organisational Approach to IS Strategy-Making. In M.J. Earl (ed.) *Information Management: The Organizational Dimension.* Oxford University Press, Oxford.

Edwards, C., Ward, J., and Bytheway, A. (1995). *The Essence of Information Systems.* 2nd ed. Prentice Hall, London.

Farbey, B., Land, F., and Targett, D. (1993). *How to assess your IT investment: a study of methods and practice.* Butterworth Heinemann, Oxford.

Galliers, R.D., Merali, Y., and Spearing, L. (1994). Coping with information technology? How British executives perceive the key information systems management issues in the mid 1990s. *Journal of Information Technology,* 9(3).

Hochstrasser, B. (1993). Quality engineering: a new framework applied to justifying and prioritising

IT investment. *European Journal of Information Systems*, 2(3): 211-223.

Hochstrasser, B. and Griffiths, C. (1991) *Controlling IT Investment: Strategy and Management.* Chapman & Hall, London.

Kaplan, R.S. and Norton, D.P. (1996). Using the balanced scorecard as a strategic management system. *Harvard Business Review*, Jan-Feb 1996: 75-85.

Kempis, R. and Ringbeck, J. (1999). Do IT Smart: Seven Rules for Superior Information technology Performance. Free Press, New York.

King, J. L. (1996). Where are the payoffs from computerization?: technology, learning and organizational change. In Kling, R. (Ed.), *Computerization and Controversy: value conflicts and social choices*, 2nd Ed., Academic Press, San Diego, California.

Laudon, K.C. and Laudon, J.P. (2000). *Management Information Systems: Organization and Technology in the Networked Enterprise.* 6th ed. Prentice Hall, N.J.

Lillrank, P., Lehtovaara, M., Holopainen, S., and Sippa, S. (1998). A pre-study report: the impact of information and communication technologies (ICT) on business performance - a constructive empirical study and philosophical enquiry: http://www.interactive.hut.fi/projects/ict-bp/pre_re/ict-julk.htm.

McKague, A. (1998). Time has Come to Deal with Issue of IT Failures. *Computing Canada*, June 8.

Olve, N.G., Roy, J. and Wetter, M. (1999) *Performance Drivers: A Practical Guide to Using the Balanced Scorecard.* Wiley, Chichester.

Peppard, J. (1993). *IT Strategy for Business.* Pitman Publishing, London.

Peters, G. (1994). Evaluating your computer investment strategy. In L. Willcocks (ed.) *Information Management: The Evaluation of Information Systems Investments.* Chapman and Hall, London.

Peters, G. (1996). From strategy to implementation: identifying and managing benefits of IT investments. In L. Willcocks (ed.) *Investing in Information Systems: Evaluation and Management.* Chapman and Hall, London.

Phillips, L.D. (1989). Gaining Corporate Commitment to Change. Working Paper, Decision Analysis Unit, London School of Economics, Houghton Street, London WC2A 2AE.

Rai, A., Patnayakuni, R., and Patnayakuni, N. (1997). Technology investment and business performance. *Communications of the ACM*, 40(7): 89-97.

Remenyi, D., Sherwood-Smith, M., and White, T. (1997). *Achieving maximum value from information systems: a process approach.* Wiley, Chichester.

Remenyi, D.S.J., Money, A., and Twite, A. (1993). *A Guide to Measuring and Managing IT Benefits.* 2nd ed. NCC Blackwell, Oxford.

Serle, J. (1994). Is IT good for business? *Public Finance,* 7 January, 20.

Smyrk, J. (1994). If only a tape measure would do ..., *Management Information Systems*, June, 48-50.

Sprague, R.H. and McNurlin, B.C. (1993). *Information Systems Management in Practice.* 3rd ed. Prentice Hall, Englewood Cliffs.

Strassmann, P.A. (1997). *The Squandered Computer; Evaluating the Business Alignment of Information Technologies.* Information Economics Press, New Canaan, Connecticut.

Sutherland, F. (1994). Some current practices in the evaluation of IT benefits in South African organisations, *Proceedings of the 1st European Conference on IT Investment Evaluation*, Henley, September, 27-43

Thorp, J. (1998). *The Information Paradox: Realizing the Benefits of Information Technology.* McGraw Hill, Toronto.

Tozer, E.E. (1996). *Strategic IS/IT Planning.* Butterworth Heinemann, Boston.

Ward, J. and Griffiths, P. (1996) *Strategic Planning for Information Systems.* 2nd ed. Wiley, Chichester.

Ward, J., Taylor, P., and Bond, P. (1996). Evaluation and realisation of IS/IT benefits: an empirical study of current practice. *European Journal of Information Systems*, 4(1996): 214-225.

Weill, P. (1990). Strategic investment in information technology: an empirical study. *Information Age*, 12(3): 141-147.

Weill, P and Olsen, M.H. (1989). Managing investment in information technology: mini case examples and implications. *MIS Quarterly*, March, 3-17.

Willcocks, L. (1992). Evaluating Information Technology Investments: Research Findings and Reappraisal. *Journal of Information Systems* (1992)2: 243-268.

Willcocks, L. (1994). Introduction: of capital importance. In L. Willcocks (ed.) *Information Management*. Chapman and Hall, London.

Willcocks, L. (1996). Introduction: beyond the IT productivity paradox. In L.Willcocks (ed.) *Investing in Information Systems: Evaluation and Management*. Chapman and Hall, London.

Willcocks, L. and Lester, S. (1997). Assessing IT productivity: any way out of the labyrinth? In Willcocks, L., Feeny, D.F. & Islei, G. (Eds.) *Managing IT as a Strategic Resource*. McGraw-Hill, London.

Wilson, T.D. (1989). The Implementation of Information Systems Strategies in UK Companies: Aims and Barriers to Success. *International Journal of Information Management*, 9.

Part II:

IT Evaluation Research and Methods

Chapter IV

A Review of Research Issues in Evaluation of Information Systems

Vassilis Serafeimidis
KPMG Consulting[1], London, UK and
University of Surrey, Guildford, UK

INTRODUCTION

Information technology (IT) and information systems (IS) have become an organizational necessity in order to support routine data processing operations, initiatives for competitive advantage, business transformation exercises in products, organizational structures, work-roles, and patterns of relationships between organizations. IS are critical components of business, taking part in increasingly complex organizational changes, redefining whole markets and industries, as well as the strategies of the firms that compete within them (e.g., increasing focus on the use of the Internet). As information becomes embedded in organizations, in their products and services and in their relationships with partners and customers, IS cannot be separated from human intellect, culture, philosophy and social organizational structures.

The cost of IT has plummeted dramatically since the 1960s, while its potentials have increased, generating enormous investment and increasing the pace of IT adoption by organizations. According to the Gartner Group, in 1998 the average IS budget was 4.17% of the organizational revenue. This trend is expected to continue, as most organizations have passed 'unharmed' the Millennium landmark and attempt to conquer the arena of *eBusiness* (KPMG Consulting, 2000).

Organizational resources are expected to be invested in anticipation of the highest future gains. The initial enthusiasm for IT, during the 1970s and 1980s, has been overtaken by a sense of pragmatism in the 1990s. Senior management seeks solid justification of the business value of IT, IS departments have been questioned concerning their contribution to business success, and system developers and business users have been forced to become much more familiar with both IT"s potential and constraints.

It has been realized that successful IT outcomes do not occur by default; they are highly uncertain and in order to achieve organizational success, IS has to be managed effectively and be considered broadly. This issue is very important, particularly today, where IT outcomes refer to an 'ecosystem' of networked partners. The additional difficulties in

Copyright © 2001, Idea Group Publishing.

identifying and measuring potential benefits and costs, deriving from current organizational practices, forced many organizations to establish management control mechanisms. Among these mechanisms are the thorough appraisal of potential IT investments and the evaluation of their deliverables.

The role that evaluation plays as an organizational process varies. It is strongly related to other management and decision making processes. The management expectation from IS evaluation is about establishing by quantitative and/or qualitative means the worth of IT to the organization (Farbey et al., 1993; Willcocks, 1994) and IT's contribution to the organizational growth. This can be achieved by effective IS evaluation which ranks alternatives (Clemons, 1991) and forms a central part of a complex and incremental planning, decision-making and control (diagnosis) process (Blackler and Brown, 1985; Hawgood and Land, 1988). Evaluation is then a crucial feedback function (Angell and Smithson, 1991; Baker, 1995), which helps the organization learn (Earl, 1989; Hirschheim and Smithson, 1987; Farbey et al., 1993; Walsham, 1993) and thereby reduces the uncertainty of decisions. This feedback helps trace and understand the underlying factors leading to the success or otherwise of an IT investment. In many cases (Farbey et al., 1995; Gregory and Jackson, 1992; Powell, 1992a) evaluation is a mechanism for gaining commitment and, in highly politically influenced environments, for legitimization and in some other occasions is a mechanism for exploration and discovery (Serafeimidis, 1997).

This chapter discusses the role of evaluation in the management of IS investments and its gradually increasing importance as part of the IS governance. It reviews extensively the related literature (published during the last decade) across conceptual and operational dimensions. The following section outlines a framework for discussion which provides a structure for the review. Then the authors examine the traditional formal/rational evaluation approaches referring to the technical and financial focused developments. The interpretive research epistemology has been adopted by many researchers to provide more valid alternatives to IS evaluation. These alternatives are presented in detail and critically discussed in detail in this chapter.

A FRAMEWORK FOR ANALYSIS

Information systems evaluation is a highly complicated phenomenon. In order to examine the extensive literature in the area and clearly identify the problems, it is essential to have a conceptual framework for discussion. The framework adopted here is based on a definitional approach, starting from a working definition of evaluation, including the assumptions underlying the evaluation approaches and methodologies.

However, despite the huge number of articles written in the area, very few authors have taken this approach and attempted to draw the boundaries of the phenomenon under study because of its elusiveness and broadness. IS evaluation is highly subjective and context dependent, as well as covering a wide area of situations and activities. As IS have been integrated into modern organizations their evaluation has become more important than before, and also qualitatively and structurally different. Evaluation involves a large number of stakeholders both internal and external to the organization each with their own particular values and objectives. A good definition is adopted from Remenyi and Sherwood-Smith (1997).

> "Evaluation is a series of activities incorporating understanding, measurement, and assessment. It is either a conscious or tacit process which aims to establish the value of or the contribution made by a particular situation. It can also relate to the determination of the worth of an object."

This definition can be broken down into a number of elements (see below). These elements are not independent; they are closely interrelated and are determined in practice by the demands of the situation.

Purpose/Reasons → *Why?*
The subject → *What?*
Criteria/Measurement → *Which aspects?*
Time frame → *When?*
People → *Who?*
Methodologies/Tools → *How?*

The Purpose/Reasons of Evaluation

Here the focus is on the range of different purposes that IS evaluation can serve and the related variety of outcome/deliverable. For example, summative evaluation (Legge, 1984) is concerned with the *ex ante* selection of one course of action, or design, from a number of available alternatives (including 'Go - No Go' decisions). On the other hand, formative evaluation is concerned with *ex post* feedback by evaluating an existing system, or one recently developed/implemented. Here the concern is with a rational approach to resource allocation and the achievement of predefined objectives. Other purposes of evaluation are much more political, as it has been used to gain commitment before starting a project or as a disengagement device for developers at the end of a project.

What Is Being Evaluated

Here the interest is in the subject of the evaluation exercise, i.e. the entity that is being evaluated. This is concerned with drawing boundaries around the evaluation. In this regard, the evaluation could focus on the technology alone, the technology plus the information provided, a particular system (including or excluding the people involved), a new way of doing things based on the system, or the achievement of some business objectives using the system/information (such as a new product or business transformation).

The Criteria/Variables Investigated

Any evaluation involves the measurement of certain variables and the comparison of these measurements against certain criteria. These could be technical measures (e.g., response time, function points), financial measures (e.g., costs), measures of system or information quality (DeLone and McLean, 1992), user satisfaction (Ives et al., 1983), or some other form of impact measurement. Usually the criteria are derived from the purpose of the evaluation but influential stakeholders may be able to impose particular variables and criteria. Clearly the decision of what to measure and what constitutes an acceptable level of performance on each criterion affects the evaluation considerably (e.g., methodology used).

The Time Frame Followed

Evaluation can be performed at various points within the systems life cycle. The most common points are related to the feasibility and the post-implementation stages. During the early stages costs and benefits are estimated and the potential IS investment is justified in terms of its likely contribution. Possible alternatives are compared and ranked before final selections are made. On the other hand, after the system's implementation a post-event action aims to judge the success achieved in comparison with the planned objectives. For IS projects this action is usually ignored or focuses on limited issues (such as technical performance and costs occurred).

The People Involved

Various authors (e.g., Hirschheim and Smithson, 1988; Symons, 1990; Walsham, 1993) have emphasized the subjectivity and social nature of evaluation and thus the people aspects are of considerable importance. These aspects should refer not only to those people doing the evaluation but also to other stakeholders that are affected in one way or another (victims or beneficiaries according to Guba and Lincoln, 1989). Even stakeholders who are not directly involved may still have considerable political influence on the evaluation (e.g., senior managers). Typically there are hidden agendas as well as formal objectives. It should also be noted here that everyone involved tends to make informal assessments regardless of any formal evaluation exercise.

The Methodologies and Tools Used

This is concerned with the how of evaluation: the procedures including methodologies, techniques and tools used. The rational aspect of evaluation implies that a well-defined methodological framework is essential. Much of the literature in the 1970s and 1980s was concerned with this aspect of evaluation, which accounts for the wide range of methodologies available today (e.g., cost benefit analysis, ROI, information economic). Theoretically, at least, decisions related to how the evaluation is to be carried out should flow from the elements of the evaluation described above.

Underlying Assumptions of IS Evaluation

The adopted framework classifies research based on its underlying assumptions and follows broadly the three streams (zones) of Hirschheim and Smithson (1988) and Smithson and Hirschheim (1998): efficiency; effectiveness; and understanding. Studies in the efficiency zone are built upon rational/objective principles regarding the nature of the IS and its evaluation, and they try to judge the achievement of well-determined goals. In terms of effectiveness, evaluation attempts to measure usage or utilization of the IS outcomes. The understanding zone refers to highly interpretive oriented studies, which seek understanding of the integrated role of technology within its rich organizational context.

Nature of the Study

Another dimension of the literature review refers to the nature of the research studies. On a conceptual level the complicated nature of evaluation led researchers to search for solutions in reference disciplines such as finance, economics, engineering, education and sociology, where evaluation has been studied before. The majority of these studies focus on the development of concepts and frameworks for IS evaluation utilizing previous knowledge and experience. However, the importance of practice encouraged many researchers to search for evidence and solutions in organizational practices. These studies have attempted to develop new concepts, theories and understanding based on empirical evidence. The above distinction appears to be a trend in the review literature (see for example DeLone and McLean, 1992).

Most of the above issues are indirectly determined by the organizational context and culture, the role of IS and their evaluation, and the maturity of concepts and structures developed around them. Another important aspect to note from the above discussion is the breadth, diversity and richness of each element which will be used below to critically examine the literature.

THE TRADITIONAL RATIONAL VIEWS OF IS EVALUATION

The Technical/Functional Stream

Purpose/Reasons

The focus of technical/functional evaluation is around the technical (IT) components. The evaluation is concerned with the relationship between inputs and outputs. Its main focus is on efficiency in terms of the technical performance and the control of resources. Further interests lie around increased and improved capacity, fewer errors, greater reliability, manpower savings, etc. and also software performance developments and quality.

Subject and criteria

One of the earliest contributions of IT was the automation of clerical tasks and the consequent reduction of operating costs. IT is used both to automate manual manufacturing tasks and support back-office processes such as payroll and accounting in order to improve transaction efficiency. In the early years, this was the primary and only focus of IT. Information systems were assumed to be an objective, external force with deterministic impacts on organizations. Although today the variety of IT roles is much more extensive, simple automation applications are still in place.

Time frame

The driving force for the timing of evaluation is the traditional system's development life cycle. Therefore, two discrete phases of evaluation are perceived. Firstly, a feasibility study is conducted in an attempt to estimate costs and potential efficiency contributions. Secondly, post-implementation evaluation is used as a disengagement device for the project team and as an instrument to measure the achievement of the initial goals.

These actions implicitly assume that IS impacts are short-term (Hopwood, 1983) and they are delivered with the completion of the system's implementation. These assumptions are only relevant for simple automating applications (e.g., payroll systems, office support systems). Although the expected contribution of such an IS is direct, high uncertainty and risk characterize the IS implementation (e.g., Angell and Smithson, 1991; Charette, 1989). The spiral model has been proposed by Boehm (1988) as a complementary development of the traditional life cycle model with a particular emphasis on regular assessment of risk factors.

People

As the primary focus of evaluation is around the technical system, IT departments and IT experts used to dominate the evaluation. Technical evaluation then becomes an internal organizational matter and the participation of business users is limited to the approval of proposals and the acceptance of the system's functionality.

Methodologies/Tools

The emphasis here is on reliability and cost-efficiency. Demand for systems quality control (e.g., on-line response, reliability) led to extensive research into software metrics (see, for example, Fenton, 1991; Gilb and Finzi, 1988). The total quality management (TQM) movement reinforced this and provided a number of ways of defining and measuring quality. Some of the techniques provide useful ways of maintaining a continuous evaluation of the quality of both the development process (e.g., ISO 9000/9001, BS 5750) and the information system as a product (Finkelstein, 1989; Land and Tweedie, 1992; Zultner, 1993).

Regarding cost-efficiency, experience from the evaluation of civil engineering projects and their cost-related issues influenced many researchers in the IS field, according to Powell (1992b). Those cost estimations are perceived as predictions, which therefore should not be expected to be accurate in the accounting sense. Thus, approaches aiming at accurate cost predictions were developed, including parametric cost and effort estimation techniques based on a statistical analysis of past project data, e.g., COCOMO model (Boehm, 1981) and SLIM model of differential equations (Putnam, 1980). Alternative methods are based on a measure of the complexity of the project such as function point analysis (e.g., Albrecht and Gaffney, 1983). A number of tools were developed (e.g., METRICS, ESTIMACS, OLIVER, SLIM SIMON), to collect and process the vast amount of data required (Fenton, 1991; Lederer and Prasad, 1993).

Empirical evidence

Despite the dramatic shift in the last decade towards a concern with business values and organizational objectives, technical criteria are still widely used in practice (DeLone and McLean, 1992). Bacon (1994) found that 79% of his survey participants used technical/system requirements at the primary level of their decision processes. However, these requirements were only applied on average to 25% of the organizations' projects. In Lederer and Prasad's (1993) survey of IS managers, 43% stated that system development cost estimation is the most important issue. Nevertheless, in Willcocks and Lester's (1994) survey, the lack of any technical criteria used as a basis for evaluation is rather surprising.

The technical/functional view focuses on the quality and the costs of software development which in some cases are translated into savings, but neither of them addresses the important dimension of the business benefits (including intangibles, strategic and their associated risks) nor do they consider the social and organizational impacts of an information system. The accuracy and usefulness of these models and tools is still debatable (see, for example, Heemstra, 1992; Kemerer, 1987; Lederer and Prasad, 1993). However, this may be the only area where practitioners extensively use methods and tools developed in academia. Although much research interest has shifted away from the earlier preoccupation with IS efficiency, it is still a necessary (but not sufficient) aspect of IS evaluation in practice. This is especially the case for high volume transaction processing systems, as well as the network and database components that form important element of a modern organization's infrastructure.

The Economic/Financial Stream

Purpose/Reasons

The contribution of IT became less transparent and, considering its importance, IT has started to be viewed as a capital investment and not just as an operational cost. This means a move from considering IS as a set of hardware and software components to viewing it as an investment or a business facilitation project. Therefore, the focus of IS evaluation has shifted from the performance of the IS *per se* to the quality of its outputs (e.g., information) and their utilization (e.g., customer satisfaction, creation of business value). Here the evaluation of IT investments is based on an organizational analysis which emphasizes the achievement of predetermined outcomes as a measure of effectiveness (e.g., critical success factors, business objectives). In contrast to the technical/functional measures, the importance here is on the relationship of outputs to outcomes (e.g., more information available leads to better decision making).

> "There has been a shift in focus from operations - making the technology work and keeping it working, to impact - understanding, predicting and influencing the micro and macro-level effects of IT on individuals, organizations and the economic system, and especially on how to maximize the return on IT investments. These trends have raised issues concerning the value added by IT and the cost of providing IT resources." (Bakos and Kemerer, 1992 p.366)

Subject and criteria

The constant improvements in technology shifted the interest of IS evaluation research more towards the business and human aspects (e.g., human computer interfaces). IT started to facilitate the more effective management and control of the organization (informate, Zuboff, 1988) by improving the personal productivity and effectiveness of managers and their support staff (e.g., management information systems, decision support systems). Effectiveness is gauged in terms of the outcomes of given activities or services measuring information quality, user satisfaction, etc. The search for the business value of IT investments has driven many researchers to formulate approaches based on finance and economic principles. Accounting and economic measures are used to express the hard quantifiable contribution of IT to organizations, the impacts on individual's productivity and to national economies. The methods found here are still widely seen as objective (Hirschheim and Smithson, 1988; Powell, 1992b) based on similar principles as the technical/functional approaches.

Investment decisions, traditionally, are associated with high uncertainty and risk. Future uncertainty is responsible for inaccurate predictions of the investment returns, and consequently with the effectiveness of IS, and in most cases of the investment's costs. As the IS outputs become more strategic and indirect, the study (analysis and management) and breadth of risks have become more significant. Risks related to efficiency usually refer to the development issues. Within the economic evaluation stream, the study of risks considers additional multiple risk factors associated with the ability of the IS to deliver business value (Clemons 1991; Coleman and Jamieson 1994; Kumar, 1996; Parker et al., 1988). The studies range from simple economic sensitivity analysis to more advanced frameworks such as those proposed by Birch and McEvoy (1995), Ward (1992; 1994), Willcocks and Margetts (1994).

Time frame

The structured systems development life cycle is also here the dominant time frame for evaluation actions. In *ex ante* terms management is most concerned with investigating the broader context from where business opportunities and constraints for IT investments will derive. Furthermore, the scope of the strategy and high-level goals, as well as potential alternatives and their costs, benefits and associated risks are examined. However, in the later stages of analysis and design, the concerns are more detailed, involving more precise specifications of the project's aims. In *ex post* evaluation the aim is to identify the actual costs incurred and (financial) benefits achieved. Then, the problem becomes much more one of measurement, of determining the precise impact of the system on the business processes and objectives in line with the initial requirements.

People

IS evaluation is broader and stretches beyond the technical experts, becoming much more a business matter. The involvement of users of the system and/or its outputs has become more common. In addition, economists, accountants and finance experts are involved, and more business managers participate in the evaluation exercises. Methodologies such as information economics (Parker et al., 1988) attempt to bridge the gap between business and systems.

Approaches originating in economics

A number of studies use economics as a reference discipline to investigate the value of IT (in micro and macro environments), the value of information and the real cost of providing IT resources. However, the theoretical robustness of these studies has been questioned (Kauffman and Weill, 1989; Mukhopadhyay, 1993). Bakos and Kemerer (1992) use a classification based on six major areas of economic theory (information economics; production economics; economic models of organizational performance; industrial economics; institutional economics; and macroeconomics) to review the existing studies. Mukhopadhyay (1993) distinguishes between the strategic impact of IT and the productivity impacts, while Hitt and Brynjolfsson (1996) also consider the consumer value. Gurbaxani and Whang (1991) analyze the impacts of IS on organizations by determining the effects of IT on the cost structure of the organization, and examining, from the perspective of agency theory and transaction cost economics (Ciborra, 1987), how these effects result in changes to various attributes of the firm. Dos Santos (1994), Kambil et al. (1993) and Kumar (1996) discuss an approach based on the option pricing theory. This approach recognizes among other things the importance of learning, and it is highly relevant to IS infrastructure projects.

The pure economic view of IS evaluation has a lot of similarities with (Lucas and Moore, 1976), and would benefit by experience from, the evaluation of research and development programs such as OECD (1990 & 1992), Capron (1992) and Krull (1993).

Various empirical studies attempted to investigate the relationship between IT investments and productivity, with mixed results both in the manufacturing sector (Hochstrasser and Griffiths, 1991; Loveman, 1988; Roach, 1989; Weill, 1992) and the service industries (Francalanci and Galal, 1998; Hackett, 1990; Roach, 1991; Strassmann, 1990). The lack of evidence of positive relationships (Barua et al., 1991; Breshnahan, 1986; Brynjolfsson and Hitt, 1993), success rates as low as 30%-40% (Hochstrasser and Griffiths, 1991; Willcocks and Lester, 1994) and productivity rates thirteen times less than with traditional capital investments (Landauer, 1995) make up what is commonly known as the IT-related productivity paradox (the lack of a strong positive relationship between IT expenditure and improved productivity). A number of authors (Brynjolfsson, 1993; Farbey et al., 1993; Henderson and Venkatraman, 1993; Landauer, 1995; Scott Morton, 1991) attribute this to: measurement issues (wrong measures and errors), benefit-related problems (time lag, redistribution, invisibility), IS mismanagement (poor business/IT alignment, limited system utilization) and human-related issues (slow learning and adjustment, reluctance, limited usability).

Approaches originating in finance

Another set of studies, derived from accounting and finance, supports the view of many organizations that an IS project is a classical capital investment. They focus on quantification of costs and benefits in monetary terms and try to predict the time value of money in order to support decision making (see, Powell, 1992b; Sassone, 1988; Shoval and Lugasi, 1987). Among the most popular views (see, Ballantine et al., 1995 - 72%; Willcocks and Lester, 1994 - 96%) is cost benefit analysis (CBA).

"Everybody does cost benefit analysis, however, most of them are fictional not only for the benefits side but also for the costs." (Grindley, 1991; Scott Morton, 1991)

The first complete methodology that tries to view the contribution of an IS beyond the simple financial benefits but in terms of business value is information economics (Parker et al., 1988). The risk factors influencing the delivery of the business value are also identified and considered one as well as their representative stakeholders. The method has attracted the interest of practitioners (see, Coleman and Jamieson, 1994; Wiseman, 1994).

Regarding other financial techniques, Bacon (1994), suggests that 75% of the organizations use some form of discounted cash flow (DCF) in selecting their IS projects. The most popular techniques are the net present value (NPV) and the internal rate of return (IRR) especially for large projects. From the non-DCF methods, the majority of organizations use the payback (PBK) method as a financial criterion for project justification (Bacon, 1994; Ballantine and Stray, 1999; Ballantine et al., 1995; Currie, 1995) regardless of its implicit short-term orientation and its theoretical shortcomings. A number of computer-based tools, such as spreadsheets, support the finance-driven activities (e.g., NPV, IRR) of the evaluation function.

Approaches originating in behavioral science

Another set of studies uses principles of behavioral science (organizational behavior and cognitive science) to investigate the impact of IT on the individual's satisfaction and decision making behavior, and finally the impacts on the organization. In these cases IS effectiveness is measured by the use of the information provided by an IS expressed in terms of the user satisfaction and the quality of decisions made.

Strassmann's (1985) return on management (ROM) tries to determine the impact of IT on management productivity. Although it is a meaningful and easy to calculate statistic, it does not take into account the contextual factors influencing management productivity. An extensive study of Ives et al. (1983) developed an instrument for user satisfaction measurement which has been used and improved by Baroudi and Orlikowski (1988). Remenyi et al. (1991) propose the use of gap analysis and factor analysis as user satisfaction measurements for the information systems effectiveness. Their approach is supported by a computer-based tool. Value analysis (Keen, 1981) was introduced with the particular idea of focusing on the highly intangible benefits side of decision support systems.

A SEARCH FOR INTERPRETIVE ALTERNATIVES TO IS EVALUATION

The growing importance of IT and its greater integration with organizational functions and structures led to the need for more sophisticated and/or holistic approaches to study IS within their context. The general evolution of IS research philosophies from positivistic to the interpretive paradigm has strongly influenced evaluation research (e.g., Walsham, 1993). The formal-rational approaches emphasize the technology aspects at the expense of the organizational and social aspects. Therefore, the worth of an IS, is sought in the system's performance and financial profitability. A move towards the interactionist role of the technology with the organizational structures, culture and stakeholders has shifted IS evaluation to the analysis and understanding of the social and subjective nature of the phenomenon. Such interpretive thinking to evaluation would facilitate a richer understanding (Hirschheim and Smithson, 1988; Walsham, 1993) which would provide information for action. A wide range of alternative proposals is available, building upon experience from other disciplines (e.g., education), providing meta-methodologies, seeking for new measurements, etc.

Contingency Models

Having to deal with a spectrum of different IS projects with a wide range of impacts, a number of studies argue that evaluation should be driven be the type of information system to be evaluated (e.g., Berghout, 1997; Ward, 1994; Willcocks, 1992). Therefore, researchers propose the use of a contingency model which would assist in the understanding of the role and impacts (e.g., benefits, costs, risks, performance measures, criteria) of the particular case in hand. This argument is further supported by the unsuccessful search for one evaluation methodology rich enough to satisfy the diversity of evaluation requirements. Any such classification attempts to assist and guide the way that applications need to be handled, the identification of their impact and the methods by which they can be evaluated.

There are many ways of classifying information systems (e.g., based on users, systems versus applications), but for investment appraisal purposes the role they play in the business and the contribution they are expected to make should be the key parameters for such classification. Many such classifications have been proposed. Thus, Parker et al. (1988), in information economics, suggest three types of applications: substitutive, complementary and innovative, based on the benefits accrued and how they can be quantified to help justify the investment. Ward (1994) expanded these ideas further.

Farbey et al. (1993) argue that the type of application and the type of objective ought to be the most influential factors in the choice of an evaluation method. They claim that projects should be classified based on their expected returns and the levels of uncertainty associated with them. Based on these principles they suggest the benefits evaluation ladder of IS projects.

Gregory and Jackson's (1991) proposal for a contingency model is based on the characteristics of the group of people participating in the evaluation. Two dimensions are important: the orientation to evaluation (qualitative versus quantitative) and the variety of the group (ranging from low to high).

Peters (1994; 1996) suggests an investment map to relate IS investment projects to the expected benefits and to the organizational/business needs. He starts with a classification of the expected benefits based on their degree of tangibility, and he continues with the investment orientation in terms of infrastructure, business operations, and market influence. Mapping all the company's characteristics under these labels would assist in the effective representation of IS plans and strategies. Additionally, the maps allow the comparison of investment proposals and assess their contribution to provide benefits and their orientation to the business value chain.

Versatility in Evaluation

Besides the traditional reasons for evaluation, there is a stream of research studies, which emphasize a number of new purposes. These new roles are more holistic in terms of their organizational contribution, stretching beyond the simple technical, financial or economic issues. IS evaluation views the whole organizational change program and investigating its multiple effects (e.g., technical, financial, social) within and outside the organization. Evaluation could thus be a fundamental component of understanding and interpreting an IS, organizational learning, business and IS strategy formulation, an organizational discourse instrument and a safeguard mechanism for preventing and mitigating IS failures.

The above arguments are supported extensively by Symons (1990) who looks for a deeper understanding of the perspectives of individuals and groups of stakeholders to reveal human intuition, judgement and understanding of the politics involved. She bases her analysis on the multiple perspectives of evaluation and uses the principles of content, context and process of organizational change (Pettigrew, 1985a; 1985b) associated with the introduction of an IS. Walsham (1993) aims for the involvement of a variety of stakeholders, with an extensive focus on learning and understanding, in order to generate involvement and commitment.

Serafeimidis (1997) investigates the institutional properties of IS evaluation. He studies the primary organizational roles (strategists for evaluation and evaluators) and the institutional characteristics of four possible evaluation orientations (control, social learning, sense making and exploratory).

Evaluation as a feedback mechanism aims to facilitate learning. As Zuboff (1991) argues, "Learning is the new form of labor and the ultimate source of value added." This assigns a more cognitive role to evaluation. Evaluation facilitates a deeper understanding of the interaction between the technology and the underlying organizational processes within a particular organizational context and facilitates a dialectic process (Symons, 1993) which will generate motivation, commitment and knowledge.

Criteria and Measurement

In both the functional and the financial approaches the content (e.g., benefits, costs, risks) is assumed to be well-defined, direct and short term, and the measures used are relatively straightforward (e.g., Hopwood, 1983). A rational relationship between cause and effect is maintained. The changing role of IT means that these content elements have changed considerably (e.g., intangible benefits, highly uncertain). The traditional evaluation approaches are unable to cope with these changes. A survey by the Nolan Norton Institute (1998) suggests that 40% of all companies do not measure IT costs and benefits as part of their IT decision making process. On this basis two streams of proposals have been identified: those suggesting alternative measures and those which focus on improving the reliability of the existing ones.

Regarding alternative measures, Bakos and Kemerer (1992) argue that it is more relevant to examine intermediate outputs based on time, quality, cost and flexibility, which are more directly linked to IT than profit or other financial measures of performance. Organizational learning and organizational commitment are other (less tangible) intermediate measures (Scott, 1994). Nonfinancial measures proposed by Carlson and McNurlin (1992) include: the level of IT resources used, business efficiency and quality management. Likewise, Parker (1996) argues for the use of output-based added value, economic value added, market value added as extensions to the initial proposals of value linking, value acceleration, value restructuring and innovation (Parker et al., 1988). Further attempts in the area include Mooney et al.'s (1995) process oriented framework which focuses on intermediate business processes generating insight into the creation of business value through IT.

The major characteristics of the existing measures (technical and financial) are the tangibility of inputs and outcomes and the assignment of meaningful figures. These requirements restrict their relevance to many IS outcomes. However, sometimes a decision can be made rationally and analytically, even when it cannot be made numerically (Clemons, 1991).

Kaplan (1986) argues that instead of putting monetary tags on the intangibles, managers should estimate how large these benefits must be to justify the proposed investment. Usually these benefits should be expressed in terms of corporate goals and objectives. This view is also shared by practitioners. There are a number of attempts reported in the literature (e.g., Coleman and Jamieson, 1994; Whiting et al., 1993) to formulate methodologies and tools able to deal adequately with the intangible issues. Other researchers (e.g., Cronk and Fitzgerald, 1999) are focusing on the determination of the business value in quantitative and qualitative terms prior to any evaluation attempts.

Extending the Traditional Life Cycle Through Benefits Management

The use of the traditional systems development life cycle imposes a number of milestones to IS evaluation which are one-dimensional and linear as they are entirely related to the progress of the IS development. Their concern focuses on the utilization of resources and the achievement of predetermined objectives.

A stream of research attempts to extend conceptually and operationally the traditional systems development life cycle to a model which is driven by the needs of evaluation. Particular attention is paid to a continuous evaluation mechanism for maximizing the benefits delivered rather than just deciding where to invest initially (e.g., feasibility studies). This suggests an increased concern for managing the benefits after implementation and during the operational life, which was often neglected in the past.

The most significant developments in this area have been made by Schäfer (1988), Ward et al. (1996), Ward and Griffiths (1996) and Willcocks (1996). Schäfer's (1988) Functional Analysis of Office Requirements (FAOR) method incorporates a benefit analysis framework which is approached as a continuous process. Ward et al.'s proposal considers the whole process by which IS benefits are identified, appraised, realized and evaluated. Their approach is concerned with "the evaluation of the outcome of investment in IS/IT, rather than appraisal of the project of investment in IS/IT" (Ward et al., 1996 p.214). This proposal focuses on a limited part of the traditional IS development life cycle without really integrating the concepts of benefits management with the other stages. For Willcocks (1996 p.5) an evaluation life cycle is a countermeasure to the productivity paradox "... which attempts to bring together a rich and diverse set of ideas, methods, and practices ... and point them in the direction of an integrated approach to the systems lifetime." Willcocks correctly attempts to incorporate principles of IS development, evaluation and IS management together. However, the final model introduces only a simple, strategic front-end to the traditional IS development life cycle.

Stakeholders in the Evaluation Judgement

In the center of an interpretive evaluation philosophy are the social actors. The identification of the stakeholders and their stakes is an underlying principle which covers developments discussed above including new roles of evaluation, an extended life cycle and new evaluation methodologies (e.g., experimental).

In the heart of Cyert and March's (1963) theory is the formulation of various coalitions around a goal. Gregory and Jackson (1992) introduced the idea of the 'evaluation party' as the group of stakeholders involved and/or affected in an evaluation situation. The 'evaluation party' is a dynamic coalition that depends on the type of investment, the time that evaluation is performed, the then current context, etc. Gregory and Jackson (1991) argue that the variations of the characteristics of the 'evaluation party' (i.e. its approach to evaluation, the group and its variety) should be determinant factors in the evaluation approach adopted.

The deeper understanding of the perspectives of individuals and groups is necessary to reveal human intuition, judgement and understanding of the politics involved. Serafeimidis (1997) investigated the structural and behavioral aspects of particular members of the evaluation party and mapped them across the social actions of the participants to guide successful evaluation exercises.

The identification of stakeholder groups and their concerns is a key principle in the fourth generation evaluation of Guba and Lincoln (1989). Farbey et al. (1993) have developed an evaluation stakeholder's map to identify the people involved in an IS evaluation exercise. However, this map does not play, as would be expected, an explicit role in their further development of an evaluation meta-methodology (Farbey et al., 1992).

Ward et al. (1996) investigate the role of stakeholders in their benefits management process model. The focus of their model is on the identification and management of stakeholders who would act as the levers or as beneficiaries of the IS benefits. Initially, benefit owners are identified, and responsibilities are allocated to them to deliver the benefits. In a second stage additional stakeholders who can affect the delivery are investigated. This is particularly important as IT facilitates changes of organizational roles and tasks (which highly affect the people). These changes contribute to the success (or not) of the IS. Also stakeholders can spot unexpected benefits and assist in their realization. A similar thinking has been applied by Remenyi and Sherwood-Smith (1999) their continuous participative evaluation approach.

New Methodological Developments

Assuming a uniform context across organizations and a highly static context within an organization, the hunt for the best methodology was understandable. However, today's diversity and instability of contexts implies that IS evaluation methodologies should be flexible enough to adapt to the new requirements and they should become context sensitive. Particular methodologies are suitable to particular contexts, specific points in time and refer to particular audiences (e.g., accountants). The range of circumstances is so wide that no single methodology can cope. Therefore, the search for a universally accepted evaluation methodology becomes a fiction (Farbey et al., 1993; Symons and Walsham, 1991). In addition little or no thinking took place regarding how evaluation methodologies would be integrated with other decision making processes, who would benefit from using them, and what organizational changes were required (e.g., skills of users). Furthermore, the dominant systems development life cycle presupposes the compatibility of evaluation methodologies with the requirements of IS development methodologies. Thus, their focus becomes clearly one point in time (e.g., *ex ante*, *ex post*).

Because of the problems with the existing evaluation methodologies, researchers have firstly proposed the use of meta-methodologies which attempt to utilize the existing plethora of methods according to a contingency approach. Secondly, there is a development of new paradigms which address primarily the underlying principle of understanding.

Meta-methodologies

There is a clear need for contingent evaluation approaches in order to deal with the range of circumstances encountered. This implies that IT projects, as well as their contexts have certain characteristics which influence the choice of a suitable evaluation method. Similarly, every evaluation methodology or technique has characteristics which point to the set of circumstances in which it could be applied more successfully.

Farbey et al. (1993) argue that IS projects should be classified based on five groups of factors which influence evaluation: the role of the evaluation, the decision environment, the system characteristics, the organizational characteristics, and the cause and effect relationships between an investment and its benefits. Farbey's et al. (1992) proposal uses those five characteristics as key drivers to classify the project under investigation on a set of two-dimensional matrices. The same dimensions are used to classify, on a sixth matrix, eleven of the most popular existing evaluation methods. In a follow up publication (Farbey et al., 1994) they add one more dimension: the portfolio position of the project.

Willcocks and Lester (1994) using similar reasoning, based on their above classification propose a 'cost-benefit map' which drives the selection of the most suitable method. On the other hand, Hares and Royle (1994) propose a structured framework which utilizes the existing financial appraisal methodologies. They argue that the drawbacks of the individual financial methods can only be overcome by using multiple complementary methods.

Building Understanding Methods

The need to understand the plurality and multiplicity of stakeholders' views provides an opportunity for the use of experimental and exploratory methods. Such methods could facilitate the elicitation of the subjective informal and formal views and political agendas.

Experimental methods (e.g., simulation, prototyping, game-playing) attempt to assist decisions by judgement based on empirical evidence. Nissen (1994) introduces a methodology called virtual process measurement, which uses simulation in the measurement of IT value.

Farbey et al. (1993; 1994; 1995) discuss a number of methods which emphasize the process of obtaining agreement through exploration, mutual learning and negotiation. In methods such as multi-objective multi-criteria and furthermore in art criticism, accreditation, adversarial methods (borrowed from evaluation research, House, 1983), numbers are less important for the stakeholders involved than a thorough understanding of the issues, threats and opportunities.

Avgerou (1995) argues that new evaluation methods are needed to clarify and/or support stakeholders' views. She suggests the use of soft systems methodology (SSM) and fourth generation evaluation as diagnostic tools for further understanding and negotiation.

CONCLUSIONS

This chapter has reviewed research on IS evaluation on the basis of a framework of analysis which addressed the why, what, which aspects, when, who and how dimensions. Three different streams of IS evaluation theoretical and practical developments have been identified; the technical/functional, the economic/financial and the interpretive.

The functional and economic streams promote a logical rationalistic philosophy which searches for the efficiency and effectiveness of an information system in technical and business terms. Developments under these areas have attracted attention for a long period of time as they have addressed necessary questions regarding the performance and the financial aspects of the technical components and their investment returns. These two modes, although necessary and complementary, suffer a number of deficiencies. Their limitations include the:

- limited consideration of the organizational context,
- narrow purposes deriving from the formal/rational paradigm,
- lack of consideration of the new content elements and relevant measures,
- confined and fragmented time horizon,
- neglect of human aspects of evaluation,

- narrow methodological focus.

As a response to the limitations identified above, alternative evaluation approaches based on the interpretive paradigm have been proposed. The more interpretive approaches come closer to the social and real nature of IS and evaluation, but they have not managed to provide a completely 'rich picture' of the phenomenon. Moreover, these attempts lack empirical evidence for their applicability and wide acceptance by practitioners. A summary of the major issues addressed in each stream is presented in Table 1 below.

Table 1 - Literature review summary

	Technical stream	Economic stream	Interpretive alternatives
Purpose/ Reasons	• Technical performance (e.g. quality) • Control of resources (e.g. costs)	• Quality & utilization of IS outputs (e.g. accuracy of information)	• Context-sensitive (i.e. contingent, emergent) • Understanding of social actions • Organizational learning
The subject of evaluation Criteria and measurement	• IT system • Automate - Cost reduction	• IS outputs • Informate - Productivity - Business value - User satisfaction • Uncertainty/Risks	• Broad portfolios of processes and systems • Intermediate relevant measures (e.g. more reliable systems)
Time frame	• *Ex ante* and *ex post* in relation to the systems development life cycle	• *Ex ante* & *ex post* in relation to the systems development life cycle	• Continuous benefits management
People	• IT experts	• IT experts • Finance experts • Business managers	• 'Evaluation party' including internal and external stakeholders
Methodologies /Tools	• Quality-related (e.g. TQM, software metrics) • Cost-related (e.g. COCOMO, function point analysis)	• Economic oriented (e.g. agency theory) • Finance oriented (e.g. CBA, SESAME, DCF, IRR) • Behavioral science driven (e.g. ROM, value analysis)	• Meta-methodologies • Contemporary methods (experimental and exploratory)
Underlying assumptions	• Efficiency	• Effectiveness	• Understanding

It is evident that further research in the area is necessary to enhance the qualities of interpretive evaluation approaches and provide them with the necessary practicalities.

REFERENCES

Albrecht, A. and Gaffney, J. (1983). Software Function, source lines of code, and development effort prediction. *IEEE Transactions on Software Engineering*, SE-9 (6), 639-647.

Angell, I.O. and Smithson, S. (1991). *Information systems management - Opportunities and risks*. MacMillan, London.

Avgerou, C. (1995). Evaluating Information Systems by Consultation and Negotiation. *International Journal of Information Management*, 15 (6), 427-436.

Bacon, C.J. (1994). Why companies invest in information technology. In *Information Management. The evaluation of information systems investments*, (ed. L. Willcocks), Chapman & Hall, London, 31-47.

Baker, B. (1995). The role of feedback in assessing information systems planning effectiveness. *Journal of Strategic Information Systems*, 4 (1), 61-80.

Bakos, J.Y. and Kemerer, C.F. (1992). Recent applications of economic theory in Information Technology research. *Decision Support Systems*, 8, 365-386.

Ballantine, J., Galliers, R. and Stray, S.J. (1995). The use and importance of financial appraisal techniques in the IS/IT investment decision-making process - resent UK evidence. *Project Appraisal*, 10 (4), 233-241.

Ballantine, J. And Stray, S. (1999). Information systems and other capital investments: evaluation practices compared. *Journal of Logistics and Information Management*, 12 (1-2), 78-93.

Baroudi, J.J. and Orlikowski, W.J. (1988). A Short Form Measure of User Information Satisfaction: A Psychometric Evaluation and Notes on Use. *Journal of Management Information Systems*, 4 (4), 45-59.

Barua, A., Kriebel, C.H. and Mukhopadhyay, T. (1991). *Information Technologies and Business Value: An Analytic and Empirical Investigation*. Working Paper, Graduate School, Carnegie Mellon University, May.

Berghout, E. (1997). *Evaluation of information systems proposals: Design of a decision support method*. E.W.Berghout, Rotterdam.

Birch, D.G.W. and McEvoy, N.A. (1995). Structured risk analysis for information systems. In *Hard Money - Soft Outcomes*, (eds B. Farbey, F. Land and D. Targett), Alfred Waller, Oxon, 29-51.

Blackler, F. and Brown, C. (1985). Evaluation and the impact of Information Technologies on People in Organizations. *Human Relations*, 38 (3), 213-231.

Boehm, B. (1981). *Software Engineering Economics*. Prentice Hall, New Jersey.

Boehm, B. (1988). A spiral model of software development and enhancement. *IEEE Computer*, May, 61-72.

Brenshahan, T.F. (1986). Measuring the spillovers from technical advance: Mainframe computers in financial services. *The American Economic Review*, 76 (4), 742-755.

Brynjolfsson, E. (1993). The productivity paradox of information technology. *Communications of the ACM*, 36 (12), December, 67-77.

Brynjolfsson, E. and Hitt, L. (1993). *New Evidence on the Returns to Information Systems*. Working Paper 3571-93, MIT Sloan School of Management, October.

Capron, H. (Ed.) (1992). *Proceedings of the Workshop on Quantitative Evaluation of the Impact of R&D Programmes*. January 23-24, Commission of the European Communities, Brussels.

Carlson, W.M. and McNurlin, B.C. (1992). *Uncovering the Information Technology Payoffs*. I/S Analyzer Special Report, United Communications Group, Rockville, Maryland.

Charette, R.N. (1989). *Software Engineering Risk Analysis and Management*. New York, NY. McGraw Hill.

Ciborra, C. (1987). Research agenda for a transaction cost approach to information systems. In *Critical issues in information systems research* (eds R.J. Boland and R. Hirschheim), John Wiley & Sons, Chichester, 253-274.

Clemons, E.K. (1991) Evaluation of Strategic Investments in Information Technology. *Communications of the ACM*, 34 (1), 23-36.

Coleman, T. and Jamieson, M. (1994). Beyond return on investment. In *Information Management. The evaluation of information systems investment,* (ed. L. Willcocks), Chapman & Hall, London, 189-205.

Cronk M.C. and Fitzgeralnd, E.P. (1999). Understanding IS business value: derivation of dimensions, *Journal of Logistics and Information Management*, 12 (1-2), 40-49.

Currie, W. (1995). *Management Strategy for IT. An International Perspective*. Pitman Publishing, London.

Cyert, RM. and March, J.G. (1963). *A Behavioural Theory of the Firm*. Prentice Hall, Engelwood Cliffs, New Jersey.

DeLone, W.H. and McLean, E.R. (1992). Information Systems Success: The Quest for the Dependent Variable. *Information Systems Research*, 3 (1), 60-95.

Dos Santos, B.L. (1994). Assessing the value of strategic information technology investments. In *Information Management. The evaluation of information systems investment,* (ed. L. Willcocks), Chapman & Hall, London, 133-148.

Earl, M. (1989). *Management Strategies for Information Technology*. Prentice Hall, London.

Farbey, B., Land, F. and Targett, D. (1992). Evaluating investments in IT. *Journal of Information Technology*, 7, 109-122.

Farbey, B., Land, F. and Targett, D. (1993). *How to Assess your IT Investment. A study of Methods and Practice*. Butterworth Heinemann, Oxford.

Farbey, B., Land, F. and Targett, D. (1994). Matching an IT project with an appropriate method of evaluation: a research note on 'Evaluating investments in IT'. *Journal of Information Technology*, 9, 239-243.

Farbey, B., Land, F. and Targett, D. (eds) (1995). *Hard Money - Soft Outcomes. Evaluating and Managing the IT Investment*. Alfred Waller Ltd, Oxon.

Fenton, N.E. (1991). *Software Metrics. A Rigorous Approach*. Chapman & Hall, London.

Finkelstein, C. (1989). *An introduction to information engineering: from strategic planning to information system*. Addison Wesley, Reading MA.

Francalanci, C. and Galal, H. (1998). Aligning IT investments and workforce composition: the impact of diversification in life insurance companies. *European Journal of Information Systems*, 7 (3), pp175-184.

Gilb, T. and Finzi, S. (1988). *Principles of Software Engineering Management*. Addison Wesley, Wokingham.

Gregory, A.J. and Jackson, M.C. (1991). Evaluating organizations: a systems and contingency approach. *Systems Practice*, 5 (1), 37-60.

Gregory, A.J. and Jackson, M.C. (1992). Evaluation Methodologies: A System for Use. *Journal of Operational Research Society*, 43 (1), 19-28.

Grindley, K. (1991). *Managing IT at Board Level. The Hidden Agenda Exposed*, Pitman Publishing, London.

Guba, E.G. and Lincoln, Y.S. (1989). *Fourth Generation Evaluation*. Sage Publications, Newbury Park, California.

Gurbaxani, V. and Whang, S. (1991). The Impact of Information Systems on Organizations and Markets. *ACM Communications*, 34 (1), 59-73.

Hackett, G. (1990). Investment in technology: the service sector sinkhole?, *Sloan Management Review*, Winter, 97-103.

Hares, J. and Royle, D. (1994). *Measuring the Value of Information Technology*. John Wiley & Sons, Chichester.

Hawgood, J. and Land, F. (1988). A Multivalent Approach to Information Systems Assessment. In *Information Systems Assessment: Issues and Challenges*, (eds N. Bjorn-Andersen and G.B. Davis), North Holland, Amsterdam, 103-124.

Heemstra, F.J. (1992). Software cost Estimation. *Information and Software Technology*, 34 (10), 627-639.

Henderson, J.C. and Venkatraman, N. (1993). Strategic alignment: Leveraging information technology for transforming organizations. *IBM Systems Journal*, 32 (1), 4-16.

Hirschheim, R. and Smithson, S. (1987). Information Systems Evaluation: Myth and Reality. In *Information Analysis: Selected Readings*, (ed. R. Galliers), Addison Wesley, Sydney, 367-380.

Hirschheim, R. and Smithson, S. (1988). A critical analysis of information systems evaluation. In *Information Systems Assessment: Issues and Challenges*, (eds N. Bjorn-Andersen and G.B. Davis), North Holland, Amsterdam, 17-37.

Hitt, L.M. and Brynjolfsson, E. (1996). Productivity, Business Profitability, and Consumer Surplus: Three *Different* Measures of Information Technology Value. *MIS Quarterly*, June, 121-142.

Hochstrasser, B. and Griffiths, C. (1991). *Controlling IT Investment. Strategy and Management*. Chapman & Hall, London.

Hopwood, A.G. (1983). Evaluating the Real Benefits. In *New Office Technology: Human and Organizational Aspects*, (eds H.J. Otway and M. Peltu), Frances Pinter, London, 37-50.

House, E.R. (ed.) (1983). *Philosophy of Evaluation*. Jossey-Bass, San Francisco, CA..

Ives, B., Olson, M. and Baroudi, J.J. (1983) The Measurement of User Information Satisfaction. *Communications of the ACM*, 26 (10), 785-793.

Kambil, A., Henderson, J. and Mohsenzadeh, H. (1993). Strategic Management of Information Technology Investments: An Options Perspective. In *Strategic Information Technology Management: Perspectives on Organizational Growth and Competitive Advantage*, (eds R.D. Banker, R.J. Kauffman and M.A. Mahmood), Idea Group Publishing, Harrisburg, PA, 161-178.

Kaplan, R.S. (1986). Must CIM be Justified on Faith Alone?, *Harvard Business Review*, March-April, 87-95.

Kauffman, R.J. and Weill, P. (1989). An Evaluative Framework for Research on the Performance Effects of Information Technology Investment. In *Proceedings of the Tenth International Conference on Information Systems*, Boston, MA, December, 377-388.

Keen, P. (1981). Value Analysis: Justifying Decision Support Systems. *MIS Quarterly*, 5 (1), 1-16.

Kemerer, C. (1987). An empirical validation of software cost estimation models. *Communications of the ACM*, 30 (5), 416-429.

KPMG Consulting (2000). *eBusiness and Beyond: Board Level Drivers and Doubts*, Research Report, January 2000, London.

Krull, W. (1993). *Guidelines for the Evaluation of European Community R&D Programmes. Recommendations.* European Commission, Luxembourg.

Kumar, R.L. (1996). A Not on Project Risk and Option Values of Investments in Information Technology. *Journal of Management Information Systems*, 13 (1), 187-193.

Land, F. and Tweedie, R. (1992). *Preparing for Information Technology Implementation: Adding a TQM Structure.* Working Paper, Bond University, Gold Coast, Queensland.

Landauer, T.K. (1995). *The Trouble with Computers. Usefulness, Usability, and Productivity.* The MIT Press, Cambridge, MA.

Lederer, A.L. and Prasad, J. (1993). Information systems software cost estimating: a current assessment. *Journal of Information Technology*, 8, 22-33.

Legge, K. (1984). *Evaluating Planned Organizational Change.* Academic Press, London.

Loveman, G.W. (1988). *An assessment of the productivity impact of Information Technologies.* Working paper 88-054, Management in the 1990s, Sloan School of Management, MIT.

Lucas, H.C. and Moore, J.R. (1976). A Multiple Criterion Scoring Approach to Information System Project Selection. *Information*, 19 (1), 1-12.

Mooney, J.G., Gurbaxani, V. and Kraemer, K.L. (1995). A Process Oriented Framework for Assessing the Business Value of Information Technology. In *Proceedings of the Sixteenth International Conference on Information Systems,* (eds J.I. DeGross, G. Ariav, C. Beath, R. Hoyer and C. Kemerer), December 10-13, Amsterdam, The Netherlands, 17-27.

Mukhopadhyay, T. (1993). Assessing the Economic Impacts of Electronic Data Interchange Technology. In *Strategic Information Technology Management: Perspectives on Organizational Growth and Competitive Advantage*, (eds R.D. Banker, R.J. Kauffman and M.A. Mahmood), Idea Group Publishing, Harrisburg, PA, 241-296.

Nissen, M.E. (1994). Valuing IT through virtual process measurement. In *Proceedings of the Fifteenth International Conference on Information Systems,* (eds J.I. DeGross, S.L. Huff and M.C. Munro), December 14-17, Vancouver, Canada, 309-323.

Nolan Norton Institute (1998). *IT Governance and management. Research Memorandum*, Nolan Norton & Co, June 1998.

OECD (1990). *Principles for the Evaluation of Programmes Promoting Science, Technology and Innovation.* Committee for Scientific and Technological Policy, Paris.

OECD (1992). *National Programmes for Supporting Local Initiatives: Content and Evaluation.* OECD, Paris.

Parker, M.M. (1996). *Strategic Transformation and Information Technology. Paradigms for Performing While Transforming.* Prentice Hall, Upper Saddle River, NJ.

Parker, M.M., Benson, R.J. and Trainor, H.E. (1988). Information Economics: Linking Business Performance to Information Technology. Prentice-Hall, NJ.

Peters, G. (1994). Evaluating your computer investment strategy. In *Information Management. The evaluation of information systems investment*, (ed. L. Willcocks), Chapman & Hall, London, 99-112.

Peters, G. (1996). From strategy to implementation: identifying and managing benefits of IT investments. In *Investing in Information Systems. Evaluation and Management*, (ed. L. Willcocks), Chapman & Hall, London, 225-240.

Pettigrew, A.M. (1985a). *The Awakening Giant: Continuity and Change in ICI*. Blackwell, Oxford.

Pettigrew, A.M. (1985b). Contextualist Research and the Study of Organizational Change Processes. In *Research Methods in Information Systems*, (eds E. Mumford, R. Hirschheim, G. Fitzgerald and T. Wood-Harper), Elsevier Science Publishers, North Holland, Amsterdam, 53-78.

Powell, P. (1992a). Information Technology and Business Strategy: A Synthesis of the Case for Reverse Causality. In *Proceedings of the Thirteenth International Conference on Information Systems,* (eds J.I. DeGross, J.D. Becker and J.J. Elam), December 13-16, Dallas, Texas, 71-80.

Powell, P. (1992b). Information Technology Evaluation: Is It Different?. *Journal of Operational Research Society*, 43 (1), 29-42.

Putnam, L.N. (1980). *Software Cost Estimating and Life Cycle Control*. IEEE Computer Science Press, London.

Remenyi, D., Money, A. and Twite, A. (1991). *A Guide to Measure and Managing IT Benefits*. NCC Blackwell, Manchester & Oxford.

Remenyi, D. and Sherwood-Smith, M. (1997). *Achieving Maximum Value from Information Systems*, John Wiley, NY.

Remenyi, D. and Sherwood-Smith, M. (1999). Maximising information systems value by continuous participative evaluation. *Journal of Logistics and Information Management*, 12 (1-2), 14-31.

Roach, S.S. (1989). America's white-collar productivity dilemma. *Manufacture Engineering*, August, 104.

Roach, S.S. (1991). Services under siege - The restructuring imperative. *Harvard Business Review*, September-October, 82-92.

Sassone, P.G. (1988). A survey of cost-benefit methodologies for information systems. *Project Appraisal*, 3 (2), 73-84.

Schäfer, G. (1988). *Functional Analysis of Office Requirements*. John Wiley & Sons, Chichester, England.

Scott, J. (1994). The link between organizational learning and the business value of information technology. In *Proceedings of the Fifteenth International Conference on Information Systems. Doctoral Consortium*, (eds R.D. Galliers, S. Rivard and Y. Ward), December 12-14, Vancouver Island, Canada.

Scott Morton, M.S. (ed.) (1991). *The corporation of the 1990s. Information Technology and Organizational Transformation*. Oxford University Press, New York.

Serafeimidis, V. (1997). *Information Systems Investment Evaluation: Conceptual and Operational Explorations*, Unpublished PhD thesis, London School of Economics and Political Science, UK.

Shoval, P. and Lugasi, Y. (1987). Models for computer system evaluation and selection. *Information and Management*, 12, 117-129.

Smithson, S. and Hirschheim, R. (1998). Analysing information systems evaluation: another look at an old problem, *European Journal of Information Systems*, 7 (3), pp158-174.

Strassmann, P. (1985). *Information Payoff*. Free Press, New York.

Strassmann, P. (1990). *The Business Value of Computers*. Information Economics Press, New Cannan, Conn.

Symons, V.J. (1990). *Evaluation of Information Systems: Multiple Perspectives*. Unpublished PhD thesis, University of Cambridge.

Symons, V.J. (1993). Evaluation and the failure of control: information systems development in the processing company. *Accounting, Management and Information Technology*, 3 (1), 51-76.

Symons, V.J. and Walsham, G. (1991). The evaluation of Information Systems: a critique. In *The Economics of Information Systems and Software,* (ed. R. Veryard), Butterworth Heinemann Ltd, Oxford, 71-88.

Walsham, G. (1993). *Interpreting Information Systems in Organizations*. John Wiley & Sons, Chichester, England.

Ward, J. and Griffiths, P. (1996). *Strategic Planning for Information Systems*. 2nd Edition, John Wiley & Sons, Chichester.

Ward, J., Taylor, P. and Bond, P. (1996). Evaluation and realisation of IS/IT benefits: an empirical study of current practice. *European Journal of Information Systems*, 4 (4), 214-225.

Ward, J.M. (1992) *Assessing and Managing the Risks of IT Investments*. Report SWP 24/92, Cranfield School of Management, Cranfield.

Ward, J.M. (1994). A portfolio approach to evaluating information systems investment and setting priorities. In *Information Management. The evaluation of information systems investment*, (ed. L. Willcocks), Chapman & Hall, London, 81-96.

Weill, P. (1992). The Relationship Between Investment in Information Technology and Firm Performance: A Study of the Valve Manufacturing Sector. *Information Systems Research*, 3 (4), 307-333.

Whiting, R.E., Davies, N.J. and Knul, M. (1993). Investment appraisal for IT systems. *BT Technology Journal*, 11 (2), 193-211.

Willcocks, L. (1992). Evaluating Information Technology investments: research findings and reappraisal. *Journal of Information Systems*, 2 (4), 243-268.

Willcocks, L. (1994). Introduction: of capital importance. In *Information Management. The evaluation of information systems investment,* (ed. L. Willcocks), Chapman & Hall, London, 1-27.

Willcocks, L. (1996). Introduction: beyond the IT productivity paradox. In *Investing in Information Systems. Evaluation and Management*, (ed. L. Willcocks), Chapman & Hall, London, 1-12.

Willcocks, L. and Lester, S. (1994). Evaluating the feasibility of information systems investments recent UK evidence and new approaches. In *Information Management. The evaluation of information systems investment*, (ed. L. Willcocks), Chapman & Hall, London, 49-75.

Willcocks, L. and Margetts, H. (1994). Risk and information systems: developing the analysis. In *Information Management. The evaluation of information systems investments*, (ed. L. Willcocks), Chapman & Hall, London, 207-227.

Wiseman, D. (1994). Information economics: a practical approach to valuing information systems. In *Information Management. The evaluation of information systems investments*, (ed. L. Willcocks), Chapman & Hall, London, 171-187.

Zuboff, S. (1988). *In the age of the smart machine: The future of work and power*. Basic Books, New York.

Zuboff, S. (1991). Informate the Enterprise: An Agenda for the 21st Century. *National Forum*, Summer, 3-7.

Zultner, R.A. (1993). TQM for technical teams. *Communications of the ACM*, 36 (10), 78-91.

ENDNOTE

1 The statements and opinions in this chapter are in all respects those of the author and do not represent the views of KPMG Consulting.

Chapter V

Methodologies for IT Investment Evaluation: A Review and Assessment

Egon Berghout
Delft University of Technology, The Netherlands

Theo-Jan Renkema
Eindhoven University of Technology, The Netherlands

INTRODUCTION

The evaluation of information technology (IT) investments has been a recognised problem area for the last four decades, but has recently been fuelled by rising IT budgets, intangible benefits and considerable risks and gained renewed interest of both management and academics. IT investments already constitute a large and increasing portion of the capital expenditures of many organizations, and are bound to absorb a large part of future funding of new business initiatives. However, for virtually all firms, it is difficult to evaluate the business contribution of an IT investment to current operations or corporate strategy. Consequently, there is a great call for methods and techniques that can be of help in evaluating IT investments, preferably at the proposal and decision-making stages.

The contribution of this chapter to the problem area is twofold. First, the different concepts, which are used in evaluation are discussed and more narrowly defined. When speaking about IT investments, concepts are used that originate from different disciplines. In many cases there is not much agreement on the precise meaning of the different concepts used. However, a common language is a prerequisite for the successful communication between the different organizational stakeholders in evaluation. In addition to this, the chapter reviews the current methods for IT investment evaluation and puts them into a frame of reference. All too often new methods and guidelines for investment evaluation are introduced, without building on the extensive body of knowledge that is already incorporated in the available methods. Four basic approaches are discerned: the financial approach, the multi-criteria approach, the ratio approach and the portfolio approach. These approaches are subsequently compared on a number of characteristics on the basis of methods that serve as examples for the different approaches. The chapter concludes with a review of key limitations of evaluations, suggestions on how to improve evaluation practice and recommendations for future research. This chapter draws on earlier work as published in Renkema and Berghout (1997), Berghout (1997), and Renkema (1996; 2000).

Copyright © 2001, Idea Group Publishing.

BACKGROUND

Investments for developing and implementing information systems (IS) are large and increasing. They constitute up to 50 percent of the capital expenditures of large organizations (Earl, 1989; Davenport and Short, 1990; Keen, 1991). Information systems are not only used in administrative and decision making tasks but are changing the shape of production processes (e.g. embedded software, workflow management, or ERP packages) and enable the development of new products and services (e.g. chip card services, e-commerce). The many empirical studies carried out in the 1990s show that organizations have several problems with the evaluation of proposals for IT investments (e.g. Hochstrasser and Griffiths, 1990; Bacon, 1992; Farbey et al., 1992; Yan Tam, 1992; Willcocks and Lester, 1993). A number of causes can be identified. Because information systems are often for a great extent integrated in the organization, it is difficult to establish the boundaries of the system. For instance, which user costs of a new electronic mail system should be considered in an investment proposal? Another possible cause is the ongoing dispute on the relevant decision criteria. How should, for example, long term impacts of an IT investment be incorporated? An example of this is the contribution of a database management system to the realisation of corporate data infrastructure in an organization.

A plethora of methods and techniques have been proposed to assist in the evaluation of IT investment proposals. A number of researchers have identified over sixty methods, that all aim to be of help in the evaluation of IT investment proposals (Swinkels and Irsel, 1992; Berghout and Renkema, 1997). An overview of these methods is given in the Appendix. Already in 1961 the International Federation of Information Processing devoted its first conference to evaluation issues (Frielink, 1961) and in 1968, Joslin, wrote a book on computer selection (Joslin, 1968). The purpose of this chapter is to improve insight in the current methods for the evaluation of IT investments, with a focus on evaluation in the proposal stage, and to assess their main strengths and weaknesses. For the moment this is the maximum that can be strived for, as research that has validated evaluation methods is hardly available. General prescriptions about the use of which method in which circumstances should be applied can not be given. Current research is still focusing on finding the essential evaluation criteria, the circumstances in which these should be used and the inclusion of the criteria in the evaluation process (Berhout, 1999; Renkema 2000).

TERMINOLOGY

Necessity

In order to be able to compare methods for the evaluation of IT investment proposals, one should avoid misinterpretations about the different concepts used. Also in evaluation practice the communication between stakeholders in the evaluation process can be improved by the use of a common language. Often when discussing IT investments notions are used, e.g., costs and benefits, without being certain that everyone means the same thing. This section discusses and defines the concepts that are used in evaluation and in the remainder of the hapter.

Definitions

A distinction is made between *financial* and *nonfinancial* impacts. Financial impacts are the impacts which are expressed in monetary terms. Nonfinancial impacts are not expressed in monetary terms. The latter category is referred to as *contribution*. An *impact* is defined as an event that arises from the introduction of the information system, starting with the decision to go ahead with the investment. Financial and non-financial impacts together determine the

Figure 1: Definitions

Definitions		Positive	Negative
Financial and non-financial	Business Value	Benefits	Sacrifices
Financial	Profitability (profits or losses)	Yieldings	Costs
	Return	Earnings	Expenditures
Non-financial	Contribution	Positive	Negative

(business) value of an information system. *Benefits* refer to all positive impacts of an IT investment and *sacrifices* to all negative impacts.

With respect to financial impacts, a further distinction is made between *profitability* and *return*. The *return* is determined by cash flow evaluation. Positive, incoming cash flows are *earnings* and negative, outgoing cash flows are *expenditures*. The *profitability* in terms of profits (positive) or losses (negative) is defined as the accounting registration of *yieldings* and *costs*. A sound financial evaluation of a proposed investment is based upon an analysis of the return and not on the profitability (Brealy and Myers, 1988; Fox et al., 1990). Figure 1 gives an overview of the defined concepts.

In several methods *risk* is included as a separate criterion. Risk is seen as a measure of uncertainty with respect to a specific impact of an investment. This uncertainty can for example be expressed in terms of the chances that the expected expenditures will be higher or the expected earnings will be lower. This implies that risk refers to every impact of an IT investment. Scarce (financial and nonfinancial) resources are allocated to an *investment* on a certain decision moment, and these are expected to yield a positive result during several years. By doing this, financial assets are translated into real assets. An investment has a certain lifetime. If it is decided to stop allocating resources to an investment, before the planned investment lifetime is over, the investment has become a *divestment*. Investments which are done after the initial decision stage, and which require separate justification, are referred to as *maintenance investments*. An *information system* is defined as the result of an IT investment and refers to all components that together provide the necessary information. These components are *hardware* and *software*, *people* and *procedures* with which they work, and the *data* that are processed by the system (Brussaard, 1993).

A REVIEW OF EVALUATION METHODS

Different authorities have given an overview of the available methods for the evaluation of IT investment projects, although often from different perspectives (see e.g. Carlson and McNurlin, 1989; Powell, 1992; Farbey et al., 1993). More than 65 discrete methods are listed in the Appendix, all with the purpose to assist organizations with their IT investment decisions. The existing methods are structured according to the type of method and a number of key characteristics, by means of discussing ten typical methods. The different available financial methods are not analyzed in detail in Figure 2, since the characteristics of these methods have many similarities.

Types of methods

Apart from methods that limit themselves to a purely financial appraisal of investment projects, three main nonfinancial approaches can be discerned. These are the multi-criteria

Figure 2: Methods for IT evaluation

Characteristics		Type of method							
		Financial methods	Multi-criteria methods		Ratio methods		Portfolio methods		
			Information Economics	SIESTA	Return on Management	IT-Assessment	Bedell's method	Invsetment Portfolio	Investment Mapping
Objects of the method	Breadth	Project-level	Project-level	Project-level	Organization-level	Project en organization-level	Project en organization-level	Project-level	Project en organization-level
	Type of application area	Business investments	IT investments	IT investments	Business investments	IT investments	IT investments	IT investments	IT investments
Evaluation criteria	Financial	Earnings & expenditures	Earnings & expenditures (average accounting rate of return)	Unclear	Own measure	Yieldings and costs	Implicit assessment, specification of expenditures required	Earnings & expenditures (NPV)	Earnings & expenditures (IRR)
	Non-financial	None	4 business criteria 1 technological criterion	7 business criteria 6 technological criteria	Unclear	Different business	Quality and importance IT domain	Business domain & 3 investment separately	3 type of benefits and orientations
	Risks	Deduction from expectations or coverage through adjusted discount rate	1 business risk 4 technological risks	4 business risks 8 technological risks	None	None	None	Deduction from expectations, spread can be specified	Spread can be specified
Support of the decision-making process		None	Discussed examples and mentions stakeholder groups	None	None	None	Maximal appraisal is once every year, mentions top mgt., users, IT staff	Discusses responsibilities, addresses role of business mgt, IT mgt. en projectmgt.	None
Measurement scale		Ratio and interval	Ordinal	Ordinal	Interval	Several scales	Ordinal	Ordinal and interval	Ordinal and interval

The financial approach

Methods with a financial approach to investment evaluation only consider impacts which can be monetary valued. Traditionally they are prescribed for the justification and selection of all corporate investment proposals, and focus on the incoming and outgoing cash flows as a result of the investment made (see the standard accounting and capital budgeting texts, e.g., Brealey and Myers, 1988; Weston and Copeland, 1991). A calculation of yieldings and costs, as an estimation of the profitability of an investment, is not considered as being correct for evaluating the financial impacts of an investment proposal. When appraising a project in financial terms the purpose is to evaluate what ex-ante (i.e. before the investment is actually implemented) the financial return is, as a consequence of the earnings and expenditures which result from the investment. Both are cash flows, as becomes clear when the terms cash proceeds and cash outlays are used (Bierman and Smidt, 1984). A payment to an external supplier for computing equipment is an example of an expenditure and a reduction of staffing levels is an example of earnings. Cost and yieldings are essentially accounting-based terms, which are meant to report on and account for ex-post profits or losses, and therefore open to many difficult to trace accounting decisions. Lumby (1991) therefore speaks of costs and yieldings as reporting devices and of expenditures and earnings as decision-making devices.

The multi-criteria approach

Apart from financial impacts, an IT investment will generally have nonfinancial impacts, i.e. positive or negative impacts that can not or not easily be expressed in monetary terms. Because of the differences between financial and nonfinancial impacts it is difficult to compare the different impacts on an equal basis. This however is a prerequisite for a comprehensive evaluation of an investment proposal and adequate prioritization of different proposals. Methods from the multi-criteria approach solve this problem by creating one single measure for each investment. Multi-criteria methods are used in many decision making problems and are well known for their strength in combining quantitative and qualitative decision-making criteria. A good theoretical treatment of multi-criteria methods, applied in the realm of investments in advanced production technologies is given by Canada and Sullivan (1989). Different variants of multi-criteria methods exist, however, normally the method works as follows. Before using a multi-criteria method, a number of goals or decision criteria have to be designed. Subsequently, scores have to be assigned to each criterion for each alternative considered. Also the relative importance of each alternative should be established, by means of weights. The final score of an alternative is calculated by multiplying the scores on the different decision criteria with the assigned weights and adding or multiplying these.

The ratio approach

Economic researchers have already for a long time been paying special attention is to the possibilities to compare organizational effectiveness by means of ratios. Several ratios have been proposed to assist in IT investment evaluation (Butler Cox, 1990; Farbey et al., 1992). Examples of meaningful financial ratios are: IT expenditures against total turnover and all yieldings that can be attributed to IT investments against total profits. Ratios do not necessarily take only financial figures into account. IT expenditures can for instance be related to the total number of employees or to some output measure (e.g. products or services).

The portfolio approach

Portfolios (or grids) are a well-known decision-making tool in the management literature. A portfolio used in many strategic analyses, is the Growth Share portfolio for the positioning of product families of the Boston Consulting Group, which distinguishes between wild cats, stars, cash cows and dogs. In a portfolio several investment proposals are plotted against (sometimes aggregated) decision-making criteria. Portfolio methods combine the comprehensiveness of multi-criteria methods with the graphic opportunities of portfolios. The number of evaluation criteria is generally less than in multi-criteria methods, however, the result may often be more informative.

Key Characteristics of Methods

Apart from the type of method, the many available IT investment appraisal methods also differ in terms of their characteristics. This chapter distinguishes between four key characteristics to identify the main features of the different methods to be discussed: objects of the method, evaluation criteria of the method, support of the decision-making process and the measurement scale of the method. These characteristics have the following meaning:

a. Objects of the method

The objects of the method define into what extent a method limits itself to certain type of IT investments. Discerned are the *breadth* of the method and the *type* of investments for which the method can be used.

a.1 Breadth of the method

With respect to the breadth of a method a distinction is made between project level assessment (discrete investment proposals) and organizational level assessment (the method takes a higher level than the project level into account, for instance by looking at the IT intensity or automation degree of certain departments).

a.2 Type of application area

A method can also limit itself to a certain application area. A distinction is made between methods that were designed specifically for IT investment appraisal and methods which are use for all types of business investments.

b. Evaluation criteria of the method

The evaluation criteria are the aspects that are addressed in the decision whether to go ahead with the proposed investment and to set priorities between competing projects. A distinction is made between financial impacts, nonfinancial impacts and risks. With respect to the financial impacts it is important to know whether the return or the profitability is evaluated.

c. Support of the decision-making process

Support of the decision-making process refers to the extent into which a method indicates or prescribes how this should be used in evaluation practice. Complaints with respect to the evaluation of investment proposals often have to do with how difficult it is to make the possible benefits more tangible. A method might for instance suggest possible ways to identify the benefits of the investments, in addition to giving mere evaluation criteria. Furthermore, support could be given regarding persons to be involved in the evaluation process and their responsibilities, data collection on the right level of detail, and the frequency of evaluations after the proposal stage.

d. Measurement scale of the method

The outcomes of an evaluation method can be measured on the following four measurement scales, with increasing meaning:
- *Nominal scale*: measurement units are used to classify on the basis of uniformity. This is done by defining labels (for instance the purpose of the investment: value improvement, cost containment, risk reduction, etc., or simply type 1, 2 and 3 investments);
- *Ordinal scale*: measurement units are also used to represent a certain order. An example is to order all investment proposals from good to bad. The differences in this order indicate that a certain proposal is better or worse than another proposal, but not how much better or worse;
- *Interval scale*: also the differences (or intervals) between measurement units have a real meaning. This implies that one can not only speak of better or worse, but also of this much better and this much worse. A clear definition of a unity of distance is required to be able to do this (a well-known example is the thermometer unit Celsius degrees). Intervals are defined in terms of a number of measurement units, which leads to a measurement rod. When ranking investment projects on a scale from 0 to 100 interval units, the difference between the scores 2 and 8 is for instance three times more than between 10 and 12.
- *Ratio scale*: which in fact is an interval scale with a real, absolute zero (e.g. weight, length, contents or money). Only one object measured can be given the number zero. When using a ratio scale, also the relation between two measurement units has practical meaning. A project with an estimated financial result of $100,000 not only is worth $50,000 more than an project with a result of $50,000, but also has a double value.

METHODS AND THEIR CHARACTERISTICS

In the previous section a review of the type of methods available for IT investment appraisal and their distinctive characteristics was given. In this section a number of methods are described in more detail. These methods serve as examples of the particular types of methods. The requirements for discussing a method are:
- The method should be well documented and accessible for further analysis;
- The method should have a clearly defined structure. This means that a method consisting of merely heuristic guidelines will not be discussed;
- The method should be characteristic of the type of method reviewed.

Financial Methods

Payback period

The payback period is the period between the moment that an IT investment gets funded and the moment that the total sum of the investment is recovered through the net incoming cash flows. If a proposed project for instance requires a investment of $1 million and realizes cost savings of $500000 a year, the payback time is 2 years. The investing organization decides on a time period within which the sum must be recovered; if it is less or equal than the calculated payback period then it is justified to invest in the proposed project.

Average accounting rate of return

When calculating the accounting rate of return, the first step is to estimate the financial return of an investment for each year of the projected lifetime. This is then divided by the lifetime of the project. By dividing this then by the initial investment sum, the remaining ratio is the return on investment. A related financial indicator, which sometimes is also referred to as the average accounting rate of return, is calculated by deducting the average yearly depreciation from the average yearly net cash flow and divided by the average investment sum during the investment's lifetime.

Net present value

The starting point in the net present value (NPV) method is the opportunity cost of capital. This rate is used as the discount rate to calculate the net present value. If this value is larger than zero, it is considered to be wise to go ahead with the investment. The higher the net present value, the higher the project priority should be. By using the discounting technique, earnings and expenditures, which do not occur at the same moment in time, become comparable. To illustrate this: if one for instance puts $100 in a savings account, this will be worth $105 when receiving an interest of 5%. Consequently, this is also valid the other way around, if one has to pay $105 in a year, this liability is worth $100 now. By applying this method, all earnings and expenditures can be made today's money.

The internal rate of return

The internal rate of return (IRR) is the threshold at which, after discounting the incoming and outgoing cash flows, the net present value equals zero. If this threshold exceeds the opportunity cost of capital, the investment is considered to be worthwhile. So, instead of using a fixed rate of 5%, as in the net present value method, the internal rate of return for earnings and expenditures is calculated and checked whether this exceeds the 5%. This method normally requires substantial calculations.

The latter two methods, often referred to as Discounted Cash Flow (DCF) methods, are from a financial perspective considered to be the superior financial methods, as they take the time value of money into account. This means that if the moment of receipt of cash flow is further into the future, the value of these cash flows will be less (as decision makers normally have an aversion to risk).

Multi-criteria Methods

Information Economics

In the field of evaluating IT investment proposals, the Information Economics multi-criteria method of Parker et al. (1988, 1989) is one of the best known methods. A substantial number of firms have already applied this method and many consulting firms have developed their own variants (see Wiseman, 1992). The first criterion of the Information Economics method gives a financial appraisal of a proposed investment. Parker et al. call this the enhanced return on investment (ROI). This ROI not only considers cash flows arising from cost reduction and cost avoidance, however, additional techniques are provided to estimate incoming cash flows:
- Value linking: additional cash flows that accrue to other departments;
- Value acceleration: additional cash flows due to reduced time scale for operations;

- Value restructuring: additional cash flows through restructuring work and improved job productivity;
- Innovation valuation: additional cash flows arising from the innovating aspects of the investment (e.g. competitive advantage).

With respect to nonfinancial impacts and risks, Information Economics makes a distinction between the business domain and the technology domain. The technology domain offers IT opportunities to the business domain, while the business domain focus is on optimal deployment of IT and pays the technology domain for use of resources. In the two domains several criteria are discerned (see Figure 3). To summarize, the total appraisal of an IT investments proposal takes place in three steps, covering financial, business and technological criteria, both positive and negative.

More recently, the concept of the Balanced Scorecard, as introduced by Kaplan and Norton (1992) has received a lot of attention by business and financial managers. This method can be seen as and extension of the traditional accounting-based financial methods, and has also been applied for IT investment assessment and measurement (Willcocks and Lester, 1994; Zee, 1996; Grembergen, 1999). The balanced scorecard has been developed in order to get a broader perspective in management accounting metrics than available by using strictly financial methods. A division is made between a financial perspective, a customer perspective, an internal perspective and an innovation and learning perspective. This method is, therefore, similar to the Information Economics method, except that the technology domain is missing. In fact the Gartner Group even advises to add technological criteria to the balanced scorecard (and make this method identical to the Information Economics method).

Figure 3: Information economics method (Parker et al., 1988)

Appraisal criteria	Meaning
Business Domain	
Strategic match	The extent into which the investment matches the strategic business goals
Competitive advantage	The extent into which the investment contributes to an improvement of positioning in the market (e.g. changes in industry structure, improvements of competitive positioning in the industry)
Management information	The extent into which the investment will inform management on core activities of the firm
Competitive response	The extent into which not investing implies a risk; a timely investment contribute to strategic advantage
Organizational risk	The extent into which new competencies are required
Technology Domain	
Strategic information systems architecture	The extent into which the investment matches the IT plan and the required integration of IT applications
Definitional uncertainty	The extent into which user requirements can be clearly defined
Technical uncertainty	The extent into which new technical skills, hardware and software are required
Infrastructure risk	The extent into which the investment requires additional infrastructure investments and the IT department is capable to support the proposed system

Methodologies for IT Investment Evaluation 87

SIESTA (Strategic Investment Evaluation and Selection Tool Amsterdam)

The SIESTA method probably is one of the most comprehensive multi-criteria methods (Van Irsel et al. 1992). The method has more than 20 criteria and is supported by questionnaires and software. The appraisal criteria are deduced from a model, in which a distinction is made between the business and the technological domain and three levels of decision-making are discerned. Benefit and risk criteria are deduced from the extent into which the different elements of the model fit the total model (inspired by the Strategic Alignment Model). Figure 4 illustrates the structure of the model.

Figure 4: SIESTA method (Van Irsel et al., 1992)

Ratio Methods

Return on Management method

A ratio approach which attracted a lot of attention is the Return on Management (ROM) method of Strassmann (1990), see also Van Nievelt (1992). The method presupposes that in today's information economy management has become the scarce resource and that right type of management defines the extent into which business value is derived from IT deployment. In the ROM method the value added by management is related to the costs of management. Analysis with the ROM method is supported by the MPIT database that contains company data of about 300 companies over several years. This database can be used for complete organizational diagnosis or for analyzing the impacts of specific investments. Unfortunately the database is not for public use, but has a commercial trademark. Figure 5 gives the formulas to calculate ROM.

Figure 5: Return on management method (Van Nievelt, 1992)

$$ROM = \frac{yieldings - full\ operating\ costs}{total\ costs - full\ operating\ costs}$$

$$= \frac{value\ added\ by\ management}{full\ cost\ of\ management}$$

$$= 1 + \frac{economic\ profit\ before\ taxes}{full\ cost\ of\ management}$$

IT assessment method

The consulting firm Nolan Norton has designed a method, IT assessment for the evaluation of information technology effectiveness from a strategic point of view (Zee and Koot, 1989; Janssen *et al.*, 1993). An important part of the method focuses on the analysis of financial and nonfinancial ratios. The ratios are subsequently compared with benchmarks; average values that were collected through research in other organizations. These benchmarks are not for public use. The ratios are also used for a historical analysis of the organization and its use of IT. Used this way, the ratio's can be of help in decision-making on new strategic initiatives.

Portfolio Methods

Bedell's method

The first portfolio method that is described is Bedell's portfolio method (Bedell, 1985; Van Reeken, 1992). In this method, subsequently, three questions are answered:
- Should the organization invest in IT applications?
- In which activities should the organization invest?
- Which IT applications should be developed?

The central premise of Bedell's method is that a balance is required between quality and importance of information systems. This is also the basis upon which the answers to the three questions are sought. IT investments are more appropriate if the relation between the perceived quality of the systems and the importance of information systems is worse. In this case information systems are more important, if they support important activities and if the activities are more important to the organization. Before the three questions can be answered and calculations are made, several assessments have to be performed. These assessments concern:
- The importance of an activity to the organization;
- The importance of IT-based support to the activities;
- The quality of the IT support in terms of effectiveness, efficiency and timing.

In a more detailed investment analysis the above three questions are answered by the involved stakeholders: senior management, user management and IT specialists. The prioritization of investment proposals is carried out by calculating the contribution of each IT system and by plotting three portfolios. The contribution of an IT application is defined as the importance of the system multiplied with the improvement of quality after development. To evaluate the business value of the investment, a Project-Return index can be calculated, by relating the contribution of the IT system to the development costs.

Investment Portfolio

In the second portfolio method, the Investment Portfolio, IT investment proposals are evaluated on three criteria simultaneously (Berghout and Meertens, 1992; Berghout, 1997; Renkema and Berghout, 1999). These three criteria are:
- The contribution to the business domain;
- The contribution to the IT domain;
- The financial return.

The portfolio serves as a framework to make the preferences of the different stakeholders explicit and debatable. Important stakeholders taken into account are senior management, IT management and project management of the evaluated project. An example of an Investment Portfolio is illustrated in Figure 6.

The size of the Net Present Value (NPV) of an investment proposal is plotted in the portfolio by means of circle. The larger the circle, the higher the expected NPV. The contribution to the business domain focuses on the long term benefits, leading to an improvement of the products or services of the organization. The authors suggest that the criteria of the Information Economics method can be used for this. The contribution to the IT domain is assessed by criteria such as: conformance to technology standards, market acceptance of the used technologies and continuity of the suppliers. In addition to evaluating a single IT investment proposal, the Investment Portfolio can be used to compare and prioritize several investment projects. The method also offers a risk and sensitivity analysis by varying the size of the circle and by changing the position of a circle.

Figure 6: The investment portfolio (Berghout and Meertens, 1992)

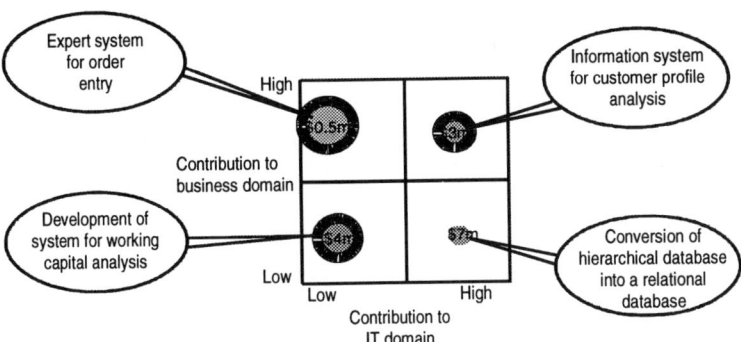

Investment Mapping

Peters (1988) developed the Investment Map, in which investment proposals are plotted against two main evaluation criteria: the investment orientation and the benefits of the investment. The investment orientation is broken up into infrastructure, business operations and market influencing. The benefits are broken up into enhancing productivity, risk minimization and business expansion. These categories partly overlap. Figure 7 gives an illustration of the Investment Map. The position of an investment proposal on the two axes is determined by a score on the evaluation criteria, which can be input to more detailed evaluation at the individual project level (Peters, 1990). A portfolio that has been visualized makes the IT investment strategy more explicit and debatable. The Investment Map can also be used to investigate the alignment of the IT investment strategy and the business strategy. To do this, a distinction is made between a quality driven or a cost leadership strategy. Additionally, it is possible to do a competitor analysis by plotting the strategies of the main competitors in the portfolio.

ASSESSMENT OF EVALUATION METHODS

Figure 7: Investment mapping (Peters, 1988)

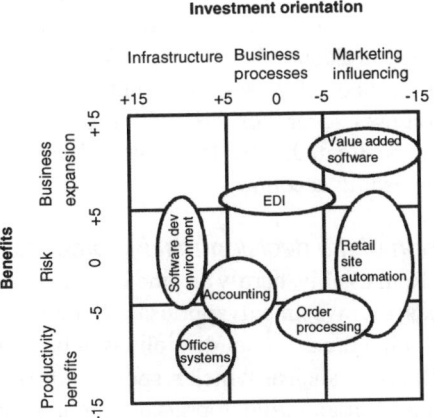

Previously we discussed that the methods described in this chapter serve as examples and have been assessed using a number of assessment criteria. In what follows a summary of this assessment will be given. An overview of the assessment was given in Figure 2.

Objects of the method

The discussed methods do not limit themselves much in terms concerning the objects of the investment:

- *Breadth of the method.* Almost all methods discussed are meant to support the appraisal of specific IT investment projects. The ROM method and Investment Mapping try to measure investment effects on the level of the entire firm. Bedell's method and IT Assessment additionally look at separate business processes or activities in the organization.

- *Type of application area.* Regarding the type of application area, financial methods differ from nonfinancial methods. The financial methods are used for all types of business investments. The discussed nonfinancial methods all are specifically designed for investments in IT. Only the ROM method is also meant for a more generic assessment of organizational value.

Evaluation criteria

Except for the financial methods, the methods exhibit a large variety of used evaluation criteria.

Financial criteria

Almost all nonfinancial methods to some extent take financial criteria into account when evaluating investment impacts. The Investment Portfolio and Investment Mapping look at cash flows through the net present value and the internal return respectively. Information Economics calculates the accounting rate of return. Regarding the SIESTA method, it is not clear how financial consequences are accounted for, while ROM uses its own specific financial measure. Bedell's methods implicitly uses a financial valuation when evaluating the quality criterion and requires a specification of the investment expenditures. It is assumed here that the criteria of the financial methods do not require further explanations.

Nonfinancial criteria

The nonfinancial methods take many different criteria into account to evaluate the impact of investments. All criteria concern business goals and technology related aspects. None of the methods explicitly motivates the choice for the set of criteria, or underpins this with a particular theory. Consequently, the methods differ significantly in terms of the number and type of criteria covered.

Risks

The methods include risks in several ways. In General, risks can be translated into a variation of expected outcomes with particular probabilities. Risk analysis can then, for example, be carried to calculate best and worst case situations. Both the financial methods and the Investment Portfolio calculate, explicitly, with most likely outcomes. When using Information Economics, the estimated risk is deducted from the most likely outcomes. The ROM method, IT Assessment and Bedell's method do not consider risks. The SIESTA method uses several criteria with respect to risk, however, it is not clear how these risks should be treated. The Investment Mapping method takes the spread of the possible outcomes into account.

Support of the decision-making process

Remarkably, hardly any method offers clear practical support for the decision-making process of investment appraisal. If mentioned at all, this support is often limited to mentioning the different disciplines to be represented (e.g. management, IT staff and end-users). Issues such as which responsibilities are needed, or how often an evaluation should be done after an initial appraisal, do not get much attention. A firm and prominent role is given to top management to make the final investment decision. Although the final responsibility for resource allocations lies there of course, it can be doubted whether top managers should be fully informed of the impacts of all investment initiatives.

Measurement scale

Methods that focus on a financial appraisal are by definition using a ratio and/or an interval scale. The other methods employ measurement scales rather arbitrary. Several methods calculate a final score for each investment project using ordinal scores, although an ordinal measurement scale does not allow this (e.g. Information Economics, SIESTA and Bedell's methods).

LIMITATIONS OF EXISTING EVALUATION METHODS

In the previous sections an overview was given of existing evaluation methods for IT investments by discerning types of methods and their main characteristics. This section discusses the most prominent limitations of these methods for taking adequate investment decisions (Renkema, 2000). The limitations identified here are in the following areas

Long-term and enabling nature of investments

The described methods particularly focus on decisions regarding IT applications in an end-user environment. As such, the methods seem to take existing methods for IT system development as a reference. For many of the contemporary infrastructure-oriented IT investments this orientation does not suffice, and in fact can impede a good view on the real business value of infrastructure. For getting such a view, investment appraisal should focus on the future, long-term, advantages of IT-based infrastructure.

Organizational collaborations

An implicit starting point of existing evaluation methods is that investment decisions are taken within a clearly identifiable firm or business unit. Consequently, the characteristics that investments cross the boundaries of individual organizations or organizational units is not reflected in these methods. Also the cooperation with entities in the organizational environment are hardly a point of attention. Also there is not much attention for the integration of an organizational unit within a higher cooperation level (e.g. a business unit, division or diversified corporation).

Coherence of decision-making criteria

The evaluation or decision criteria of the existing methods virtually all refer to money, technology and general business goals of profit-oriented firms. There is hardly any attention for investment impacts on e.g. organizational culture or work conditions and for political implications of proposed investments. Contemporary investments do, however, profoundly impact these aspects. This limited view on impacts may be due to the apparent lack of theoretical grounding and practical validation of methods. Consequently, the decision criteria do not have much cogency and often seem to be based on ad-hoc choices. A typical example of this is that often no difference is made between earnings/expenditures versus costs/yieldings. Although individual methods suggest a quite rigid set of criteria, with some manipulation, many methods can quite easily be translated in another method (Kusters and Renkema, 1996). The overall conclusion, therefore, is that a coherent and structured framework of decision criteria, validated in practice, which can be used to derive appropriate organization-specific value metrics does not exist.

Integration of 'hard' and 'soft' evaluation elements

An important element of business value generation from IT is the integration of hard, content-related issues with soft elements, focusing on social and political processes. Concerning these softer issues, the available arsenal of evaluation methods does not offer much support for investment decision-making. Virtually all methods focus entirely on the 'hard' side of management decisions, by suggesting a set of selection criteria. This is done at the expense of other decision-making elements, such as how and by whom this criteria should be used, and when and in what way these criteria play a role in organizational decision-making (Berghout, 1997). Walsham (1993) in his plea for a more socio-political interpretation of evaluation puts this as follows: "Much of the literature on evaluation in general, and IS evaluation in particular, takes a formal-rational view of organizations, and sees evaluation as a largely quantitative process of calculating the preferred choice and evaluating the likely cost/benefit on the basis of clearly defined criteria." Based on the organizational change theories of Pettigrew (1985), several researchers working from an interpretive research philosophy have proposed a model for investment evaluation. This consists of the following three elements (Farbey et al.1992):

- Content: the what of the evaluation. This concerns the criteria covered in decision-making, and the way these are assessed and measured;
- Process: the how of the evaluation. This concerns such aspects mutual understanding, the development of a common language and the use of learning experiences;
- Context: the who and why of the evaluation. This refers to the organizational and external environment in which decision-making takes place.

In perspective of these three elements, it can be concluded that the existing methods for IT investment evaluation give relatively too much attention to the content related, decision criteria oriented aspects of decision-making. A consequence of this is that process- and context related aspects, which are not less important, do not get sufficient attention.

CONCLUSIONS AND RECOMMENDATIONS
Conclusions

Evaluation of IT investment proposals currently is a major issue for both management and academics. This chapter has first introduced and defined the concepts used in evaluation. These definitions can be used to improve communication between the different stakeholders in the evaluation process and to gain insight in the differences between the several evaluation methods. The review and comparison of evaluation methods showed that the available methods differ in many respects and that conclusions regarding the overall quality can not be drawn. However, some general observations can be made:

- The available evaluation methods are hardly underpinned by theory. Consequently, the choice of criteria seems rather arbitrary;
- The available methods focus on the evaluation criteria, less attention is paid to the evaluation process;
- There seems to be trade-off between the inclusion of nonfinancial criteria and the ease of use of a method. Graphical tools as used in portfolio methods can be of help;
- The differences between the methods can partly be overcome by combining features of the different approaches. For instance a financial or ratio assessment combined with the nonfinancial impacts (contribution) represented in a portfolio.

Recommendations

Gaining insight into the different and interdependent aspects of evaluation is essential for improving the evaluation of IT investment proposals. It is important to have a clear understanding of the relevant evaluation criteria, of the circumstances under which they should be used and of the embedding of the criteria in the evaluation process. In our opinion this understanding can only be gained through a significantly improved theoretical underpinning and validation in practice. We are aware of the difficulties involved in validating evaluation methods in evaluation practice. A major difficulty lies in the causality between the success of investment and the use of an evaluation method. However, some suggestions can be made. The number of independent variables can be reduced by focusing research on a specific type of investment (e.g. infrastructure investments) or on a single line of business (e.g. financial services or public service). A more thorough understanding of the evaluation process in practice would also be a step forward. A prerequisite for this is the possibility to do research *in* organizations and to be allowed to publish the insights gained. Progress can only be made by communicating with each other and by exchanging ideas and insights.

An important observation is that many different interpretations are given to the concepts used in evaluation methods. A good example of this the use of the term cost benefit analysis, which in some cases refers to cash flows in some cases to costs and yieldings. A way forward definitely is a clearer definition of applied concepts. In evaluation practice it is particularly important to make clear which criteria and which aspects are evaluated to avoid misinterpretations in the course of the evaluation process. Overall, our conclusion is that although a plethora of methods can be found, the methods do not offer much support for the complex nature of today's IT investment decisions of organizations. For a value-focused support, appropriate attention should be given to the long-term and enabling nature of investments, possible intra- and interorganizational collaborations, the use of coherent set of decision-criteria, and last but not least to a better integration of content-related issues with explicit support for decision processes and political appraisal issues.

REFERENCES

Agarwal, R., Tanniru, M.R., and Dacruz, M. (1992). Knowledge-based support for combining qualitative and quantitative judgments in resource allocation decisions. *Journal of Management Information Systems*, No.1, 165-184.

Bacon, C.J. (1992). The use of decision criteria in selecting information systems/technology investments. *MIS Quarterly*, September.

Banker, R.D., Kauffman, R.J., and Mahmood, M.D. (1993). *Strategic information technology management*. Idea Group Publishing, Hershey, Pennsylvania.

Bedell, E.F. (1985). *The computer solution: strategies for success in the information age*. Dow-Jones Irwin, Homewood.

Berghout, E.W. (1997). Evaluation of information system proposals: design of a decision support method. Ph.D. Thesis Delft University of Technology, June.

Berghout, E.W., and Meertens, F.J.J. (1992). Investeringsportfolio voor het beoordelen van voorstellen van informatiesystemen ('Investment portfolio for the evaluation of IT investment proposals). *Theme issue on IT investment evaluation (in Dutch), Informatie*, 677-691.

Berghout, E.W., and Renkema, T.J.W. (1994). Beoordelen van voorstellen voor investeringen in informatiesystemen ('Evaluating information systems investment proposals'). *Research report*, Delft University of Technology.

Bouma, J.L. (1980). *Leerboek der bedrijfseconomie, deel II (Textbook on Business Economics, part II; in Dutch)*. Wassenaar, Delwel, 1980.

Brealey, R.A., and Myers, S.C. (1988). Principles of corporate finance. McGraw-Hill, New York.
Brussaard, B.K. (1993). *Organisatie van de informatievoorziening ('Organization of Information Processing').* Department of Mathematics and Computer Science, Delft University of Technology.
Buss, M.D.J. (1983). How to rank computer projects. *Harvard Business Review*, January-February.
Butler Cox Foundation (1990). Getting value from information technology. *Research report 75.*
Canada, J.R., and Sullivan, W.G. (1989). *Economic and multiattribute evaluation of advanced manufacturing systems.* Prentice Hall, Englewood Cliffs.
Carrol, J.S., and Johnson, E.J. (1990). Decision research, a field guide. *SAGE*, Newsbury Park.
Carter, W.K. (1992). To invest in new technology or not? New tools for making the decision. *Journal of Accountancy*, May, 58-62.
Clemons, E.K., and Weber, B.W. (1990). Strategic information technology investments, guidelines for decision making. *Journal of Management Information Systems*, 2, 9-28.
Davenport, Th. J., and Short, J.E. (1990). The new industrial engineering: information technology and business process redesign. *Sloan Management Review*, Summer, 11-27.
Delahaye, I., and Reeken, A.J. van (1992). Waarom investeren in welke informatiesystemen? Een vergelijkende toepassing van 'Bedell' en 'Parker/Benson/Trainor' bij De Gazet (Why invest in which information systems, a comparison of 'Bedell' and 'Parker/Benson/Trainor at The Gazet). *Theme issue on IT investment evaluation, Informatie*, 655-670.
Dos Santos, B.L. (1991). Justifying investments in new information technologies. *Journal of Management Information Systems*, nr. 4, 71-90.
Douglas, D.P., and Walsh, L. (1992). Basic principles for measuring IT value. *I/S Analyzer*, 10.
Earl, M.J. (1989). *Management strategies for information technology*. Prentice Hall, London.
Farbey, B., Land, F., and Targett, D. (1992). Evaluating investments in IT. *Journal of Information Technology*, 7, 109-122.
Farbey, B., Land, F., and Targett, D. (1993). *How to assess your IT investment, a study of methods and practice.* Butterworth-Heinemann, Oxford.
Fox, R., Kennedy, A., and Sugden, K. (1990). Decision making, a management accounting perspective. Butterworth Heinemann, Oxford.
Frielink, A.B. (ed.) (1961). *Auditing Automatic Data Processing.* Elsevier, Amsterdam.
Grembergen, W. (2000). IT balanced Scorecard. *Proceedings of the second International Symposium,* March 20–21, Antwerp, Belgium.
Harrington, J. (1991). *Organizational structure and information technology*. Prentice Hall, London.
Hochstrasser. B., and Griffiths, C. (1990). Regaining control of IT investments, a handbook for senior UK management. *Kobler Unit,* Imperial college, London.
Hochstrasser, B. (1993). Quality engineering: a new framework applied to justifying and prioritising IT investments. *European Journal of Information Systems*, 3, 211-223.
Huigen, J., and Jansen, G.S.H. (1991). De baten van informatiesystemen ('The benefits of information systems'; in Dutch'). *ESB*, 3/7.
Irsel, H.G.P., and Fluitsma, P., and Broshuis, P.N.A. (1992a). Evaluatie van IT investeringen: het afstemmen van vraag en aanbod (IT investment evaluation: aligning supply and demand). *Theme issue on IT investment evaluation, Informatie*, 716-726.
Irsel, H.G.P., and Fluitsma, P. (1992b). Het plannen en rechtvaardigen van infrastructurele IT-investeringen ('Planning and justification of infrastructural IT investments'). *Compact*, Summer, 38-48.
Ives, B., and Learmonth, G. (1984). The Information System as a Competitive Weapon. *Communications of the ACM*, 12, 1193-1201.
Joslin, E.O. (1968). *Computer selection.* Addison-Wesley, London.
Kambil, A., Henderson, J., and Mohsenzadeh, H. (1993). Strategic management of information technology investments: an options perspective. *In Banker e.a.*
Kaplan, R.S., and Norton, D.P. (1992). The balanced scorecard -measures that drive performance. *Harvard Business Review*, 71-79, January-February.

Keen, P.G.W. (1981). Value Analysis: Justifying Decision Support Systems. *MIS Quarterly*, March.
Keen, P.G.W. (1991). *Shaping the future: Business design through information technology*, Harvard Business School Press, ISBN 0-87584-237-2.
King, J.L., and Schrems, E.L. (1978). Cost-Benefit Analysis in information systems development and operation. *Computing Surveys*, 10(1), 19-34.
Kleijnen, J.C.P. (1980). Computers and profits, Reading, Massachusetts, Addison-Wesley.
Kusters, R.J. and Renkema, T.J.W. (1996). Managing IT investment decisions in their organizational context, in: *Proceedings of the third European Conference on IT investment evaluation,,* Brown. A.; Remenyi, D., (eds.) University of Bath, UK.
Lincoln, T.J. (1986). Do computer systems really pay-off? *Information and Management*, 11, 25-34.
Lincoln, T.J. (1990). *Managing Information Systems for Profit*, John Wiley & Sons, Chicester.
Lincoln, T.J. and Shorrock, D. (1990). Cost justifying current use of information technology. In: *Lincoln, T.J. (ed.), Managing information systems for profit*, John Wiley & Sons, Chicester.
McFarlan, F.W. and McKenney, J.L. (1983). Corporate information systems management. *The issues facing senior executives*, Dow Jones Irwin, Homewood.
McKeen, J.D. and Smith, H.A. (1991). The value of information technology: A resource view. *Proceedings of the ICIS.*
Merdith, J.R. and Hill, M.M. (1987). Justifying new manufacturing systems: A managerial approach. *Sloan Management Review*, Summer.
Nievelt, M.C.A. van (1992). Managing with information technology, a decade of wasted money? *Compact*, Summer, 15-24.
Oonincx, J.A.M. (1982). Kengetallen en financiële ratio analyse ('Ratio's and financial ratio analysis'). *Financiële Leiding en Organisatie*, January, 2370-1 - 2370-22.
Parker, M.M., Benson, R.J., and Trainor, H.E. (1988). *Information Economics, linking business performance to information technology*. Prentice-Hall, New Jersey.
Parker, M.M., Benson, R.J., and Trainor, H.E. (1989). Information Strategy and Economics. Prentice-Hall, New Jersey.
Peters, G. (1988). Evaluating your computer investment strategy. *Journal of Information Technology*, September.
Peters, G. (1989). The evaluation of information technology projects. *Ph.D. Thesis*, Brunel University.
Pettigrew, A.M. (1985). *The awakening giant: continuity and change in ICI*, Blackwell, Oxford.
Porter, M.E. (1985). Competitive advantage, creating and sustaining superior performance. The Free Press, New York.
Powell, P. (1992). Information technology evaluation, is it different? *Journal of the Operational Research Society*, 1, 29-42.
Reeken, A.J. van (1992). Investeringsselectie van informatiesystemen, de methode van Eugene Bedell (Selection of investments in information systems, Eugene Bedell's method). *Handboek BIK*, C 1030-1–1030-32.
Renkema, T.J.W. (1996). *Investeren in de informatie-infrastructuur: richtlijnen voor besluitvorming in organisaties (Investing in the IT infrastructure)*. Kluwer.
Renkema, T.J.W. (2000). The IT Value Quest: How to Capture the Business Value of IT-based Infrastructure. John Wiley & Sons, Chicester.
Renkema, T.J.W. (1993). Evaluation of investments in information technology, preliminary research findings. *Proceedings of the fourth European Software Cost Modelling Meeting*, Bristol.
Renkema, Th., and Berghout, E. (1999). *Investeringsbeoordeling van IT-projecten: een methodische benadering' (A methodological approach to IT investment appraisal)*. 2nd Revised edition, Information Management Institute, B.V., Rotterdam.
Renkema, T.J.W., and Berghout, E.W. (1997). Methodologies for information system investment evaluation at the proposal stage: A comparatative view. *Information and Software Technology*, Elsevier Science Ltd, Oxford, 39(1), 1-13.

Rockart, J.F. (1989). *Chief executives define their own data needs*. Harvard Business Review, March-April, 81-93.

Saaty, T.L. (1988). *The analytic hierarchy process*. McGraw-Hill, New York.

Sassone, P.G. (1988). A survey of cost-benefit methodologies for information systems. *Project Appraisal*, 2, 73-84.

Sassone, P.G. and Schaffer, W.A. (1978). *Cost-Benefit Analysis: A Handbook*. Academic Press, New York.

Schaefer, G. (ed.) (1988). Functional analysis of office requirements: A multiperspective approach. John Wiley & Sons, Chicester.

Silk, D.J. (1990). Managing IS benefits for the 1990s. *Journal of Information Technology*, 185-193.

Spraque, R.H. and Carlson, E.D. (1982). *Building effective decision support systems*. Prentice-Hall, Englewood Cliffs.

Strassmann, P.A. (1990). Information Payoff. Free Press, New York.

Swinkels, G.J.P. and Irsel, H.G.P. van (1992). Investeren in informatietechnologie, take IT or leave IT (Investments in information technology, take IT or leave IT; in Dutch). *Compact*, Summer, 3-13.

Symons, V.J. (1991). A review of information system evaluation: content, context, and process. *European Journal of Information Systems*, 1, 205-212.

Udo, G. and Guimaraes, T. (1992). Improving project selection with a socio-technical approach. *IRMA Conference Proceedings*, 204-213.

Vaid-Raizada, V.K. (1983). Incorporation of intangibles in computer selection decisions. *Journal of Systems Management*, 30-36, November.

Ward, J.M. (1990). A portfolio approach to evaluating information systems investments and setting priorities. Journal of Information Technology, 222-231.

Willcocks, L. (1992). Evaluating information technology investments: research findings and reappraisal. *Journal of Information Systems*, 2, 243-268.

Willcocks, L. and Lester, S. (1993). Evaluating the feasibility of information technology investments. *Research report RDP93/1*, Oxford Institute of Information Management.

Wiseman, C. (1985). *Strategy and computers, Information systems as competitive weapons*. Dow Jones Irwin, Homewood.

Wissema, J.G. (1985). *An introduction to capital investment selection*. Francis Pinter, London.

Wolstenholme, E.F. et al. (1992). The design, application and evaluation of a system dynamics based methodology for the assessment of computerised information systems. *European Journal of Information Systems*, 5, 341-350.

Yan Tam, K. (1992). Capital budgeting in information system development. *Information and Management*, 23, 345-357.

Zee, H.T.M. van der (1996). In search for the value of information technology. *Ph.D. Thesis*, Tilburg University, May.

Zee, H.T.M. van der and Koot, W.J.D. (1989). I/T-assessment, een kwalitatieve en kwantitatieve evaluatie van de informatieverzorging vanuit een strategisch perspectief ('I/T assessment, a quantative and qualitative evaluation of information processing from a strategic perspective'). *Informatie*, 11, 805-900.

Zmud, R.W. (1983). *Information Systems in Organizations*. Scott, Foresman.

APPENDIX A: METHODS FOR THE EVALUATION OF IT INVESTMENT PROPOSALS

This appendix is based upon research at the universities of Amsterdam, Delft and Eindhoven, all in The Netherlands. Although this research has been carried out with the utmost care, the review can not be exhaustive. Almost daily new methods are published and consultants often use a well-considered method, which is however not published because of a possible competitive advantage. Furthermore, several methods combine features of other methods.

For some methods, not the original source is given, but a reference to articles or books in which the method is mentioned or reviewed. The list of references is also not exhaustive, however, gives the references with the best description of the method. Furthermore, not all methods are specifically designed for the evaluation of IT investment proposals.

METHOD:	REFERENCES:
Accounting rate of return:	Bacon, 1992.
Analytic hierarchy process:	Saaty, 1980; in: Carter, 1992.
Application benchmark technique:	in: Powell, 1992.
Application transfer team:	in: Lincoln, 1990.
Automatic value points:	in: Lincoln, 1990.
Balanced scorecard:	Kaplan and Norton 1992, in: Douglass and Walsh 1992.
Bayesian analysis:	Kleijnen, 1980.
Bedell's method:	Bedell, 1985; in: van Reeken, 1992a, 1992b.
Buss's method:	Buss, 1983.
Benefits-risk portfolio:	McFarlan and McKenney, 1983; in: van Irsel and Swinkels, 1992.
Benefit assessment grid:	Huigen and Jansen, 1991.
Breakeven analysis:	Sassone, 1988.
Boundary value:	in: Farbey et al., 1992, 1993.
Cost benefit analysis:	King and Schrems, 1978; Sassone and Schaffer, 1978.
Cost benefit ratio:	Yan Tam, 1992.
Cost displacement/avoidance:	in: Sassone, 1988.
Cost effectiveness analysis:	in: Sassone, 1988.
Cost-value technique:	Joslin, 1977.
Cost revenue analysis:	in: Farbey et al. 1992.
Critical success factors:	Rockart, 1979.
Customer resource life cycle:	Ives and Learmonth, 1984; in: Hochstrasser and Griffiths, 1990.
Decision analysis:	in: Sassone, 1988; in: Powell, 1992.
Delphi evidence:	in Powell, 1992.
Executive Planning for Data Processing:	in: Lincoln, 1990.
Functional Analysis of Office Requirements:	Schaeffer et al., 1988.
Gameplaying:	in: Farbey et al., 1992.
Hedonic wage model:	in: Sassone, 1988.
Information Economics:	Parker et al. 1988, 1989.
Investment mapping:	Peters, 1988, 1989.
Investment portfolio:	Berghout and Meertens, 1992.
Information systems investment strategies:	in: Lincoln, 1990.
Knowledge based system for IS evaluation:	Agarwal et al., 1992.
MIS utilisation technique:	in: Powell, 1991.
Multi-objective, multi-criteria methods:	in: Farbey et al. , 1992; Vaid-Raizada, 1983.
Option theory:	Dos Santos, 1991; Kambil et al., 1993.
Potential problem analysis:	in: Powell, 1992.
Profitability index:	Bacon, 1992.
Process quality management:	in: Lincoln, 1990.
Quality engineering:	Hochstrasser, 1993.
Return on investment:	Brealey and Myers, 1988; Farbey et .al, 1992.
Return on management:	Strassmann, 1990; van Nievelt, 1992.
Requirements-costing technique:	Joslin, 1977.
Schumann's method:	in: van Irsel and Swinkels, 1992.
SESAME:	Lincoln, 1986; Lincoln and Shorrock, 1990.
Seven milestone approach:	Silk, 1990.
Strategic application search:	in: Lincoln, 1990.
Strategic option generator:	Wiseman, 1985.
Systems investment methodology:	in: Lincoln, 1990.
Simulation:	Kleijnen, 1980; in: Farbey et al., 1992; in: Powell, 1992.
Socio-technical project selection:	Udo and Guimaraes, 1992.
Satisfaction and priority survey:	in: Lincoln, 1990.
Structural models:	in: Sassone, 1988.
System dynamics analysis:	Wolstenhome et al., 1992.
Systems measurement:	Spraque en Carlson, 1982; in: Powell, 1992.
Time savings times salary:	in: Sassone, 1988.
User utility function assessment technique:	in: Powell, 1992.
Value analysis:	Keen, 1981.
Value chain analysis:	Porter, 1985.
Ward's portfolio analysis:	Ward, 1990.
Wissema's method:	Wissema, 1983.
Zero based budgeting:	in: Zmud, 1987.

Part III:

Alternative Ways of Traditional IT Evaluation

Chapter VI

The Institutional Dimensions of Information Systems Evaluation

Vassilis Serafeimidis
KPMG Consulting[1], London, UK and University of Surrey, UK

INTRODUCTION

Information systems evaluation is embedded in many social and organisational processes, and thus is a particularly complex decision process. Evaluation happens in many ways (e.g. formally, informally), uses diverse criteria (e.g. financial, technical, social), follows rigorous methodologies or gut feelings and often becomes a political instrument which influences the balance of organisational power.

The existing literature (Ballantine et al., 1995; Farbey et al. 1992; Willcocks and Lester 1994; Ward et al. 1996) identifies noticeable gaps between academic theories, commercially available methodologies, and actual evaluation practice within organisations. Such gaps between theory and practice are not unusual and they have been reported in other research areas. Hsia (1993 p.14) for example argues: "The truth is that most companies have two sets of practices; one real, the other recommended." In other words there are the formal evaluation practices promoted by organisational rules and structures, the informal practices implemented by stakeholders involved, and finally the academic recommendations which in many cases recognise the delicate nature of evaluation and suggest more interpretive considerations.

The better theories tend to emphasise the complexity and richness of the evaluation problem situation or context while the available methodologies tend to oversimplify the process through cookbooks that focus on the more measurable aspects of the outcome of IT/IS investment. Meanwhile, the actual use of such methodologies in practice is often largely determined by the subjective views of individual stakeholders facing a combination of business, organisational and technological pressures.

The reasons for the apparent gaps, I believe, are related to the institutional dimensions of IT/IS evaluation as an organisational process and their limited understanding. The limited consideration of the organisational/institutional context where evaluation is integrated (e.g. the system's development life cycle, the IS management practices) and furthermore, the limited study/understanding of the stakeholders' behaviour (the socialisation) lead to differences (or mismatches) between theory and practice. The lack of appropriate cultural and structural foundations (e.g. organisational learning and maturity, training) could also explain the crisis of utilisation (Legge 1984) of the evaluation approaches.

Copyright © 2001, Idea Group Publishing.

The above aspects will be the focus of discussion in this chapter. Furthermore, this chapter discusses a series of concepts to be considered in order to build an alternative way of the IT evaluation. The proposed approach focuses on the role of individuals and the roles they play in their organisational context in relation to evaluation. The following section provides a short review of the available approaches in IS evaluation and attempts to justify the rationale for the institutional suggestions. It is followed by my recommendations regarding the various components that need to be considered within the organisational framework of IS evaluation.

WAYS OF APPROACHING IT EVALUATION

Like much IS research, the study of IS evaluation used to be dominated by a positivistic and scientific paradigm (see, for example, Lee et al., 1997; Walsham, 1995a). The traditional (formal-rational or functionalist) conception sees evaluation as an external judgement of an information system that is treated as if it existed in isolation from its human and organisational components and effects. It also places excessive emphasis on the technological and accounting aspects at the expense of the organisational and social aspects. In so doing it neglects the organisational context and process of IS development and its content, elements that are critical to the successful application of IT in support of the business. In general, more attention has been focused over the years on prescribing how to carry out evaluations (with project-driven and cost-focused accounting frameworks) rather than analysing and understanding their role, interactions, effects and organisational impacts (Hirschheim and Smithson, 1988; Smithson and Hirschheim, 1998).

Formal/rational approaches have been challenged for their internal validity and their external ability to generalise in other areas of social research (Legge, 1984). They contribute to one piece of the picture but are not rich enough to describe the complex impacts within organisations. Thus they cannot encompass the uncertainties, risks and context dependencies concerning the value of IS investments to a business that is undergoing often considerable organisational change. Evaluation is a socially embedded process in which formal procedures entwine with the informal assessments by which actors make sense of their situation.

Many authors (e.g., Hirschheim and Smithson, 1988; Iivari, 1988; Walsham, 1993; Smithson and Hirschheim, 1998) argue that IS evaluation would be improved by using an interpretive epistemology. This stance offers a framework for analysis that assists in the understanding and assessment of the meanings assigned by individuals to evaluation phenomena. Interpretive evaluation aims to involve a wide variety of stakeholder groups and to focus on a discourse for learning and understanding. Such designs are driven mainly by the determination of the content according to the organisational context and they are organised around the concerns, issues and actions of the stakeholders. The suitability of interpretive epistemology to the study of IS evaluation is supported further by Symons (1993 p.74) who argues that:

> "Interpretive methodologies of evaluation actively analyse the experience of organisational reality, focusing on stakeholder interests and perspectives. ... They increase the effectiveness of organisational activity by ensuring a match between elements of organisational change and the dynamics of organisational culture and politics."

However, research (e.g., Ballantine et al., 1995; Farbey et al., 1992; Willcocks and Lester 1994; Ward et al., 1996) suggests that most of the interpretive approaches find limited use in practice and they remain among the academic community. The main reasons for that limited success (Serafeimidis, 1997) are around the integration of the interpretive approaches with the existing organisational practices and institutions. IS evaluation should be perceived as an organisation matter and not simply as an IT management process. The following section discusses a number of tools that can assist managers to establish interpretive IS evaluation practices.

A RICH ORGANISATIONAL FRAMEWORK FOR IS EVALUATION ADOPTION
Soft and Social Cognitive Perspectives
Evaluation Stakeholder Analysis

The dominant organisational nature of IS evaluation brings into the foreground the human actor, a stakeholder, of an evaluation exercise. An organisation-wide participatory stakeholder analysis prior to any evaluation is an essential step to the formulation of the evaluation party (Gregory and Jackson, 1992; Guba and Lincoln, 1989). The members of the evaluation party will have the opportunity to express their views and attempt to influence the evaluation decisions on the basis of their interests and backgrounds. Current research on information systems evaluation is usually restricted in stating the importance of evaluation stakeholders, even though the need to involve certain types of stakeholders in IS decision making has been emphasised in the literature for some time. For example, Mumford (Mumford and Weir, 1979) was one of the pioneers in supporting the involvement of end users as a factor of effective IS development and implementation, using implicitly the stakeholder concept in this area. Yet, stakeholder identification in IS research is still limited to some obvious interested parties (e.g. users, developers, managers) and thus fails to understand the potential range of stakeholder roles in an evaluation situation (Pouloudi and Whitley, 1997). In particular, because of the emphasis on general strategy or development, little guidance or attention is given to stakeholders of IS evaluation. However, stakeholders play a key role in the evaluation process, not least because it is the different stakeholders who determine whether a system has been successfully developed and implemented. This important role of stakeholders has been articulated in previous research on IS failures (Lyytinen and Hirschheim, 1987; Lyytinen, 1988) arguing that failure is contingent on the capability of an IS to meet the expectations of different stakeholders.

However, the identification of the evaluation stakeholders is not trivial. The existing literature in the area lacks systematic guidelines to assist in their identification and classification. What is usually available refers to generic lists (e.g., Gilbert et al., 1988) with questionable general applicability.

A useful starting point in a attempt to define the evaluation stakeholders would be the application of the following principles (originated in the work of Ginzberg and Zmud, 1988):
- The adopted view of evaluation (e.g., functional, economic, social quantitative, qualitative).
- The relative position of the stakeholder to the organisation (e.g., internal, external, seniority level).
- The relationship of the stakeholder to the evaluation (e.g., an agent who designs an evaluation exercise, supervises it, provides information, applies a methodology, receives the results).
- The depth of impact of evaluation to the person (e.g., direct, indirect, informal, beneficiary, victim).
- The level of aggregation (e.g., individuals, groups, industry—variety; Gregory and Jackson, 1991).

102 Serafeimidis

Farbey et al. (1993) have developed a stakeholders map suitable for evaluation situations. Ward et al. (1996) have also developed the idea of stakeholder analysis with practical suggestions on how to determine the stakeholders in IT benefits management.

Evaluation schema

Cognitive psychology provides us with the necessary structures to reveal and understand the individual's knowledge, views, emotions and perceptions. The idea of a 'schema' (Bartlett 1932) has been used as an individual's interpretation of their reality. It is a network of general knowledge based on previous experience which provides predictive and explanatory power for understanding (Norman, 1983). A more flexible and evolving notion is that of mental models (Craik, 1943) which are dynamically constructed, as creations of the moment, by activating stored schemas.

Giddens (1984) uses the similar term interpretative schemes as a modality of shared cognitive structures for understanding and as an initiative for actions. Beach (1990) uses the idea of schema, called "image" as a major component to develop an alternative to the classical decision making theories. His proposed "Image Theory" focuses upon individuals' three different types of images, representing the values, goals and plans of a decision maker. The knowledge stored in those images is framed dynamically based on the existing context.

Orlikowski and Gash (1994) develop the concept of "technological frames" in order to investigate cognitions and values of stakeholders (in particular those of technologists and users) of an IS and to explain and anticipate actions and meanings. Three domains characterise the interpretations of individuals captured in a frame: the technology itself; the technology strategy; and technology in use. Orlikowski and Gash's ideas have been used and enhanced further in order to understand the broader set of organisational changes needed (for IS experts and business managers) to accommodate new IT (Gallivan, 1995).

Likewise, I introduce the concept of evaluation schema as the individual stakeholder's views of the evaluation and/or the IS under evaluation (see Figure 1). The evaluation schema should play a key role as an elicitation and interpretation mechanism for understanding the interests of the individuals and the groups that they form (e.g., evaluation party).

The members of the evaluation party have individual interpretations of an evaluation situation depending on their previous experiences, knowledge, personal goals and emotions (Norman, 1980). They also have a set of core beliefs in common. It is not unusual that many groups of stakeholders share common schemas when there is a significant overlap of cognitive categories (properties) and content. On the other hand schemas can be so different (in content or structure) as to lead to conflict, and furthermore, to a negotiation to deal with it. In any case evaluation schemas are seem as the basis for communication (Giddens, 1979) by diffusing formal and informal conceptions.

The evaluation schemas of the same stakeholder may vary according to the time and context. Schemas are very rich and therefore need to be framed so knowledge that is associated with the time and the context is brought to bear in that context and endow the situation

Figure 1: The notion of the 'evaluation schema'

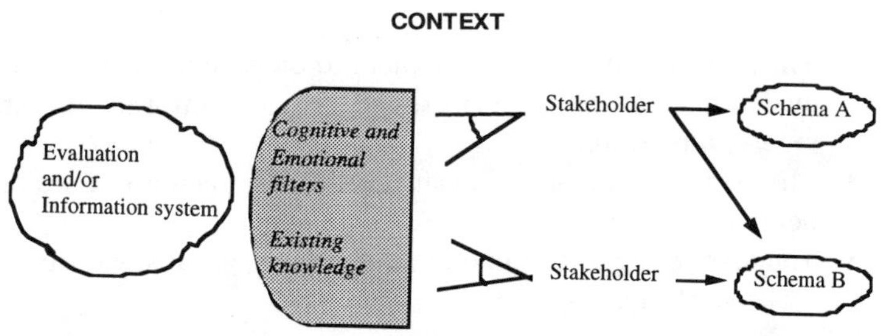

with meaning. For instance, the views of the project manager regarding the expected project benefits may vary considerably between different stages of the IS development lifecycle. During the early stages of the lifecycle, while the IS context is not well-defined, the benefits are characterised by uncertainty and vagueness. When the system reaches its operational stage a clearer picture can be drawn.

The concept of schema is a process rather than a variance theory (Orlikowski and Gash, 1994). It is then particularly valuable in examining the changes associated with the implementation of a change over time (e.g., IS-related change). Evaluation schemas could be used to track changes in the meanings stakeholders ascribe to an IS and its evaluation over time, thus providing a way of investigating the processes and outcomes of organisational and evaluation-related changes. In other words, the construct of an evaluation schema could be additionally useful as a complementary element to the heart of contextualism (Pettigrew, 1985) (see Table 1). The elements of contextualism could assist in framing the knowledge carried by the individual's schema. The four layers are strongly linked together.

The evaluation schema as a cognitive structure is very powerful and important to the understanding and explanation of IS evaluation. However, in practical terms, as with all similar cognitive models (e.g., images, mental models), the evaluation schema experiences a number of problems which constrain its use (e.g., Johnson-Laird, 1989; Norman, 1983). These include:

- The lack of a complete definition and theories.
- The theories that invoke models as a representation of knowledge are also incomplete.
- People's abilities to run their models are severely limited.
- Models are unstable, unscientific and parsimonious.
- Depending on the domain they use, theorists emphasise either on their content and its representation (e.g., cognitive maps, Eden, 1992) or on their structure and the process of their construction.
- Discovering and modelling (a model's structure and content) are difficult tasks which require a lot of experience.

Table 1: The Element of Contextualism

Context; Context is concerned with the multi-level identification of the various systems (e.g. social, political, economic) and structures (e.g. organisational processes) within which the organisation is located. Various stakeholder groups, both internal and external, should be identified together with the processes and tasks with which they are involved.
Content; The content refers to the values and criteria to be considered and what should be measured. It is here where it is particularly important to look beyond the narrow quantification of costs and benefits to an analysis of the opportunities presented by IT, together with the potential constraints and risks in its application. These include the linkage to organisational goals and a consideration of the implementation process.
Process; The process layer is concerned with the way in which evaluation is carried out (the techniques and methods used), and furthermore its relationship to IS planning, decision making, systems development and project management methods and techniques. It is very important to establish a means of communication with all the stakeholders involved in order to achieve organisational and individual learning.
History; A historical understanding of all the above conceptual elements is necessary because IT-related changes and their evaluation (including learning from failures) evolve over time and, at any particular point, present a series of constraints and opportunities shaped by the previous history.

Evaluation Roles

The above discussion indicates that although evaluation schemas are a powerful construct there are still practical difficulties which restrict their applicability. Thus, I suggest to approach and understand the actions of these individuals from the perspective of their organisational roles. A social and/or organisational role is associated with a set of behaviours appropriate and expected for a given position or status (Biddle, 1979; Brown, 1970; Fincham and Rhodes, 1988; Levinson, 1971; Merton, 1971). For Parsons (1951) roles belong to the social system, can be explained through role expectations that are held by participants and are supported by sanctions. Mintzberg (1973, p.54), in his attempt to identify the roles of a manager, defines the term as:

"an organised set of behaviours belonging to an identifiable office or position. Individual personality may affect how a role is performed, but not that it is performed. Thus, actors, managers, and others play roles that are predetermined, although individuals may interpret them in different ways."

In the centre of this argument are the structurally given demands (e.g. norms, expectations, responsibilities) and the behaviour of an individual acting out a role (e.g. individual role-conceptions and individual role-performance). Schema are at the heart of people's behaviour, consequently influencing their actions in specific context and time dimensions. In addition, schema are critical in the interpretation of actions by other individuals, as reactions are based on the meanings given to situations with which the schema are associated. As Simon (1991 pp.126-127) argues:

"Roles tell organisation members how to reason about the problems and decision that face them: where to look for appropriate and legitimate information premises and goal (evaluative) premises, and what techniques to use in processing these premises."

The current research on IS evaluation does not go further than the identification of the evaluation stakeholders, and does not consider the potential range of roles these individuals play in an evaluation situation. Thus, it cannot explain associations, interactions and conflicts. An organisational stakeholder has one or more positions (e.g. finance director, IT developer, IS director). This stakeholder can perform a number of roles associated with his position. For example, the finance director acts as a user in relation to the IT developer and as the financial policy maker in the eyes of the company's shareholders. Furthermore, he has a different role in relation to the IS director.

There are many evaluation roles. Focusing in particular on the principle that distinguishes the relationship of the stakeholder to the evaluation, a number of individual roles can be determined. These could include the "strategist for evaluation", the "evaluator", the "champion", and the "sponsor". The first two types of roles are in compliance with the objectives of this chapter, which include the understanding and explanation of the interplay between evaluation concepts and actions. Furthermore, the investigation of the organisational integration of evaluation approaches primarily involves those who perceive, implement and apply such a change, the strategists and the evaluators.

The stakeholder playing the role of the champion is the person who conceives the project idea and makes the project run. Champions are usually politically involved in the evaluation exercise and their influences are neither overt nor predictable (e.g. what approach is used, what results are expected). A sponsor provides the necessary protection and resources. These roles have been discussed in the literature (e.g. Hirschheim and Klein (1989), Markus (1994), Walsham (1993)) with regard to IS development projects and so it was decided to focus on strategists and evaluators. A first assumption is that the roles of a strategist and an evaluator are separated and are performed by different stakeholders. The primary characteristics of these roles are discussed below.

The strategist for evaluation is involved in the production, implementation and institutionalisation of evaluation norms, structures and methods. The tasks of a strategist, in general terms (Mintzberg, 1994), include the analysis of the situation, the production and reproduction of normative values (Walsham, 1993), in the maintenance and change of power relations. In a more unconventional way, a strategist can encourage and facilitate others to question conventional wisdom (see academics in this respect), and increase awareness.

The organisational status and power of the strategist (should) support his actions. A strategist usually has the authority to make commitments. In the cases of education and social programmes the above role is played by the government, which sets standards, measures and rules. In IS terms there is no clear person responsible, i.e., internal to the organisation, who carries the role of a strategist for evaluation. Externally, this role can be assigned to the various academic researchers or consultants who attempt to develop and promote alternative IS evaluation approaches.

Unlike the case of education where the role of the 'evaluator' is well-defined as a profession (House, 1993), in information systems each stakeholder from the evaluation party potentially can act as evaluator. Evaluators cannot be considered as isolated from the evaluation action. They are recipients of the changes they initiate and therefore, either beneficiaries or victims themselves (Guba and Lincoln, 1989).

The evaluated can be viewed and interpreted by his audience in various ways. In the case of a preordinate (formal/rational) evaluation exercise based largely on technical and economic criteria, the role of a formal evaluator can be taken to include not only quantitative and other assessments, but also a ritual element of demonstrating management competence (Walsham, 1993). The evaluator, in contrast with the strategist, is closer to the sources of the soft and informal information.

In the case of a responsive or interpretive evaluation design, the role of an IS evaluator is as a facilitator of the evaluation discourse amongst a wide variety of stakeholder groups. The evaluator in this context can be seen as a collaborator, learner and teacher, reality-shaper, and change agent (Walsham, 1993). Furthermore, the evaluator could be viewed as an enactor of meaning involved with organisation making (e.g. the creation and maintenance of systems of shared meaning that facilitate organised action) and also as a moral agent concerned with norms, values and power relations (Walsham, 1993).

In the case of dialectic evaluation, the evaluator is responsible for preparing and supporting an effective negotiation process by securing the participation of interested agents (Avgerou, 1995; Guba and Lincoln, 1989). As Avgerou (1995 p.436) argues further "... the evaluator is an organiser and facilitator of an interactionist evaluation process, rather than an objective assessor."

The Organisational Orientations of IS Evaluation

The organisational context (both internal and external) strongly influence the role that evaluation is expected to play (Serafeimidis and Smithson, 1999; Walsham, 1985b; Willcocks and Lester, 1994). This includes the organisational characteristics (e.g. culture, norms), structural limitations, the organisational expectations from evaluation and external requirements (e.g. academic literature). All these factors restrict and enable its impacts and contributions. Considering this relationship four different orientations of IS evaluation can be identified.

Previous literature in the area of strategic decision making (Earl and Hopwood, 1987; Harrison, 1995; Thompson, 1967) identifies four generic types of decisions which occur by computation, compromise, judgement and inspiration. Similar analyses have been used in the area of evaluation by Farbey et al. (1992 and 1994) and Hellstern (1986). The underlying principles of these classifications refer to the beliefs about causation (means and processes) and the clarity/predictability of the objectives of the decision making process (and IS in our case). The elements of uncertainty and the amount of knowledge available are critical for both dimensions as determinants of success.

Table 2: Four Orientations for IS Evaluation

		Impact on Organisation of Information Systems	
		Tactical	Strategic
Perception of Objectives	Consensus/ Clarity	Control	Social learning
	No consensus/ Ambiguity	Sense-making	Exploratory

Incorporating the dimensions/principles from these authors (e.g., Earl and Hopwood, 1987; Farbey et al., 1992; 1994) I propose four possible orientations for IS evaluation (see Table 2). This proposal recognises the importance of the context in the perception and performance of evaluation. It further considers the need for contingency approaches to evaluation due to the diversity of contexts and organisational changes attempted by IS, the clarity and uncertainty of the expected outcomes (which are related to the type of IS) and the different needs along the time horizon.

The classification defines four different evaluation sub-contexts based on two dimensions. The horizontal axis refers to the impact on the organisation of the proposed investment. The relationship with the organisational context is then evident. The expectations posed could range from tactical to more strategic. The type of expectations also reflects the uncertainty of predicting the effects (end results) of evaluation actions and their further impact on the context. The vertical axis refers to the clarity and attainability of the objectives of information systems and their evaluation. It ranges from consensus to no consensus addressing the issue of multiple interests and the plurality of impacts. This axis further depends on the nature of IS and its perceived contributions. The identification of objectives leads to the determination of relevant evaluation criteria. All these are subject to contextual factors. Each orientation does not exclude any other but complements it.

Control evaluation, for example, refers to the cases where IS are used to automate or support well-defined processes (e.g., stock control, accounting systems). The expected outcomes of the investment, usually quantitative, are rather certain, and there is an organisational consensus around them. Therefore, evaluation has clear objectives to achieve. The previous experience available and the narrow scope (tactical) makes evaluation contributions predictable and certain. Control evaluation is often experienced as a mechanism to achieve resource utilisation. It is usually implemented primarily by financial control evaluation tools.

In cases where the objectives of the investment are not clear or predictable (e.g., a decision support system, a groupware system) *sense-making* evaluation attempts to reach consensus.

Social learning evaluation contributes to decreasing uncertainty of strategic changes. The expected objectives are usually clear (e.g., increase customer base through telephone banking) but there is uncertainty of their achievement. The need for flexibility and emergent practices leads to the necessary increase of experience. Learning evaluation can facilitate a single-loop learning within the life cycle and/or a double-loop as a learning culture. Social learning attempts to increase the knowledge capital of the organisations in areas such as the management of IT investments, the application of evaluation approaches.

Exploratory evaluation is needed when the social learning faces a lack of consensus in terms of the objectives and/or the sense-making cannot deal with the strategic nature of the change and its uncertainty (e.g., major outsourcing decisions, just-in-time manufacturing systems). The exploratory orientation may will initiate a paradigm change when necessary. Examples can include the need for a more relevant approach stimulated exploration.

THE ORGANISATIONAL INTEGRATION OF IS EVALUATION

The organisational integration of IS evaluation is likely to be specific to its orientation. A number of critical success factors for the integration of each orientation are proposed in the next four paragraphs.

Control evaluation needs to be incorporated in harmony with other management processes such as financial investment appraisal, IS planning, IS project management and IS development. It is also necessary to support formal communication channels. Control evaluation assumes a rational organisational behaviour where the ritual imposed by a formal evaluation methodology can operate as a countermeasure for strategic actions. However, political influences and power games always interfere as individuals and/or groups try to promote their own interests. The allocation of the legitimate power (authority) and how the evaluation approach affects social influences are also important. Control evaluation is the most commonly held idea and it has proved a useful contribution to IS management. For all their faults, formal/control evaluation approaches remain part of organisational nature and are unlikely to disappear as they represent organisational stability. However, control evaluation is likely to be limited if used alone today because of its restricted focus.

Sense-making seeks consensus and commitment among stakeholders. It utilises communication to expose informal interests and assist in agreement of a common set of objectives. In order for evaluation to play its sense-making role successfully, a common language should be established and maintained. This can be facilitated by an accepted (formal) method. In addition, it might be necessary for special facilitators (e.g., Account managers) to be introduced to assist in the dissemination of the multiple views. Conflicts and broad interests should be exposed and then negotiated, and those affected by the situation should be actively involved. However, a broad coverage does not necessarily guarantee equal representation as power and political games can shadow them. Therefore, honesty is expected.

Social learning evaluation requires a trigger to operate and knowledge to be stored and become available to relevant stakeholders at the right time. It should be facilitated by formal processes as well as informal reasoning. Learning can be achieved by many ways including sense-making and experimentation. The knowledge stored in organisational memory in order to be useful needs to be accessible. Furthermore, learning evaluation requires a conceptual foundation (e.g., skills, other experience/knowledge) which could be acquired by prior training and/or education (e.g., in the case of a new evaluation approach introduced).

Exploratory evaluation questions the validity of existing paradigms and leads to innovation and organisational renewal. It shapes its context by radically challenging its assumptions, principles and norms. It can develop knowledge which assists in dealing with new requirements and uncertainty. Exploratory evaluation requires more than traditional structures and methods; it needs the power to promote institutional and culture arrangements. Special attention should be paid to discursive actions which should be regulated as they involve political and power underpinnings. Individuals should be empowered by skills, learning capabilities and knowledge to operate and perform their roles.

CONCLUSION

IS evaluation will continue to be an important topic in the decision making of IT investments. Particularly today and in the near future the increasing interlink between business and IT in the electronic economy requires a different type of evaluation. There is a need for integrated evaluation of business ideas/solutions enabled by IT. This makes more emergent the fact that the organisational dimensions - i.e. people's roles are understood and considered.

This chapter has suggested a number of basic principles for managing the organisational changes that information systems evaluation stimulates. The main principle remains the fact that different evaluation orientations require different treatments in terms of their integration with existing organisational properties. The complexity and degree of difficulty depends on the distance between the current and future state.

The identification of the stakeholders involved in the evaluation exercise is the first step. Their mental models (or evaluation schema) are the foundations which determine their behaviour. The schema are formed within the context where the stakeholders operate and they are subject to the process, content and history aspects. For reasons outlined in an earlier section it is difficult to analyse the schema. Therefore, the focus should lie on the organisational roles the individuals perform again within their context, process, content and history settings. Two types of roles are recognised in relationship to IS evaluation - the strategist and the evaluator. Their specifics are determined by the overall role evaluation plays organisationally (in other words by its orientation).

REFERENCES

Avgerou, C. (1995) Evaluating information systems by consultation and negotiation. *International Journal of Information Management*, 15 (6), 427-436.

Ballantine, J., Galliers, R. and Stray, S.J. (1995) The use and importance of financial appraisal techniques in the IS/IT investment decision-making process - recent UK evidence. *Project Appraisal*, 10 (4), 233-241.

Bartlett, F. (1932) *Remembering: A study in experimental and social psychology.* Cambridge University Press, London.

Beach, L.R. (1990) *Image theory: Decision making in personal and organizational contexts.* John Wiley & Sons, Chichester, England.

Biddle, B.J. (1979) *Role theory. Expectations, Identities, and Behaviors.* Academic Press, New York, NY.

Brown, J.A.C. (1970) The Social Psychology of Industry. Penguin Books Ltd, Middlesex, England.

Craik, K.J.W. (1943) *The nature of Explanation.* Cambridge University Press, Cambridge.

Earl, M. and Hopwood, A. (1987) From Management Information to Information Management. In Towards Strategic Information Systems, (E.K. Somogyi and R.D. Galliers), Tunbridge Wells, Kent and Cambridge, MA, 100-112.

Eden, C. (1992) On the nature of cognitive maps. *Journal of Management Studies*, 29 (3), 261-265.

Farbey, B., Land, F. and Targett, D. (1992) Evaluating investments in IT. *Journal of Information Technology*, 7, 109-122.

Farbey, B., Land, F. and Targett, D. (1993) *How to Assess your IT Investment: A study of Methods and Practice.* Butterworth Heinemann, Oxford.

Farbey, B., Land, F. and Targett, D. (1994). Matching an IT project with an appropriate method of evaluation: a research note on Evaluating investments in IT. *Journal of Information Technology*, 9, 239-243.

Fincham, R. and Rhodes, P.S. (1988) The individual, work, and organization. Weidenfeld and Nicolson, London.

Gallivan, M.J. (1995) Contradictions among Stakeholder Assessments of a Radical Change Initiative: A Cognitive Frames Analysis. In: W.J. Orlikowski, G. Walsham, M.R. Jones and J.I. DeGross (Eds.), *Information Technology and Changes in Organizational Work*, Chapman & Hall, London, 107-130.

Giddens, A. (1979) *Central Problems in Social Theory: Action, Structure and Contradiction in Social Analysis*. University of California Press, Berkeley, CA.

Giddens, A. (1984) *The Constitution of Society: Outline of the Theory of Structure*. University of California Press, Berkeley, CA.

Gilbert, D.R., Hartman, E., Mauriel, J.J. and Freeman, R.E. (1988) *A Logic for Strategy*. Ballinger Publishing Company, Cambridge, AM.

Ginzberg, M.J. and Zmud, R.W. (1988) Evolving criteria for information systems assessment. In *Information Systems Assessment: Issues and Challenges*, N. Bjorn-Andersen and G.B. Davis (Eds.), North Holland, Amsterdam, 41-55.

Gregory, A.J. and Jackson, M.C. (1991) Evaluating organizations: A systems and contingency approach. *Systems Practice*, 5 (1), 37-60.

Gregory, A.J. and Jackson, M.C. (1992) Evaluation Methodologies: A System for Use. *Journal of Operational Research Society*, 43 (1), 19-28.

Guba, E.G. and Lincoln, Y.S. (1989) *Fourth Generation Evaluation*. Sage, Newbury Park, California.

Harrison, E.F. (1995) *The Managerial Decision-Making Process*. Houghton Mifflin Company, Boston MA.

Hellstern, G-M. (1986) Assessing Evaluation Research. In Guidance, Control and Evaluation in the Public Sector, F.X. Kaufmann, G. Majone and V. Ostrom (Eds.), Walter de Gruyter, Berlin and New York, 279-312.

Hirschheim, R. and Klein, H.K. (1989) Four Paradigms of Information Systems Development. *Communications of the ACM*, 32 (10), 1199-1216.

Hirschheim, R. and Smithson, S. (1988) A critical analysis of information systems evaluation. In *Information Systems Assessment: Issues and Challenges*, N. Bjorn-Andersen and G.B. Davis (Eds.), North Holland, Amsterdam, 17-37.

House, E.R. (1993) *Professional Evaluation. Social Impact and Political Consequences*. Sage Publications, Newbury Park, CA.

Hsia P. (1993) Learning to Put Lessons into Practice, *IEEE Software*, September, 14-18.

Iivari, J. (1988) Assessing IS design methodologies as methods of IS assessment. In *Information Systems Assessment: Issues and Challenges*, N. Bjorn-Andersen and G.B. Davis (Eds.), North Holland, Amsterdam, 59-78.

Johnson-Laird, P.N. (1989) Mental Models. In *Foundations of Cognitive Science*, (ed. M.I. Posner), The MIT Press, Cambridge, MA, 467-499.

Lee, A.S., Liebenau, J. and DeGross, J.I. (eds) (1997) *Information Systems and Qualitative Research*. Chapman and Hall, London.

Legge, K. (1984) *Evaluating Planned Organisational Change*. Academic Press, London.

Levinson, D.J. (1971) Role, Personality, and Social Structure. In Sociological Theory: A Book of Readings (eds L.A.Coser and B.Rosenberg), The Macmillan Company, New York, 297-310.

Lyytinen, K. (1988) Stakeholders, IS failures and soft systems methodology: An assessment. *Journal of Applied Systems Analysis*, 15, 61-81.

Lyytinen, K. and Hirschheim, R. (1987) Information Systems Failures - a Survey and Classification of the Empirical Literature. In *Oxford Surveys in Information Technology* (pp. 257-309). Oxford: Oxford University Press.

Markus, M.L. (1994) Power, Politics, and MIS Implementation. In *Strategic Information Management*, R.D. Galliers and B.S.H. Baker (Eds.), Butterworth-Heinemann, Oxford, 297-328.

Merton, R.K. (1971) Role-Sets. In Sociological Perspectives, K.Thompson and J.Tunstall (Eds.), Penguin Books, Middlesex, England, 209-219.

Mintzberg, H. (1973) *The nature of managerial work*. Harper Collins Publishers, New York, NY.

Mintzberg, H. (1994) *The Rise and Fall of Strategic Management*. Prentice Hall, Hertfordshire.

Mumford, E. and Weir, M. (1979). *Computer systems in work design, the ETHICS method: Effective Technical and Human Implementation of Computer Systems: a work design exercise book for individuals and groups*. London: Associated Business Press.

Norman, D.A. (1980) Twelve issues for Cognitive Science. *Cognitive Science*, 4 (1), 1-32.

Norman, D.A. (1983) Some Observations on Mental Models. In *Mental Models*, D. Gentner and A.L. Stevens (Eds.), Lawrence Erlbaum Associates, Hillsdale, NJ, 7-14.

Orlikowski, W.J. and Gash, D.C. (1994) Technological Frames: Making Sense of Information Technology in Organizations. *ACM Transactions on Information Systems*, 12 (2), 174-207.

Parsons, T. (1951) *The social system*. The Free Press, Glencoe, IL.

Pettigrew, A.M. (1985) Contextualist Research and the Study of Organisational Change Processes. In *Research Methods in Information Systems*, E. Mumford, R. Hirschheim, G. Fitzgerald and T. Wood-Harper (Eds.), Elsevier Science Publishers, North Holland, Amsterdam, 53-78.

Pouloudi, A. and Whitley, E. A. (1997). Stakeholder Identification in Interorganizational Systems: Gaining Insights for Drug Use Management Systems. *European Journal of Information Systems*, 6 (1), 1-14.

Serafeimidis, V. (1997) *Information Systems Investment Evaluation: Conceptual and Operational Explorations*, PhD thesis, London School of Economics and Political Science.

Serafeimidis, V. and Smithson, S. (1999) Rethinking the Methodological Aspects of Information Systems Investment Evaluation, *Journal of Logistics and Information Management*, Special issue on Investment Decision Making of IT/IS, 12 (1-2), 94-107.

Simon, H.A. (1991) Bounded Rationality and Organizational Learning. *Organization Science*, 2, 125-139.

Smithson, S. and Hirschheim, R.A. (1998) Analysing information systems evaluation: Another look at an old problem. *European Journal of Information Systems,* 7 (3), 158-174.

Symons, V.J. (1993) Evaluation and the failure of control: information systems development in the processing company. *Accounting, Management and Information Technology*, 3 (1), 51-76.

Thompson, J.D. (1967) *Designing Organizations: A Decision-Making Approach*. McGraw Hill, New York.

Walsham, G. (1993) *Interpreting Information Systems in Organizations*. John Wiley & Sons, Chichester, England.

Walsham, G. (1995a) The Emergence of Interpretivism in IS Research. *Information Systems Research*, 6 (4), 376-394.

Walsham, G. (1995b) Interpreting case studies in IS research: nature and method. *European Journal of Information Systems*, 4 (2), 74-81.

Ward, J., Taylor, P. and Bond, P. (1996) Evaluation and realisation of IS/IT benefits: an empirical study of current practice. *European Journal of Information Systems*, 4 (4), 214-225.

Willcocks, L. and Lester, S. (1994) Evaluating the feasibility of information systems investments recent UK evidence and new approaches. In *Information Management. The evaluation of information systems investment*, L. Willcocks (Ed.), Chapman & Hall, London, 49-75.

ENDNOTE

1 The statements and opinions in this chapter are in all respects those of the author and do not represent the views of KPMG Consulting.

Chapter VII

Evaluating IS Quality: Exploration of the Role of Expectations on Stakeholders' Evaluation

Carla Wilkin, Rodney Carr and Bill Hewett
Deakin University

INTRODUCTION

IT Evaluation is essential, given that the value of investment in the IT industry is currently almost $2 trillion US. There is no doubt that an effective organisation will try to evaluate IT effectiveness, by linking performance measures with a financial perspective (i.e. a shareholders' view); an internal business perspective (i.e. company planning for excellence); a customer perspective; and the innovation and learning perspective (i.e. the means to improve and create value), in order to move consistently forward.

The last three perspectives are at times derived by using the same measures/instruments, via an interpretive approach based upon views of different tiers of stakeholders. Such an approach reflects a movement away from the more technical measures like benchmarking. Instead, IT effectiveness is evaluated in terms of the use of IT, or success of IT outcomes, through seeking to understand the effectiveness of the delivered IT application to the job performance of stakeholders. The merit of this interpretive approach is increasingly applicable to sectors like ecommerce, where it is very apparent that customers are concerned with the effectiveness of such IT applications.

With regard to IT research, the interpretive approach was initially crystallised in the Success Model formulated by DeLone and McLean (1992). Their evaluative tools were Use and User Satisfaction. However, if research in related industries is considered, it rapidly becomes apparent that evaluation of quality is a more highly regarded approach. In seeking to adapt this approach to IT, it is important to consider the key components of an IT system, for which effectiveness would be measured in terms of quality; what quality means in an IT context; and how stakeholders internally derive an evaluation of such quality.

In summary, this chapter reports on research which has produced a redefined IS Success Model, in which quality is the key to effectiveness. It also reports results of a related empirical study, which reaffirmed this IS Success Model and then investigated whether quality was better measured in terms of stakeholders' expectations for IS performance and their perceptions of actual performance, or whether measurement of perceptions alone provided sufficient understanding of IS quality/effectiveness.

Copyright © 2001, Idea Group Publishing.

BACKGROUND: QUALITY AS THE DETERMINANT OF IS SUCCESS/EFFECTIVENESS

Although DeLone and McLean's (1992) work reflected published research about delivered IS at the time of their study (1981-88), IT isn't a static phenomenon. Problems have arisen as IS has increasingly been recognised by corporate leaders as a service function. IS have moved from the mainframe era to a more decentralised approach in which computing and communication technologies merge to deliver an ubiquitous IS service over local and wide area networks. Via inter- and intra- organisational communication and information systems, where LANs, EDI and end-user computing prevail (Browning, 1994; Cattell, 1994; Drucker, 1988; Harris 1996; Phillipson, 1994; Violino and Hoffman, 1994; Ward and Griffiths, 1995), IS has become regarded as the instrument or service by which an organisation can gain or retain a comparative or competitive advantage. DeLone and McLean's model, which focused upon the stakeholders' use and feelings of satisfaction as the means to evaluate IS effectiveness, may have been relevant when IS success was so aligned to efforts by the IS department. Now the diffusion of IS within and between organisations is much wider and thus its role must be evaluated with a more business-oriented approach via stakeholders' views of IS capacity to accurately accommodate input and output data, in the performance of their jobs.

In seeking an alternative approach by which to evaluate IS success/effectiveness, it seemed pertinent to reconsider DeLone and McLean's own words. Given they used the term quality for framing the system and information components, this was the next point of consideration. Was it preferable and/or achievable to measure quality directly rather than through surrogates like use and user Satisfaction? Is there in fact a difference between satisfaction and quality? What does the term quality mean when it is used as a measure of success/effectiveness? How do stakeholders derive an internal measure of this quality/effectiveness?

DeLone and McLean's IS Success Model

Historically, in evaluating IT effectiveness, the key paradigm has been DeLone and McLean's Success Model. Despite calling this taxonomy a success model, what was claimed to be evaluated was the "output variable – IS success or MIS effectiveness" (DeLone and McLean 1992 p61). In that context, effectiveness was equated to influence and defined (following Mason 1978 p227) as the "hierarchy of events which take place at the receiving end of an information system which may be used to identify the various approaches that might be used to measure output at the influence level." Such events included receipt and evaluation of information as well as its application. The existence of an IS is fundamental to this work, but the term information system is not actually defined by DeLone and McLean, although it is consistent with their work for IS to "be defined in terms of its function and structure: it contains people, processes, data models, technology, formalised language in a cohesive structure which serves some organisational purposes or function" (von Hellens 1997 p802).

DeLone and McLean's (1992) IS Success Model (see Figure 1 below) offered a complete and coherent, yet conceptual depiction of the interdependent success components in an information system. Based upon a study of IS research and literature, they defined the evaluation of IS success in terms of six components, wherein the key for measuring effectiveness was postulated to be use and user satisfaction, with reference to the system and information so provided.

Figure 1: Interdependent success components in information systems (Source: DeLone and McLean, 1992 p87)

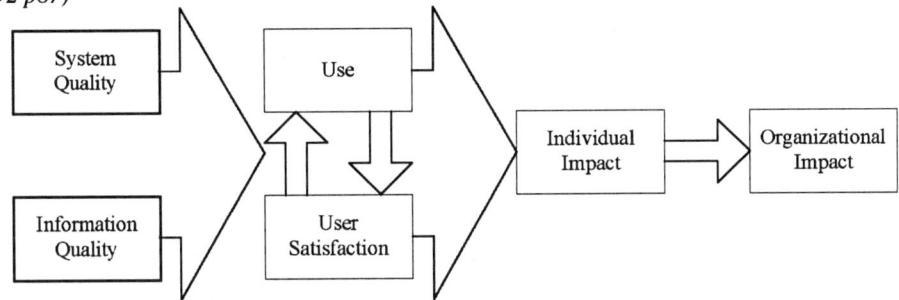

It was also worth investigating whether later research had suggested inclusion of any additional components for the IS Success Model.

Extension to the IS Success Model: Service Quality

There are two ways in which an understanding of service applies to the IS function. The first one concerns the information system itself, which is more than a technical product, for its worth lies in its capacity to serve the needs of its end-users/stakeholders. With this approach, it's the whole system which provides service to user stakeholders, for people want not merely a machine but one which serves their needs; and not merely data but information which is pertinent to their requirements.

This is a changed focus from what was accepted at the time of DeLone and McLean's study. The change is most evident from comparison of their definitions of system quality and information quality, which they derived from their review, and those which evolved as a consequence of recent work with focus groups and interviews for the empirical study reported in this article (see Table 1).

The second understanding of service concerns the role of the service or support facility. Such units, whether outsourced or in-house, are required to ensure acceptable system performance; sufficiently trained users; and technical facilities with the capability of generating the information desired. Hence it is most important to incorporate service quality as the third core component for delivered IS, because such IS departments deliver "informa-

Table 1: Comparison of Key Definitions (DeLone and McLean, 1992; Wilkin and Hewett, 1999)

Component	Meaning as Formulated by DeLone & McLean	Definition Developed by Wilkin
System Quality	Measures of the information processing system itself	A global judgement of the degree to which the technical components of delivered IS provide the quality of information and service as required by stakeholders including hardware, software, help screens and user manuals.
Information Quality	Measures of information system output	A global judgement of the degree to which these stakeholders are provided with information of excellent quality, with regard to their defined needs excluding user manuals and help screens (features of System Quality).

tion through both highly structured information systems and customised personal interactions" such that "effectiveness of an IS unit can be partially assessed by its capability to provide quality service to its users" (Pitt et al, 1995 p183). Again such a view aligned with findings in recent focus groups and interviews associated with the empirical study reported in this article.

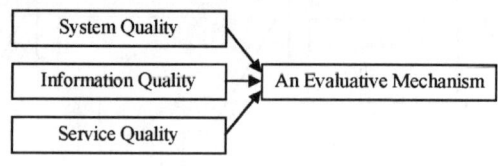

Figure 2: Service quality in the IS success model (Wilkin and Hewett 1997)

With this inclusion, the IS Success Model now begins to reshape as set out in Figure 2.

In such a context, Service Quality is defined as "[a] global judgement or attitude relating to an assessment of the level of superiority or excellence or service, provided by the IS department and support personnel" (Wilkin and Hewett 1999).

An Evaluative Mechanism: Beyond Use and User Satisfaction

At the core of their model, DeLone and McLean evaluated the effectiveness/success of the IS in terms of the use of the system by stakeholders and in terms of such stakeholders' satisfaction with the system and information so generated. They saw use and user satisfaction to be interdependent. In turn, the outcome of this process would affect individuals in the performance of their jobs and in turn the organisation's performance in achieving business needs.

A review of the literature raised four key reasons for reconsidering Use and User Satisfaction as the evaluative mechanism.

A major problem concerns the measure of use. Extensive work by Seddon and associated researchers reveals that stakeholders have confused interpretations of its meaning. Firstly, they found IS use was confused with the concept *benefits from use*. A successful system should provide benefits like helping the user to work more efficiently or produce better work, so there has developed an assumption that the more time spent with a system, the more benefits it should provide. Such an assumption ignored individuals' work rates, expertise and the degree of user friendliness in the design. Actually, in a sense, this understanding of IS use being another term for benefits from use was found to be very close to individual impact and organisational impact in DeLone and McLean's 1992 model (Seddon and Fraser, 1995).

Secondly, the term IS use was interpreted as *future use* (Seddon and Fraser, 1995; Seddon, 1997). In this sense, IS use was measuring behaviour not IS success. This related to studies by Davis (1989, 1993) and Davis et al (1989) regarding measurement scales for predicting user acceptance of information systems, which found that perceived ease of use and perceived usefulness, were good indicators of user willingness, or user satisfaction. In fact Davis' findings of a pathway ease of use –> usefulness –> usage (assuming voluntary use) supported earlier findings by Goodwin (1987), Segars and Grover (1993), and Baroudi et al (1986), that the effective function of a system depends on useability.

Finally, IS use was only measurable after the system had been used and impacted on the individual and the organisation. Thus, these two impacts and user satisfaction as the consequences of use, would be better measures of IS success (Seddon and Fraser, 1995; Melone, 1990).

Another major problem is that the model shows use and user satisfaction to be interdependent, when the literature suggests each independently reflects IS effectiveness.

For example, where use is voluntary, a measure of success is the extent of use: but when use isn't voluntary, success is concerned with users' overall degree of satisfaction (Moynihan, 1982). One empirical study concluded that user involvement in requirements gathering, system definition and implementation, led to both user information satisfaction and system usage, and that user information satisfaction led to system usage, but not the reverse (Baroudi et al 1986). This proposition of interdependency, as discussed above, was also called into question by the finding of a different pathway (assuming voluntary use) ease of use –> usefulness –> usage (Davis 1989).

Equally problematic is that a focus upon user satisfaction implies a particular view of the IS facility, related to the attitude of end users rather than their output. As such, user computer attitudes may affect user satisfaction (Igersheim 1976; Lucas 1978). Even the degree of match between the characteristics of a task and the capacity of the system itself would control the satisfactory level of information generated and hence impinge upon user satisfaction (Goodhue 1986). Other work found that where tight links existed between system usage and work, it's possible to have an effective IS facility without satisfied users (Melone 1990).

The final problem is that investigation of user satisfaction instruments reveals some confusion about what has been measured (Galletta and Lederer 1989). For example, Doll and Torkzadeh (1988) developed one such instrument that was designed to measure user satisfaction. Etezadi-Amoli and Farhoomand (1991) found this instrument to be unreliable because it measured the frequency of satisfaction and did not provide a means to assess the relative importance of each item to the respondent. Four of the factors (Information Context, Accuracy, Format and Timeliness) related to information quality and one (Ease of Use) to system quality.

So how can the effective/successful performance of the three components (System Quality, Information Quality and Service Quality) be gauged? The obvious step is to look further at satisfaction, which DeLone and McLean had argued was the key variable, and at quality, which was the evaluative term they had attached to system and information. What is the distinction, if any, between these terms?

Satisfaction and Quality

What is meant by these two terms is not always clear. At times customer satisfaction is claimed to precede quality (Parasuraman et al 1986); at other times to be its consequence (Cronin and Taylor 1992, 1994); and at other times the terms are used interchangeably (Parasuraman et al 1994).

Despite the lack of a definition of quality, DeLone and McLean's table of empirical measures implies a technical focus, using production terms like response times, resource utilisation and investment utilisation for system quality and product terms like accuracy, precision and completeness for information quality.

In an attempt to distinguish between satisfaction and quality, some have defined customer satisfaction in terms of the user's attitude or feeling as a result of a specific transaction or as a consumer's emotion related to a specific transaction (Oliver 1981). Further definitions have included "a subjective evaluation of the various consequences evaluated on a pleasant-unpleasant continuum" (Seddon 1997 p246) or as an "affective state that [was] the emotional reaction to a product or service experience" (Spreng and Mackoy 1996 p17-18). Quality was seen as a global judgement about a product's (or service's) overall excellence (Parasuraman et al 1986).

Spreng and Mackoy's (1996) empirical study investigated the proposition that satisfaction was the result of a comparison of perceptions of service received with expectations of what will happen (predictive expectations); and service quality was the result of a comparison of perceptions of service received with expectations of the service which the service provider should provide (ideal expectations or desires) (Spreng and Mackoy 1996). The study found that satisfaction and service quality were distinct constructs; that desires did affect satisfaction; that the disconfirmation of expectations (what will happen) did not significantly affect service quality; but that expectations did indirectly have a positive effect on service quality (through perceived performance) (see Figure 3).

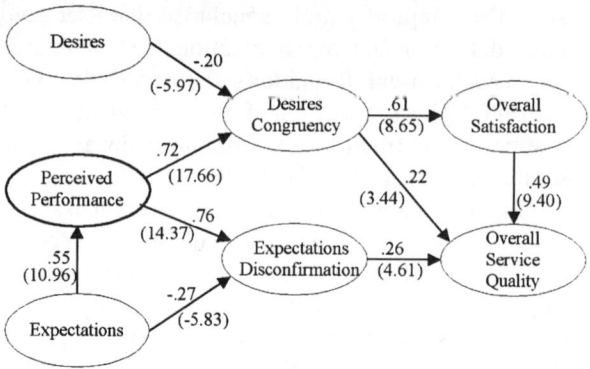

Figure 3: Relationship between satisfaction and service quality (Source: Spreng and Mackoy, 1996 p209)

The conclusion was that **satisfaction** was the result of expectations of what will happen being disconfirmed by the perceived performance and that service **quality** was derived by comparison between desires (what should happen) and perceived performance.

So how does this relate to current research? Two of their findings suggest factors that need to be considered in future research.
- As desires had an impact, "managers should not believe that merely meeting (or exceeding) predictive expectations will satisfy consumers" (Spreng and Mackoy 1996 p210).
- Expectations did influence perceptions of performance and therefore needed evaluating.

Surely if the function of IS is as a tool to gain comparative or competitive advantage, the aim would be for maximum levels of performance. Hence, the quest for IS success should be focused on a horizon beyond merely what a stakeholder may think will happen, and rather on a further horizon like what should happen. Therefore, the measurement of quality, not satisfaction, should be the focus of IS effectiveness.

The more directly an instrument can measure what stakeholders know, the less the data has to be interpreted. If the issue is user stakeholder views of the quality of the system, information or service, why use surrogates like use or user satisfaction to measure it? Why not measure in terms of quality itself?

Defining Quality

Historically, the meaning of quality has altered considerably, from conformance to product and production to specifications (Levitt, 1972; Crosby, 1979); fitness for use (Juran et al, 1974); to value (Cronin and Taylor, 1992; Garvin, 1988); and meeting and/or exceeding customers' expectations (Gronroos, 1983; 1990; Parasuraman et al., 1984; Zeithaml et al., 1990; Buzzell and Gale, 1987).

Such changes are outlined in Table 2 (see below). In this context, it is the customers/stakeholders who appear to be the driving force with their demands for higher performance requirements, faster product development and fewer defects (Kerzner, 1998; Davis and Meyer, 1998). The involvement of stakeholders in the evaluation and realisation of quality

Table 2: Changing Views of Quality (Source: Kerzner, 1998 p1042)

Past Understanding of Quality	Present Understanding of Quality
• Quality is the responsibility of blue-collar workers and direct labor employees working on the floor	• Quality is everyone's responsibility, including white-collar workers, the indirect labor force, and the overhead staff
• Quality defects should be hidden from the customers (and possibly management)	• Defects should be highlighted and brought to the surface for corrective action
• Quality problems lead to blame, faulty justification, and excuses	• Quality problems lead to cooperative solutions
• Corrections-to-quality problems should be accomplished with minimum documentation	• Documentation is essential for "lessons learned" so that mistakes are not repeated
• Increased quality will increase project costs	• Improved quality saves money and increases business
• Quality is internally focused	• Quality is customer focused
• Quality will not occur without close supervision	• People want to produce quality products
• Quality occurs during project execution	• Quality occurs at project initiation and must be planned for within the project

was best summarised by Iacocca (1988 p257), when he said, "quality doesn't have a beginning, or a middle. And it better not have an end. The quality of a product, and of the process in arriving at that product, has to go on and on to become part of every employee's mind set."

With such an understanding, there seems little distinction between quality as defined by user stakeholders and the definition of IS effectiveness as a "value judgement, made from the point of view of some stakeholders, about net benefits attributed to use of an information system" (Seddon et al., 1999 p1; Seddon et al., 1998; Grover et al., 1996). Accordingly, it would seem logical to use quality as the defining measure of IS success/effectiveness.

Redefining the IS Success Model

If the concept of quality was to be incorporated in the IS Success Model, then it should be positioned to provide the key information regarding the principal facets as they impact on both the individual and the organisation (see Figure 4). Accordingly the conceptual model which was developed here has quality as the evaluative mechanism/determinant of IS effectiveness.

Here quality is featured as the key determinant of IS success. Benefits of the model include:
- "acknowledgment of greater expertise among users and consideration of a broader audience, including customer views and organizational interests;
- the use of two key variables, expectations and perceptions, to identify underlying reasons surrounding the importance of a particular component;
- direct measurement of the key issues rather than through surrogates;
- provision of detailed information directly relevant to system, information and service quality;
- elimination of problems regarding interdependency; and
- a clear focus upon the functional effectiveness of IS, relevant to comparative and competitive advantage" (Wilkin and Hewett 1999).

With extensive growth in IT investment and concern that benefits might not be as high as expected, evaluation becomes a major concern. Moreover, given preference for the more

Figure 4: Quality as a component of an IS success model (with reference to user stakeholders)
　　　　Key:　　solid line arrows　　: variance
　　　　　　　　dotted line arrows　: influence (Source Wilkin and Hewett 1999; Wilkin 2000)

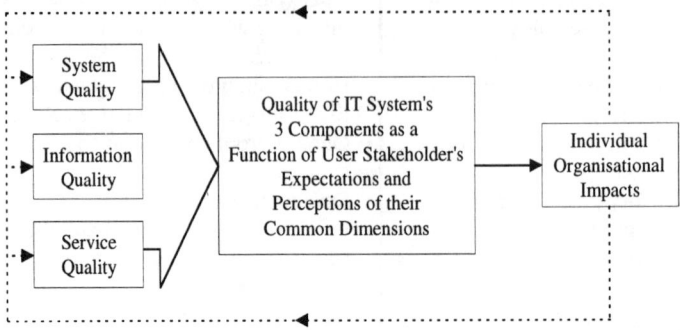

Reappraisal of expectations re future use in light of stakeholder's perceptions of quality and experiences and in light of changes to function and service requirements due to changes in system environment

customer focused definition, the next point of review is an approach by which such an evaluation can be achieved.

THE DEBATE: QUALITY = P – E

Quality has an elusive nature, and given the absence of objective measures "a useful and appropriate approach for assessing the quality of a firm's services is to measure customers' perceptions of quality. What we then need is a quantitative yardstick for gauging such perceptions" (Parasuraman et al., 1986).

Many practitioners and researchers see two key variables in the measurement of quality: perceptions and expectations. Here a measure of quality (denoted G) is derived by taking Expectations (E) away from Perceptions (P) i.e. $G = P - E$. Accordingly, the higher G, the better the level of quality, with high negative scores indicating low quality. Such an approach was considered a more sensitive measure than simply capturing the result using a single response (Perceptions only) (Parasuraman et al., 1986; Pitt et al., 1995; Wilkin and Hewett, 1997).

Three instruments evaluate quality by measurement of customer's expectations and/or perceptions: SERVQUAL, SERVPREF and SERVIT (Parasuraman et al 1986, 1991, 1994; Cronin and Taylor 1992, 1994; Wilkin and Hewett 1997). The former two are highly regarded in marketing while the latter is a derivative for use in IS. To understand the merits and problems associated with the use of these respective instruments, it is necessary to explore the terms expectations and perceptions, together with the debate which is associated with their usage.

Expectations

Expectations, variously defined as desires, wants, normative expectations, ideal standards, desired service, and hoped for service (Teas, 1994), occur at two levels, predictive expectations and desires. When used predictively, expectations relate to something that will happen and hence are linked to satisfaction, but expectations of an ideal relate to desires or something that should happen and hence are linked to evaluation of quality.

There are at least four ways people's expectations are formed: word of mouth communications, personal needs, past experience and external communications, including price (Zeithaml et al., 1990). Certain reservations should be considered as circumstances arise to cause their over-inflation or under-inflation, thereby affecting the customers' final evaluations (Olshavsky and Miller, 1972; Szajna and Scamell, 1993).

Justification for including expectations (Cronin and Taylor, 1992; 1994; Teas 1993; 1994; Van Dyke et al., 1997) centred around the insight it provided about how users formulated perceptions or how significant such users saw each dimension or statement (Parasuraman et al., 1986; Pitt et al., 1995; Kettinger and Lee, 1997; Carman, 1990). Take the example of two users. Both may rank a statement at three with regard to their perceptions of performance, but if they both expected statement one to perform at 5 ($G = 3 - 5 = -2$) and statement two at 7 ($G = 3 - 7 = -4$), then there's a distinct need to focus on statement two as an area requiring attention. As demonstrated, it was the *expectation* measurement that provided such insight.

Evident from the literature is that expectations of stakeholders are seen as essential to both understanding and achieving IS effectiveness. The concept's role in evaluation of quality/IS effectiveness has relevance, as through measurement of it as an ideal standard (the proposed means of measurement), more insight surrounding problematic areas within an organization is provided. Thus, in evaluating quality, the disconfirmation between expectations of ideal service and perceptions of reality is a less subjective, more global judgement encompassing a broader range of attributes than merely satisfaction. In this sense, measurement should be better than by use of the surrogates that other researchers have used, for example, an examination of adequacy of service (which equates to satisfaction).

A further complication concerns the different internal opinions held by different user stakeholders. Here too, unless expectations are measured as part of an evaluation of quality, a low or high perception rating could provide misleading information. Moreover, measurement of expectations provides insight regarding changes in system environment (Watson et al., 1998).

Perceptions

The perception's only measure has been proposed in both defining and evaluating quality. Believing that a measurement of service quality derived by the difference score only captured factors that were related to service quality and didn't measure customers' view of the concept itself, Cronin and Taylor (1992) proposed one which measured adequacy perceptions/importance. They argued this was the most efficient. Their justification concerned the fact that judgements of service quality and satisfaction appeared to come after evaluation of a service provider's performance.

As perceptions of delivered service are contingent upon prior expectations, they are shaped by the same four sources as expectations although additionally external communications including factors like pamphlets and advertising, are relevant. Moreover, it was found that perceptions of quality are based more heavily on perceptions of current performance than on actual service change (Bolton and Drew, 1991). However, as users are different and hence their knowledge, attitudes and methods of defining systems are different, such factors may potentially result in problems with measuring perceptions.

Support can be found for the view that a single measure of performance provides little information about a user's thoughts in relation to product features, nor does the process by which performance is converted into understandings by the consumer (Spreng et al., 1996; Oliver, 1989). Extrapolating, it can be assumed that use of expectations provides greater insight.

Dispute Regarding Expectations

Given the varying understanding of expectations, Teas (1994) argued that user interpretations would correspondingly vary.

Van Dyke et al (1997) argued that work by Teas (1993) and Boulding et al., (1993) revealed a number of different interpretations of expectations by users: a forecast or prediction, i.e., will; a measure of attribute importance (see Teas); and a classical ideal point, i.e., should. Further, they argued that it was found (Boulding et al 1993) that increased will expectations led to higher perceptions of service quality and increased should expectations led to lower perceptions. Some problems were removed when Parasuraman et al (1994) proposed expectations as a vector attribute. Here a customer's ideal point is at an infinite level, formalised with use of the word will rather than should.

Kettinger and Lee (1997) acknowledged some validity to the claim (Van Dyke et al., 1997; Teas, 1993; Boulding et al., 1993) that customers might variously interpret expectations. They used the ideal expectation format, basing this selection on material from Parasuraman et al. (1991) and Zeithaml et al. (1990).

A further issue concerned the fact that there might be confusion related to interpretation of expectations and that this confusion would be embedded in perceptions, which all researchers agreed were partly formulated from expectations (Kettinger and Lee, 1997).

Dispute Regarding the Gap Measurement

As SERVQUAL (a marketing instrument formulated by Parasuraman et al 1986) measured users' views of service quality by difference scores, i.e. its formulation was such that Service Quality (G) = Perceptions (P) – Expectations (E), criticism also related to the problems with difference scores.

Particularly influential in this regard was Peter et al (1993), who in discussing such statistical problems, cited examples of 13 researchers who investigated instruments that used difference scores. His findings revealed four key problems with such measurement.

1. Reliability. Peter et al., (1993 p658) claimed that difference scores were less reliable than their component variables such that as "the reliability of either (or both) component score decreases, the reliability of the difference score decreases." Van Dyke et al., (1997) supported the claim that as "the correlation between the component score becomes larger, the reliability of the difference score also decreases" (Peter et al., 1993 p658), a point made by Prakash and Lounsbury (1983) which they demonstrated mathematically.
2. Discriminant validity, where the term was used to describe the "degree to which measures of theoretically distinct constructs do not correlate too highly" (Peter et al., 1993 p659), is essential if the components of the measure were to have construct validity. They argued that because difference scores have lower reliability, there was an illusion that they possessed discriminant validity.
3. Spurious correlations, according to Peter et al (1993), related directly to this second discriminant validity problem. Because they felt there was a relationship between G, P and E, Peter et al. (1993) argued that the correlation between the difference scores was likely to be spurious. They felt the difference between two variables provided no more predictive or explanatory material than the two components themselves provided (i.e. G was no more useful than P and E), that often one component variable performed better than the equation (i.e., P was more accurate than G = P – E) and that the high correlation between G = P – E and P or E produced unstable parameters and misleading results.

4. Finally, there was a possible restriction of the variance of the difference score variable (Peter et al., 1993 p660). They argued that, since E was always better than P, there was a restriction on the range of scores available to those who felt that service quality was good, as opposed to the greater range of scores available to those who had lower perceptions. Thus, they reasoned that, since P would rarely equal or exceed E, users who were unhappy with service would have a greater range of difference scores than those who were happy.

Summary: Issues for Exploration in Empirical Study

With such controversy regarding the evaluation of quality in terms of perceptions and expectations, it seemed pertinent to conduct a small empirical study to look at certain key issues which related to evaluation of quality in an IS context, by user stakeholders. The principal issues were as follows:

Question 1. Whether such stakeholders structured an IS facility with the same three key components as hypothesised, namely System Quality, Information Quality and Service Quality.

Question 2. How such stakeholders then arrived at perceptions of the quality of a delivered IS function.

Question 3. Whether measurement of expectations and perceptions of quality at the same time provided a more or less accurate indication of IS Quality/Effectiveness than a perceptions only approach, as suggested by Caruana et al. (1999).

EXPLORATION OF THE ISSUE: A REAL LIFE TRIAL

In early 1999, the research work which was being undertaken, required empirical investigation of the merits of the conceptual model (Figure 4) and associated data for the formulation of an instrument by which to measure IS Quality/Effectiveness. The chosen methodology was to conduct at least four focus groups and a series of interviews with a broad range of IT professionals until consistent findings were evident.

The participants in the first two focus groups were very interested in the initial results and especially in their developed understanding of how stakeholders arrived at scaled scores for their perceptions of any one aspect of IS. Their enthusiasm and willingness encouraged this trial even though it was slightly outside the directional thrust of the general research. A few others volunteered to be involved as well.

Focus Groups/Interviews

The participants became involved by a process best described as a convenience sample, although they were not handpicked. Personal approaches to a number of firms resulted in some participants volunteering and/or being nominated by managers. Geographically, the spread was diverse, with interviews and focus groups conducted face to face in two capital cities and two regional centres, with one international contact. All were IS stakeholders who used a diverse range of delivered IS, with varying levels of seniority.

A series of four focus groups were conducted with a diverse range of stakeholders including academics, a client support officer, bookshop supervisor, national sales administration manager, and strategic development manager. Ten semi-structured interviews were conducted with a range of system developers including self-employed consultants, managers and corporate technical IS developers. Both the interviews and focus groups were aimed at ascertaining opinions about the meaning of quality; components of delivered information systems/applications; critical aspects of the system or application, information or output and support/service; and the common themes used in assessing whether these aspects are up to standard.

Two results were abundantly clear at the end of the first two focus groups (undertaken prior to the empirical study reported in this chapter), and remained a unanimous finding throughout. Firstly, all participants agreed that fundamentally a delivered IS comprised three core elements:
- System including the physical system, software and manuals.
- Information generated both in print and electronic form.
- Service/support for user stakeholders, whether outsourced or in-house.

This obviously supported aspects of the respecified conceptual model. Secondly, all agreed that evaluation of the quality of delivered IS is made by comparison of reality with ideal perceptions compared with expectations in an assessment of quality. Thus $G = P - E$ seemed to be the intuitive/internal measure, which was used in assessing IS quality whether expectations were directly assessed or not.

The Instrument

At the end of the first two focus groups, a number of participants became involved in some informed discussion about what outcomes were sought from this research. They were very interested in the debate regarding whether measurement of perceptions only, or perceptions and expectations, was the most accurate method of evaluating stakeholders' views of IS quality/effectiveness.

At this point, although the proposed three sections for an instrument appeared to be justified (system quality, information quality and service quality), the data about the dimensions were inconclusive. So the instrument used was one in which the dimensions and relevant pointers had been hypothesised from the literature and from a previous set of interviews. Hypothesised dimensions in line with the service-related literature were: tangibles, reliability, responsiveness, assurance and empathy. The statements used for this trial had been carefully refined by these earlier interviews, with categorisation and ranking according to the methodology described by Davis (1989). The only change made to this early instrument was the deletion of the tangibles dimension, because findings from the first two focus groups were that:
- although somewhat inconclusive, the tangibles dimension was ranked less significant or low;
- some statements (help screen and interface) translated from service quality and Parasuraman et al's work were linked to responsiveness;
- two statements related to portability rather than tangibles, a dimension which also ranked poorly;
- results from prior studies rated the dimension of least significance and prone to causing problems (Pitt et al., 1995; Wilkin, 1996; Wilkin and Hewett, 1997); and
- elimination of 12 statements (4 statements times 3 components) had appeal, given the instrument's length.

Thus, the hypothesised dimensions could be defined as (Table 3):

The dimensions identified for the instrument were simple enough to address only a handful of related issues at a time. The quality in each dimension is measured by presenting users with descriptions of the issues involved and by asking them to provide assessments of the magnitude of any problems perceived. By combining these individual assessments, measures of quality in each dimension are obtained (and thus system, information, service and overall IS quality).

Table 3. Definition of Dimensions (Note: The tangibles dimension was deleted, but a definition of the dimension is provided here for reader understanding)

		Dimensions				
		Tangibles	Reliability	Responsiveness	Assurance	Empathy
Definition	Service Quality Parasuraman, Zeithaml & Berry & adopted by Wilkin	Appearance of physical facilities, equipment, personnel & communication material	Ability to perform the promised service dependably & accurately	Willingness to help customers and provide prompt service	Knowledge and courtesy of employees and their ability to convey trust and confidence	Caring, individual attention the firm provides its customers
	System Quality Wilkin	Appearance of user interfaces, useability of system documentation and convenience of the physical components of the system	Ability of the system to function reliably & efficiently, and produce accurately the requested results when promised & consistency of the technology	The flexibility and integration the system allows including ease of response to users' commands and ease of selection of system features in a timely manner	The capacity of the system to inspire trust and confidence in its reliability and security	The system's consideration of the specific needs of individuals such that stakeholders feel supported & comfortable when using the system
	Information Quality Wilkin	Appearance of reports or information generated in a manner which is readable, modern looking, appealing, attractive, and formatted	The extent to which the information is complete, dependable, delivered as promised and accurate on a consistent basis.	Ease of accessibility of information	The information produced can be confidently used due to its understand-ability, comparability & validity	The information makes the user's job easier, supports key aspects of the work, and enhances but does not replace human judgement

One issue explored in this study (Question 3) is the precise way to measure the magnitude of users' assessments of problems. Take for example a specific statement from QUALIT that is used in measuring the reliability dimension of system quality.

P1. My IS systematically checks and identifies errors

P1 is one of five statements in this dimension. Stakeholders provide a measure of quality of this aspect of their system, by giving a score for this statement on a standard scale "Strongly Disagree" (score = 1) to "Strongly Agree" (score = 7). How do they arrive at the score they get? Focus groups identified that in some way they compare their actual IS with some ideal system which systematically checks and identifies all errors. The ideal system would normally but not always get a perfect score of 7. Their system is compared with this, and a score is provided. The problem is that simply asking for perceptions does not necessarily force a user to make a comparison with an ideal system, and different results may occur if the issue is forced a little by actually asking for their expectations at the same time. Moreover, different users have different internal measures with respect to both perceptions and expectations.

Therefore, this trial used an instrument hereafter called QUALIT, an IS derivative of SERVQUAL, which comprised
- three components – system quality, information quality and service quality
- four dimensions for each of the three sections – reliability, responsiveness, assurance and empathy, i.e., 12 dimensions in all. Such dimensions enabled grouping of state-

ments so that key problems or issues were addressed in four or five ways to ensure that the issue was thoroughly explored.
- 18 statements for each section
  Reliability 5 statements
  Responsiveness 4 statements
  Assurance 4 statements
  Empathy 5 statements
- Each statement was phrased in two ways. One way asked for a user's perception of actual performance (Perceptions sections) and the other for user's expectations of how the ideal would be (Expectations section).

For example, an issue or problem would be addressed through a pair of statements (for Expectations, E; and Perceptions, P).

E9. The flexibility of excellent IS will enable users to complete tasks more efficiently.

P9. The flexibility of my IS enables me to complete tasks more efficiently.

The version that was distributed first comprised solely perception statements addressing all three components (system quality, information quality and service quality), while its successor contained both expectation and perception statements, addressing the same three components. QUALIT's format was straightforward using closed statements requiring the participant to make a choice amongst a given set of alternatives by circling a number between 1 and 7, strongly disagree to strongly agree. Furthermore, it contained no negatively worded items, as these generated awkwardness.

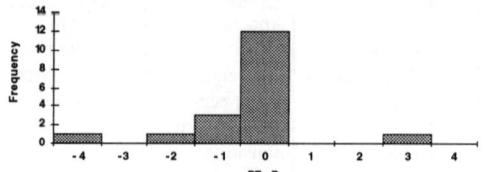

Figure 5: Distribution of respondent one's PE – P scores over all 18 system quality statements (Key: PE = Perceptions when measured with expectations and P = Perceptions only).

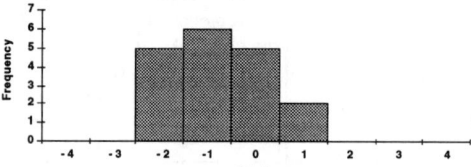

Figure 6: Distribution of respondent one's PE – P scores over all 18 information quality statements (Key: PE = Perceptions when measured with expectations and P = Perceptions only).

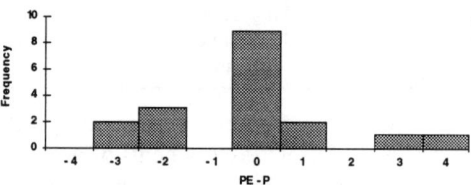

Figure 7: Distribution of respondent one's PE – P scores over all 18 service quality statements (Key: PE = Perceptions when measured with expectations and P = Perceptions only).

The Trial

Participants were all volunteers and were IS stakeholders from a wide range of occupations including administration, education, human resources, retail and professional writing sectors, who used a variety of IS. All were skilled end-users.

The initial Perceptions only version of QUALIT was administered via mail. To enhance response rates, a reminder was sent to non-respondents a week later. Two weeks after the receipt of completed questionnaires, the perceptions and expectations version was sent. Such a time lapse was deliberate for it was hoped that sufficient time had elapsed for respondents to forget specific answers to questions, but insufficient time for influential variables like training, to impact on results.

To enhance consistency, respondents were directed to complete both questionnaires with the same system in mind, a process facilitated through collection of system details in stage one, so that cover pages in stage two could be customised.

Response rates for the respective stages were relatively high, 86.66% and 76.92%, providing a total of 20 usable questionnaires. Although restrictive, the trial produced interesting findings.

4. Results

In interpreting these results, the perceptions only scores are denoted by P and the perceptions obtained when expectations were also evaluated, by PE. For each question, the difference between the scores, PE - P, provided by each respondent without expectations being measured, were determined.

Given that P and PE were on a 1 to 7 scale, it is possible for PE – P to be as large as 6 and as small as –6, but if measuring Expectations had no effect on Perceptions, values of PE – P would be close to zero. However they are not always. Figures 5, 6 and 7 show the distribution of the values of PE - P for a typical (randomly selected) respondent. For this respondent, the values of PE - P varied from a minimum of –4 on one question (on system quality) to +4 for another question (on service quality). This was typical and shows that evaluation of expectations affects measurement of perceptions for some questions and respondents.

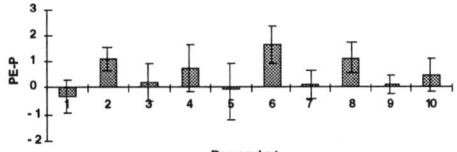

Figure 8: Respondent-by-respondent mean difference for system quality between perceptions when measured with expectations (PE) and perceptions only (P) showing ±2 Stderr

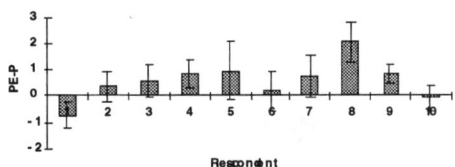

Figure 9: Respondent-by-respondent mean difference for information quality between perceptions when measured with expectations (PE) and perceptions only (P) showing ±2 Stderr

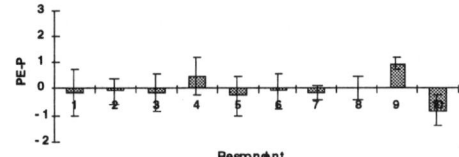

Figure 10: Respondent-by-respondent mean difference for service quality between perceptions when measured with expectations (PE) and perceptions only (P) showing ±2 Stderr

As described earlier, a tally of the Perception scores for the corresponding statements will provide a measure of quality. Again this can be done for perceptions with and without expectations being evaluated. Figures 8, 9 and 10 show the average change in the quality measurements for each participant, component-by-component (the average change is simply the average of the difference scores, PE - P, for each respondent, and is denoted as such in the charts).

It can be seen that some respondents gave much the same quality ratings whether or not expectations were determined. These are the ones where the average changes, PE – P, are close to zero. Most respondents were in this category when rating service quality for example. But it can also be seen that many respondents gave quite different ratings when expectations were determined. There is a definite trend that evaluating expectations inflates the perceptions scores, thus resulting in higher quality scores. This is most pronounced in the information quality section. The mean change in quality ratings taken over all participants reflects this (see Table 4).

The previous analysis shows that the perceptions scores are affected if expectations are evaluated at the same time. This is not particularly surprising. The most interesting and important issue is whether or not the quality ratings obtained using perceptions only are any different from those obtained by measuring the gap i.e. G = P–E. Of course a difference here means a difference up to a uniform scale factor, that is up to a linear transformation. There is nothing absolute about the ratings obtained from either the P or the P-E version.

Using the data from the PE trial, the gap was determined for each statement pair and each participant. In Figures 11, 12 and 13, these scores are plotted against the corresponding perceptions score, component by component. NOTE: to facilitate readability, the gap (derived by taking perceptions away from expectations) has been rescaled to a positive score by adding 7 and dividing by 2.

Table 4: Mean Difference Between Perceptions when Measured with Expectations (PE) and Perceptions Only (P)

Component	Mean
System Quality	0.49
Information Quality	0.56
Service Quality	-0.04

It is quite clear from Figures 11, 12 and 13 that as the perceptions only score increases, so too does the gap score. However, the relationship is far from strong and one would certainly not be able to use perceptions only as a surrogate for the gap or vice versa: it is possible to get a high perception rating and virtually any score for the gap for example. The conclusion is that, although related, the gap and perceptions measure alone are different measures of quality.

CONCLUSION

Much work has been done to improve the technical quality of delivered IS, but there is a need to focus on effectively managing or evaluating this in humanistic terms. Given that surely the most knowledgeable source of information about the effectiveness of the function of the system, the information it generates and/or support services IS delivers, is user stakeholders, then collation of their myriad views in an objective, quantitative and structured manner to match the principal facets (system quality, information quality and service quality), creates an opportunity to evaluate IS effectiveness.

What this exploration has done is provoke of a number of questions that require further investigation. Although the answers aren't clear, there's enough evidence to ask two questions.

- In an evaluation of IS quality/effectiveness, does measuring perceptions and expectations at the same time mean perceptions are more valid than when perceptions only are measured? alternatively, does measuring perceptions only result in more accurate perception scores and thus a more valid measure of IS quality/effectiveness?
- Given the deliberation about expectations and perceptions, is the gap measurement (i.e. G = P – E) a more/less

Figure 11: System quality scatter plot with regression line

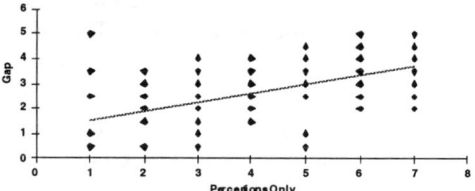

Figure 12: Information quality scatter plot with regression line

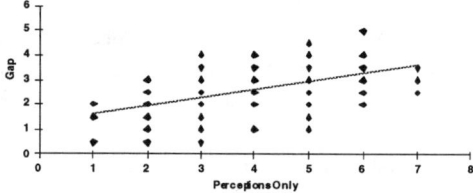

Figure 13: Service quality scatter plot with regression line

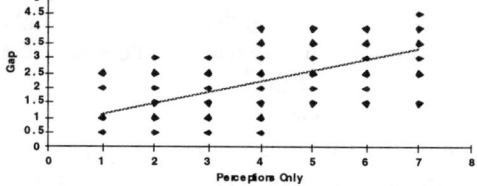

accurate measure of IS effectiveness than the perceptions only measure?

What is clear from the findings is that participants' evaluations of expectations tempered their perceptions scores to a significant unpredictable degree.

Possibly the most interesting outcome would be in ascertaining which method (perceptions or perceptions minus expectations) gives a more valid measure of quality. Exploration of this could be achieved using triangulation with interviews or an alternative questionnaire. As the effect of expectations on perceptions was only discovered after completion of this trial, triangulation such as that proposed above was not possible, as results came in so slowly that too great a time lapse could have invalidated results.

Considering the trial and results, a reasonable inference is that there really isn't any neat solution to the debate. The issue is more concerned with both benefits and problems. With that aside, given the effect of expectations on perceptions is not only indefinitive but unpredictable (not to mention the significant data expectations provides), there seems to be business value in measuring both. With the growth of the IT industry, and the increasing need for organisations to make better and better returns on investment to retain their comparative and competitive positions, there is a need to focus on such issues.

REFERENCES

Baroudi, J.J., Olson, M.H., & Ives, B. (1986). An empirical study of the impact of user involvement on system usage and information satisfaction. *Communications of the ACM, 29(3)*, 232-238.

Bolton, R.N., & Drew, J.H. (1991). A longitudinal analysis of the impact of service changes on customer attitudes. *Journal of Marketing, 55(1)*, 1-9.

Boulding, W., Kalra, A., Staelin, R., & Zeithaml, V. A. (1993). A dynamic process model of service quality: From expectations to behavioural intentions. *Journal of Marketing Research, 30(2)*, 7-27.

Browning, J. (1994). A question of communication. In P. Gray, W.R. King, E.R. McLean & H.J. Watson (Eds.), *Management of Information Systems* (2nd ed.) (pp.16-43). Fort Worth: The Dryden Press.

Buzzell, R.D., & Gale, B.T. (1987). *The PIMS principles linking strategy to performance.* New York: The Free Press.

Carman, J.M. (1990). Consumer perceptions of service quality: An assessment of the SERVQUAL dimensions. *Journal of Retailing, 66(1)*, 33-55.

Caruana, A., Ewing, M.T., & Rameseshan, B. (1999). Assessment of the three-column format SERVQUAL: An experimental approach, Forthcoming in *Journal of Business Research.*

Cattell, R.G.G. (1994). *Object data management*, Massachusetts: Addison Wesley, Reading.

Cronin, J.J., Jr., & Taylor, S.A. (1992). Measuring service quality: A reexamination and extension. *Journal of Marketing, 56(July)*, 55-68.

Cronin, J.J., Jr., & Taylor, S.A. (1994). SERVPREF versus SERVQUAL: Reconciling performance-based and perceptions-minus-expectations measurement of service quality. *Journal of Marketing, 58(January)*, 125-131.

Crosby, P.B. (1979). *Quality is free: The art of making quality certain.* New York: New American Library.

Davis, F.D. (1989). Perceived usefulness, perceived ease of use, and user acceptance of information technology. *MIS Quarterly, 13(3)*, 318-340.

Davis, F.D., Bagozzi, R.P., & Warshaw, P.R. (1989). User acceptance of computer technology: A comparison of two theoretical models. *Management Science, 35(8)*, 982-1003.

Davis, F.D. (1993). User acceptance of information technology: System characteristics, user perceptions and behavioral impacts. *International Journal of Man-Machine Studies, 38*, 475-487.

Davis, S., & Meyer, C. (1998). *Blur: The speed of change in the connected economy*, Oxford, U.K.: Capstone.

DeLone, W.H., & McLean, E.R. (1992). Information systems success: The quest for the dependent variable. *Information Systems Research, 3(1)*, 60-95.

Doll, W.J., & Torkzadeh, G. (1988). The measurement of end-user computing satisfaction. *MIS Quarterly, 12(2)*, 258-274.

Drucker, P.F. (1988). The coming of the new organisation. *Harvard Business Review*, January-February, 45-53.

Etezadi-Amoli, J., & Farhoomand, A.F. (1991). On end-user computing satisfaction. *MIS Quarterly, 15(1)*, 1-4.

Galletta, D.F., & Lederer, A.L. (1989). Some cautions on the measurement of user information satisfaction. *Decision Sciences, 20*, 419-438.

Garvin, D.A. (1988). *Managing quality: The strategic and competitive edge.* New York: The Free Press.

Goodhue, D. (1986). IS attitudes: Toward theoretical and definition clarity. In R.W. Zmud (Ed.), *ICIS 86, Proceedings of the Seventh International Conference on Information Systems* (pp. 181-194). San Diego, California, December 15-17.

Goodwin, N.C. (1987). Functionality and usability. *Communications of the ACM, 30(3)*, 229-233.

Gronroos, C. (1983). *Strategic management and marketing in the service sector*, Cambridge, MA: Marketing Science Institute.

Gronroos, C. (1990). *Service management and marketing: Managing the moments of truth in service competition.* Massachusetts: Lexington Books.

Grover, V., Jeong, S.R., & Segars, A.H. (1996). Information systems effectiveness: The construct space and patterns of application. *Information and Management, 31(4)*, 177-191.

Harris, J. (1996, October 14). "New-age 'Pcs' tipped". *The Advertiser*, p.35.

Iacocca, L. (1988). *Talking straight.* New York: Bantam Books.

Igersheim, R.H. (1976). Managerial response to an information system. In *AFIPS Conference Proceedings* (Vol. 45) (pp. 877-882). National Computer Conference.

Juran, J.M., Gryna, F.M., Jr., & Bingham, R.S. (Eds.). (1974). *Quality control handbook.* (3rd ed.). New York: McGraw-Hill.

Kerzner, H. (1998). *Project management: A systems approach to planning, scheduling, and controlling.* (6th ed.). New York: John Wiley and Sons Inc.

Kettinger, W.J., & Lee, C.C. (1997). Pragmatic perspectives on the measurement of information systems service quality. *MIS Quarterly, 21(2)*, 197-203.

Levitt, T. (1972). Production-line approach to service. *Harvard Business Review, 50(5)*, 41-52.

Lucas, H.C., Jr. (1978). Empirical evidence for a descriptive model of implementation. *MIS Quarterly, 2(2)*, 27-41.

Mason, R.O. (1978). Measuring information output: A communication systems approach. *Information and Management, 1(5)*, 219-234.

Melone, N.P. (1990). A theoretical assessment of the user satisfaction construct in information systems research. *Management Science, 36(1)*, 76-91.

Moynihan, J.A. (1982). What users want. *Datamation, 28 (4)*, 116-118.

Oliver, R.L. (1981). Measurement and evaluation of satisfaction processes in retail settings. *Journal of Retailing, 57(3)*, 25-48.

Oliver, R.L. (1989). Processing of the satisfaction response in consumption: A suggested framework and research proposition. *Journal of Consumer Satisfaction, Dissatisfaction and Complaining Behavior, 2*, 1-16

Olshavsky, R.W., & Miller, J.A. (1972). Consumer expectations, product performance and perceived product quality. *Journal of Marketing Research, IX (February)*, 19-21.

Parasuraman, A., Berry, L.L., & Zeithaml, V.A. (1991). Refinement and reassessment of the SERVQUAL scale. *Journal of Retailing, 67(4)*, 420-450.

Parasuraman, A., Zeithaml, V.A., & Berry, L.L. (1984). *A conceptual model of service quality and its implications for future research* (Report No. 84-106). Cambridge, Massachusetts: Marketing Science Institute.

Parasuraman, A., Zeithaml, V.A., & Berry, L.L. (1986). *SERVQUAL: A multiple-item scale for measuring customer perceptions of service quality* (Report No. 86-108). Cambridge, Massachusetts: Marketing Science Institute.

Parasuraman, A., Zeithaml, V.A., & Berry, L.L. (1994). Reassessment of expectations as a comparison standard in measuring service quality: Implications for further research. *Journal of Marketing, 58(January)*, 111-124.

Peter, J.P., Churchill, G.A., Jr., & Brown, T.J. (1993). Caution on the use of difference scores in consumer research. *Journal of Consumer Research, 19(4)*, 655-662..

Phillipson, G. (1994). Client/Server strategies. *Management Information Systems, 3(1)*, 73.

Pitt, L.F., Watson, R.T., & Kavan, C.B. (1995). Service quality: A measure of information systems effectiveness. *MIS Quarterly, 19(2)*, 173-187.

Prakash, V. & Lounsbury, J.W. (1983). A reliability problem in the measurement of disconfirmation of expectations. *Advances in Consumer Research, 10*, 244-249.

Seddon, P.B. (1997). A respecification and extension of the DeLone and McLean model of IS success. *Information Systems Research, 8 (3)*, 240-253.

Seddon, P.B., & Fraser, S.G. (1995). A respecification of the DeLone and McLean model of IS success. In G. Pervan & M. Newby (Eds.), *Proceedings of the 6th Austral-Asian Conference on Information Systems* (pp. 109-118), Curtain University of Technology, Perth, Australia, September 26-29.

Seddon, P.B., Staples, D.S., Patnayakuni, R., & Bowtell, M.J. (1998). The IS effectiveness matrix: The importance of stakeholder and system in measuring IS success, *ICIS 98, Proceedings of the International Conference on Information Systems*, Helsinki, Finland, December 13-15.

Seddon, P.B., Graeser, V., & Willcocks, L. (1999). *Measuring IS effectiveness: Senior management's view at the end of the 20th century.* Working Paper, University of Melbourne, Melbourne, Australia.

Segars, A.H., & Grover, V. (1993). Reexamining perceived ease of use and usefulness: A confirmatory factor analysis. *MIS Quarterly, 17(4)*, 517-525.

Spreng, R.A., & Mackoy, R.D. (1996). An empirical examination of a model of perceived service quality and satisfaction. *Journal of Retailing, 72(2)*, 201-214.

Spreng, R.A., MacKenzie, S.B., & Olshavsky, R.W. (1996). A reexamination of the determinants of customer satisfaction. *Journal of Retailing, 72(2)*, 15-32.

Szajna, B., & Scamell, R.W. (1993). The effects of information system user expectations on their performance and perceptions. *MIS Quarterly, 17(4)*, 493-516.

Teas, R.K. (1993). Expectations, performance evaluation, and consumers' perceptions of quality. *Journal of Marketing, 57 (October)*, 18-34.

Teas, R.K. (1994). Expectations as a comparison standard in measuring service quality: An assessment of a reassessment. *Journal of Marketing, 58(January)*, 132-139.

Van Dyke, T.P., Kappelman, L.A., & Prybutok, V.R. (1997). Measuring information systems service quality: Concerns on the use of the SERVQUAL questionnaire. *MIS Quarterly, 21(2)*, 195-208.

Violino, B., & Hoffman, T. (1994). From big iron to scrap metal. In P. Gray, W.R. King, E.R. McLean & H.J. Watson (Eds.), *Management of information systems* (2nd ed.) (pp.240-245). Fort Worth:The Dryden Press.

von Hellens, L.A. (1997). Information systems quality versus software quality: A discussion from a managerial, an organisational and an engineering viewpoint. *Information and Software Technology*, 39(12), 801-808.

Ward, J., & Griffiths, P. (1995). *Strategic planning for information systems* (2nd ed.). Chichester, New York:John Wiley Book.

Watson, R.T., Pitt, L.F., & Kavan, C.B. (1998). Measuring information systems service quality: Lessons from two longitudinal case studies. *MIS Quarterly, 22 (1)*, 61-79.

Wilkin, C. (1996). *Service quality as a measure of the correlation between employees' expectations for performance and their perceptions of current performance.* Unpublished honours thesis, Deakin University, Warrnambool, Australia.

Wilkin, C. (2000). *To measure the quality of information systems as a function of use.* Work in Progress on an unpublished postgraduate thesis, Deakin University, Warrnambool, Australia

Wilkin, C., & Hewett, W.G. (1997). Measuring the quality of information technology applications as a function of use. In D.J. Sutton (ed.), *ACIS 97, Proceedings of the 8th Australasian Conference on Information Systems* (pp.404-416), University of South Australia, Adelaide, Australia, September 29-October 2.

Wilkin, C., & Hewett, W. (1999). Quality in a respecification of DeLone and McLean's IS success model. In M. Khosrowpour (ed.), *Managing Information Technology Resources in Organizations in the Next Millennium, Proceedings of the 1999 Information Resources Management Association International Conference* (pp. 663-672), Hershey, Pennsylvania, USA, May 16-19.

Zeithaml, V.A., Parasuraman, A., & Berry, L.L. (1990). *Delivering quality service: Balancing customer perceptions and expectations,* New York: The Free Press.

Chapter VIII

Evaluating Evolutionary Information Systems: A Post-Modernist Perspective

Nandish V. Patel
Brunel University, UK

INTRODUCTION

Until the beginning of the 1990s, information systems (IS) were generally viewed as largely a support for business activity and were justified using cost accounting techniques. A proposed system would be developed if it could be shown that it would reduce operating costs or result in other productivity increases. No consideration was given to other benefits of an intangible or even strategic nature. As the deployment of information technology (IT) spread from operational to tactical support, the need to assess or evaluate its contribution to organisational performance and organisational reconfiguration attracted researchers' interests. Yet the same genre of cost accounting based evaluation techniques were used.

Now, as we enter the new century, IS are regarded as an essential feature of doing business, and many new kinds of businesses, such as Web-based ones, organise their business activity around IT, rather than organise the IT around the business. Executives especially regard IS as strategic tools. We are in an era of Internet-based businesses, reconfiguration of business processes with integrated IT/IS, and traditional businesses which now have to use the World Wide Web to remain viable. In this new era, the approach to IT/IS evaluation is still typically controlled using budgets and year-to-year comparisons, and by comparisons with other business costs such as human resource or production costs. With this plethora of IT/IS deployment, the actual benefits to business of introducing and *using* IS are proving inherently difficult to *measure*.

However, many of the IS in use in modern business organisations may be regarded as *evolutionary information systems* (EIS). It is argued here that EIS cannot be measured using cost-based accounting methods, or methods that seek to quantify benefits and costs in other ways. Instead, an interpretative approach is required that focuses on the subjective utility or value of IS to individuals, groups, or organisations. Such an approach is explored in this chapter.

To characterise evolutionary systems development and EIS some examples are necessary. Examples of evolutionary systems development are prototyping (Bowen, 1994) and Rapid Application Development (Pressman, 1997), amongst others. There are also developments in evolving legacy systems (Warren, 1999) that are at present not considered

Copyright © 2001, Idea Group Publishing.

in IT/IS evaluation. There is no evidence of evaluation methods that consider the improvement or enhancements made to IS through maintenance activity. The effort spent in systems maintenance, often quoted as sixty to seventy per cent of the cost of systems (Pressman, 1997), questions the value of both ex ante and ex post evaluation. Through maintenance activity it is often the case that the actual IS in operation is significantly different from the one that would have been evaluated before or after it was built. Such activity in systems development and systems usage is here termed EIS.

There are different perspectives on EIS (Land, 1982). An EIS may be a named system that is developed through time. The system changes from its inception through development to operation and final replacement. It may be regarded as the management of IT or IS over a period of time leading to maturity of systems. Finally, an EIS may be seen in a broader context in society, not solely concerned with individual systems, but with the diffusion and growth of IS through out society. A classic example of the latter is the World Wide Web

EIS can be distinguished from other IS along various dimensions, as shown in Table 1. User requirements are a critical distinguishing factor of EIS. Such systems incorporate changing user requirements. Changing user requirements requires changeable systems functionality, which is another critical distinguishing factor of EIS. Both of these factors mean that EIS are adaptable. User requirements and system functionality are normally fixed in non-evolutionary IS.

It is possible to further distinguish EIS from traditional IS with reference to their development method. Traditional IS are developed using some structured method or systems development methodology, for example Structured Systems Analysis and Design Methodology (SSADM) used in the UK by government agencies and some large companies. Yet research shows that, though a particular methodology may be named in systems development projects, it is often not adhered to but is used as a means of social defence (Wastell, 1996). An extensive literature exits detailing methods for evaluating traditional IS. It considers associated problems of quantifying individual or organisational, whether tangible or intangible, second order or third order benefits (Symons, 1991; Farbey et al., 1995; Ballantine, J. et al., 1998; Willcocks & Lester, 1999).

Most IT/IS evaluation research and practice is either done before the decision to invest or after it. This type of evaluation is suitable for methodological information system development, where a system is developed using business projects and system development methods with set budgets and time scales, and where the system is regarded as *completed*. However, it is now recognised that systems development is evolutionary, leading to IS that are classified as evolutionary systems. Examples of such systems are the World Wide Web, Internet, and extranets. Other software systems are designed to evolve too. These are usually found in process based systems development (Warboys et al., 1999).

Traditional measure-oriented evaluation techniques are not suitable for such systems. Consequently, the concept of EIS suggests that we have a class of IS that requires a radically different perspective on IT/IS evaluation. An alternative approach for evaluating EIS is proposed in this chapter as composed of interpretivism (Walsham, 1993; 1995), post

Table 1: Distinction Between Evolutionary and Non-Evolutionary IS

	Evolutionary IS	**Non-Evolutionary IS**
User Requirements	Changing, Ongoing	Established, fixed
System Functionality	Changeable	Fixed, non-changeable
Adaptability	Yes	No

modernist thinking, and situated action (Suchman, 1994). This chapter explores the problem of evaluating EIS from this perspective.

There is little or no current research into evaluating EIS. Current research and practice in IS evaluation is not suitable for systems that are classified as evolutionary. A post-modernist framework is proposed in this chapter for exploring the problem of assessing the benefits of EIS. Such evaluation is necessary but remains to be developed into a research interest in IS evaluation. Evaluation is necessary throughout the life of EIS. Though the focus of the chapter is on interpretivist evaluation, the management of benefits, risks and costs are nevertheless an important aspect of any IS evaluation.

In addressing the problem of evaluating EIS, a broad perspective of the recent developments in the IS field in terms of considering societal and human issues is taken. It is necessary first to discuss recent concepts of organisations as (business) processes, with their modern character of uncertainty and permanent organisational change. This is necessary to do because such characterisations in turn affect the conception of IS to be used in organisations. The notion of EIS is then discussed to take a deeper conceptual and philosophical view of organisational behaviour. Considering post-modernist views of society and organisations and situated action achieves this aim. Essentially, the ideas embodied in post-modernism and situated action, whether we call them by that name or not, reflect the emergence of EIS. The work of the German philosopher Martin Heideggar is introduced, especially his notion of *Dasein* (Being), as offering a philosophical explanation to justify EIS and forms a basis for an approach to the evaluation of EIS. Having thus set the conceptual ground, EIS *per se* are considered and an approach to how to evaluate them is discussed. No new *technique* is proposed, for that would essentially be contrary to the interpretivist stance adopted here, and in contradiction of Hiedeggar's concept of Being. Rather the aim is to open the debate in this increasingly pertinent area and consider the implications for further research into evaluating EIS.

BACKGROUND: BUSINESS PROCESSES AND ORGANISATIONAL CHANGE

Recent characterisations of business organisations have two prominent themes, business processes and organisational change, both of which are relevant to evaluating EIS. Hammer (1990) and Davenport (1993) argue that business processes can be changed to improve the competitive position of a company and to maximise the value that it can deliver to its customers. Changing processes, often radically, leads to improvements in performance as measured by cost, cycle time, service and techniques (Johansson et al., 1993).

The second dominant characterisation of modern companies is that they consist of permanent organisational change and uncertainty. Companies have to accept and learn to cope with uncertainty and constant change in markets and economic conditions (Handy, 1995). In manufacturing, such uncertainty and changing conditions has brought about a call for agile production techniques or systems (IMECHE, 1999). Business processes, uncertainty and change are brought together in Business Process Reengineering (BPR) in that processes, once engineered, need to be constantly changed to remain competitive (Hammer and Champey, 1997).

The role of IT and IS is significant in this modern view of organisations. Its role is to enable business processes (Moreton and Chester, 1997) and even to transform significantly organisations (Venketraman, 1991). This modern view of organisations has important bearings on IS evaluation. In it IT is affected by environmental and organisational change, and because it supports business processes that change, it too must be capable of changing or evolving.

The evaluation of IT and IS in organisations consisting of processes, uncertainty and constant unpredictable change is problematic. Since IT is used to change fundamentally business processes and change affects newly designed processes integrated with IS, quite what the benefits of the use of the technology are becomes difficult to measure. This is partly because of multiple variables such as processes changing simultaneously. Our traditional quantitative and qualitative techniques possibly are unsuitable in such an environment. For example, Aggarwal (1991) discusses procedures for assessing investments in flexible manufacturing technology, and posits that traditional capital budgeting procedures should be supplemented with strategic analysis and net present value. However, whilst such techniques may be used for assessing flexible IS too, they may not be entirely useful for evaluating the real worth of IT/IS investments.

To attempt to understand the role of IT and IS in organisations consisting of processes, uncertainty and change, and to begin to think about how EIS should be evaluated in them, requires us to take a border perspective. It is necessary to turn to the recent reconception of society as post-modern to gain a better understanding of what is happening in society and in organisations, and in that context to begin to explore how to evaluate EIS.

POST-MODERNISM, INTERPRETIVISM, AND SITUATEDNESS

Post-modernist ideas of society are relevant to the issue of evaluating EIS. Not only does post-modernism provide useful concepts to evaluate EIS, but also it is necessary to consider post modernism because it has brought about the very need for EIS.

Post-modernism has its origins in art (Bjørn-Anderson,1988). It arises from artists' concern with understanding what constitutes a "proper" piece of art, and their attempt to explore beyond traditional ideals and style. In science, the affect of post-modernism among some researchers has been to abandon the search for a grand theory that explains all physical phenomena. In the social sciences, researchers now accept that there is no one theory or no one *right* theory. Rather, they argue that all perspectives are equally acceptable. Consequently, a central theme in post-modernism is relativism. Relativism is the view that there is no objective reality that can somehow be understood and accounted for and then allows us thereby to control it. Our perception of reality is unique to us and is as valid as anybody else's in the post-modern world.

The implications of this view of society, science and organisations for IS evaluation is that such systems are not objective entities that can be measured independently of their users or contexts. Most of the evaluation techniques developed to date would be inadequate because they are based on an objective view of reality. A view in which benefits can be identified as separate from their users and quantified independently of the users' perceptions. We posit that post-modernist ideas permeate our society and organisations such that they have affected the way in which IT/IS is used in them. In particular, that post-modernism gives rise to EIS in organisations.

For the purposes of evaluating EIS, the central theme of *interpretation* and *situatedness* in post-modernist thinking are relevant. As reality itself is understood relatively, it requires the act of interpretation of reality in the situation to act in the world. Each person or group interprets phenomena individually and shares that understanding by communicating it in a social context. Consequently, in the context of post-modernism, there can be no objective evaluation. The evaluation act itself is to be thought of as an interpretation of the value of an IS to the person or group using it in the social context.

Whilst an understanding of post-modernism, interpretation, and situatedness enables us to set the context in which to think about evaluating EIS, we require a philosophical basis for thinking about them generally. In particular, the work of the German philosopher Martin Heidegger is relevant (Dreyfus, 1994). Heidegger's ontological consideration of human *Dasien (Being)* has a resonance with the notion of EIS. In systems development terms, the need for system evolution arises because new information requirements arise. This commonly observed phenomenon in IS development practice is stated in Heideggerian terms as: "Every decision...bases itself in something not mastered, something concealed, confusing; else it would never be a decision." Thus the very act of developing IS on predetermined systems requirements leaves the developer with having to tackle "...something not mastered, something concealed, confusing..."

The emergence of EIS may be explained in Heideggarian terms too. Heideggar attempts to understand how something (person or thing) is or what it means for something to *be*. For Heiddeger something *is* because of Being. Humans acting in the world do so because of and through Dasein, but they are not cognisant of it. Yet they have to act in the world to be it. It is this non-transparent being-in-the-world which gives rise for constant striving to be – or in IS terms emergent information or EIS. The method Heidegger puts forward for understanding human being is phenomenology.

Given post-modernist thinking and phenomenology, Ciborra (1997) proposes a reconception of IS based on *improvisation*. He argues that:

"A small Copernican revolution is suggested: competent actions which seem improvised are in reality deeply rooted, while structured decisions based on abstract representations and models appear to be improvised, i.e. lacking any relationship to context." p.138.

The reference to deeply rooted human actions is often supported by information that is contextually rich, and it is posited that such information is made available through EIS. There are issues that arise from viewing organisations as post-modernists. They concern the social, contextual and situated nature of organisational activity, which EIS cater for.

Suchman (1994) identifies the qualities of situatedness. The qualities are *emergence* of individual's action in a social context, action is *contextual* or *local*, it is *contingent* upon the current situation, and it is *embodied* in a physical body. Theories of social action need to remain *open*, they cannot be finalised to allow for revision, and practical information is only defined to the extent required, the rest is left *vague*. These qualities of situatedness give rise to the need for contextual information and therefore EIS. They may also be used to evaluate, in a relative sense, EIS as discussed later.

EVOLUTIONARY INFORMATION SYSTEMS

Research into evolutionary systems development and types of EIS has been prompted by conceptions of business as processes, business uncertainties and organisational change. In this section, examples of EIS development are discussed, EIS examples are provided, and characteristics that differentiate EIS from other types of IS are detailed.

Evolutionary Systems Development

Software developers have attempted to meet the challenges of the modern organisation. The challenges are conceptual and practical. Conceptually, researchers have attempted to conceive systems development as flexible or evolvable. Practically, they have proposed or implemented such methods.

There are various systems development methods that can be categorised as evolutionary systems development. These methods attempt to develop software that is appropriate to the needs of its users, and some incorporate organisational uncertainty. The methods are prototyping (Bowen, 1994), rapid application development (Martin, 1992), component based development (Jacobson et al., 1997), and the incorporation of flexibility in development methods (Fitzgerald, 1990 and Boogaard, 1994). In addition, Pressman (1997) details the incremental model, the spiral model, the component assembly model, and the concurrent development model as evolutionary software process models.

Researchers have investigated ways to evolve software processes (Lehman 1980; 1984; Conradi, 1994). Such research has been prompted by the needs of business process re-engineering and the need to model commensurate software processes (Warboys, 1994). Some programming languages facilitate flexibility, especially on the World Wide Web. For example, the recent announcements by W3C to make an extensible mark up language for the Web in the form of XML (W3C Consortium, 1999).

Patel (1999) has proposed the spiral of change model of tailorable information systems development. It is conceived to enable contextual and situated aspects of individual and organisational work to be incorporated in an interpretative way into IS development and usage. Patel and Irani (1999) suggest ways of evaluating tailorable information systems that evolve. On a conceptual level, Paul (1993) has suggested the development of living systems and proposes various development frameworks for such systems. All such developments, to varying degrees, reflect the needs of modern organisations as processes that change and which have to respond to uncertainties. Such organisations require evolutionary systems development.

Some characteristics of EIS were given in Table 1 above. Changing user requirements and systems functionality are required because the use of IS happens in dynamic organisations. The actual use of IS is intertwined with the social and business fabric of an organisation. It is largely impossible to determine what that social and business fabric is. When organisational actors know it temporally and spatially they then need appropriate IS to support their activities. For example, the development of electronic commerce not only requires certain computing hardware, which can be evaluated in monetary terms, but the actual business applications of the computing hardware will be embedded in the new business models of wired business. These business models are highly complex consisting of not only organisational but economic and social factors.

Examples of Evolutionary Information Systems

The examples of evolutionary information cited in this section are reflective of modern organisations. They enable individuals or groups to tailor information to suite their purposes, and to that extent allow interpretative use of data, information and knowledge. They also enable their users to react speedily to market demands. A defining characteristic of these systems is that they do not have the problematic phase of requirement definition as a prerequisite for design and development. Such systems, as well as meeting other individual and organisational needs, are designed to stay relevant in changing business environments and cater to unpredictable situations and future events.

For example, Pawson et al. (1995) discuss expressive systems that reduce the time to market and help tailor products and services to customers' needs, as well as to be more responsive to unexpected events. The actual system, named Kapital, is used on the derivatives floor at J.P. Morgan in New York. Traders using Kapital can choose the user interface that suites them. They can perform financial analytical calculations using mathematical models, and use real-time data from current market conditions. As stated in Pawson et al. (1995, p. 41):

> "However, what really distinguishes Kapital from other information systems is not the technology, but the fact that it does not attempt to fulfil a specified set of user requirements – at least in the conventional sense. Rather, Kapital attempts to model the very 'language' of J.P. Morgan's trading business – not only the vocabulary, but also the grammar and, arguably, the style."

The Kapital system exhibits the core characteristic of changing user requirements of EIS. The system was built without a set of predetermined user requirements, as is normal for other types of systems. As no predetermined set of user requirements in the conventional sense were available, it would not be possible to do an ex ante or ex post evaluation using traditional techniques.

Weiser (1991) presents Xerox PARC's ubiquitous computing research program, and Newman et al., (1991) identify the "Forget-me-not" application at EuroPARC. These systems capture electronically current events and occurrences in the organisation and make them available for *future* use. Such systems cannot be evaluated ex ante because their potential use is unspecified at the time of capture of data – location, time, document, a picture, a conversation. The Forget-me-not application is particularly interesting to provide a better understanding of EIS. The aim of its developers was to provide a system that users could use in some future, unpredictable time. The system could not be justified in cost-benefit terms, as the benefits were unknowable because of the unspecified future use of the system.

Patel (1999) introduces the notion of tailorable information systems, which are designed to enable users to tailor systems to particular contexts and situations. Tailorable information systems are not based on predetermined systems requirements and aim to fulfil information requirements in an emerging fashion. As with the Forget-me-not application, the benefits of tailorable information systems emerge in the future, so the decision to invest in such systems cannot be justified only in cost-benefit terms. Users, or systems developers, to suit particular situations adapt the functionality of tailorable information systems. The costs or benefits of this type of tailoring cannot be predicted when a tailorable system is installed.

Tailorable computer systems enable their users to tailor their use to individual or group needs. The Xerox Tailorable Buttons system is appropriately described by MacLean et al. (1990) as a user tailorable system. They devised simple models of users and utilised participatory design methods. Xerox Tailorable Buttons uses object-oriented design and object implementation, and provided users with user-interfaces consisting of tailorable Buttons. The system was interfaced with an e-mail system so that user-tailored systems functionality designs and implementations may be shared among users. MacLean et al (1990) state that users can tailor Xerox Buttons on different levels with different systems properties and systems consequences, ranging from simple windows customisation on a desktop interface, to complex user-programming using fifth generation languages. They cite unique and idiosyncratic uses of the system, as well as uses to support cooperative work.

A particular difficulty in evaluating tailorable computer systems that support cooperative work concerns the actual contribution they make to the achievement of task, job responsibilities or organisational objectives. A tailorable system is actually tailored and used in the context of the particular work being done at that time, consequently it is not possible to apply ex ante evaluation to such systems. Ex post evaluation techniques on the whole would not be appropriate unless they allow for contextual features to be weighed too.

The existence and use of such IS have implications for IT/IS evaluation. IS that deliver functionality in context and, even in the future, to individuals and groups' requirements raise fundamental issues concerning how we choose to evaluate IT/IS investment decisions.

ISSUES IN EVALUATING EVOLUTIONARY IS

To begin to understand how to evaluate EIS it is necessary to identify their pertinent features. This section explores such features that would have to be considered when deciding on what to evaluate and how to do it.

A central issue in developing any approach to evaluating EIS concerns requirement analysis. In EIS there is an absence of requirement analysis as practised by current developers and advocated by researchers and academics. The Kapital (Pawson et al., 1995) and "Forget-Me-Not" (Newman, 1991) systems cited above were not built with a set of predetermined requirements. Such systems' developments cause problems for the traditional IS evaluator. EIS that are built in the absence of predetermined requirements cannot be evaluated using ex ante or ex post evaluation, since there would be nothing to compare the outcome with.

Another important feature of evolutionary systems is that they are continuous and continual processes. A traditional information system is considered completed once the business project to develop it is terminated. Tailorable information systems (Patel, 1999) would not similarly be considered as completed. Users (and developers) continuously develop tailorable systems in the context of their use. It becomes problematic to decide when to evaluate systems that are being continuously developed. The techniques available to the traditional IS evaluator are unsuitable for such systems.

Evolutionary systems are interpreted entities. Individuals and groups interpret them in the sense that changes made to them are done to reflect organisational change and context. Interpretation of IS is closely tied to the concept of emergence (see Ali and Zimmer, 1998 for discussion on emergence in artificial systems). As systems developers begin to design IS that facilitate emergence, as in the case of the Kapital system, it becomes necessary to develop appropriate evaluation approaches. Emergent properties cannot be predicted and they happen in context, making ex ante evaluation unsuitable for EIS. Other features of EIS such as functionality change, relevance at a particular time, provision of contextual and situated data, and use in the future, mean that a radically alternative evaluation approach is required, as explained in the next section.

EIS are actively used and developed, they are ongoing, and provide utility to their users at the time of use. The notion of utility is an important feature of EIS, especially in the context of post-modernist thinking. Measuring the utility of such systems for knowledge workers may be difficult, because as stated by Drucker (1993, p. 24):

"One has to assume, first, that the individual human being at work knows better than anyone else what makes him or her more productive, and what is helpful or unhelpful."

Evolutionary systems development is a process of co-creation and coevolution of systems. Professional developers and users develop systems, but, significantly, the power shifts to users, because, as Drucker (1993) states, "they know better than anyone else what is required."

FUTURE TRENDS: AN APPROACH TO EVALUATING EVOLUTIONARY IS

Ex ante and ex post evaluation is not suitable for evolutionary systems development and EIS. Systems such as Kapital, Xerox Buttons, and Forget-me-not, cited above, undoubtedly contribute to individual and organisational performance, and they may even have strategic relevance. Yet it is certain that they would have been considered doubtful projects using ex ante evaluation techniques because of the unforeseen benefits. In this section the salient features of EIS discussed above are integrated with other features to suggest aspects of an approach for evaluating EIS. No specific technique is suggested. The aim is, like Bjørn-Anderson (1988) below, not to propose my way or one way, but simply to add to our communal understanding of the matter.

The approach suggested should be based around concepts that reflect post-modernism, interpretivism, situatedness, business processes, organisational uncertainty, and change. In an early paper, Bjørn-Anderson (1988), told a "number of small stories" concerning post-modernist ideas and technology assessment. In the essay, he provides an informative overview of post-modernism, and considers its influence on technology assessment and IS evaluation. He adheres to the post-modernist style and states that "…my presentation (in a true post-modernist sense) does not serve any utilitarian purpose. The value of it, if any, is in the experience it creates in the mind of the listener." (p.11)

However, he does offer a number of insights which post-modernism provides for IS evaluation. One, that no single solution should be acceptable, and that a multiplicity of perspectives should be encouraged and accommodated. Two, that pure data analysis may not reveal underlying truths or patterns, simply because there are none. Three, that the phenomenological experience in the mind of users of systems is equally as important as other evaluation criteria. Four, that we should explore other fields such as art to inspire us to use different, experimental evaluation approaches. Finally, that post modernist concepts such as recycling, reuse, patchworking, and borrowing may be valuable in IS evaluation.

In this light, what should be evaluated? Table 2 below shows a limited sample of what is currently the subject of evaluation in traditional IS and what might be evaluated in EIS in the context of post modernism. It shows that the objects of evaluation for traditional IS are objective and those for EIS are subjective. The Table is not a like-for-like listing, but merely a set of different items for traditional and evolutionary IS evaluation. Yet the question of how EIS might be evaluated still remains open. It is certain that in EIS, context and value or utility are important, so not all the objects shown in the Table for traditional IS evaluation would necessarily be evaluated in a post-modernist context. The Kapital system discussed above enables its users to react to market demands by enabling new products and services to be designed using it. Much of the data capture and analysis occurs at the time and in the context of its use. Context and utility are thus important aspects of the evaluation process.

The problematic question of how to evaluate EIS still persists. It was discussed above that the interpretations individuals and groups place on the systems they use are an important feature of EIS, and that such interpretations occur in context and in an ongoing manner. One feature of how to evaluate EIS is that formative and continuous participative evaluation would be required for such systems. In art form is considered to be a style or mode of expression, opposite of content or orderly arrangement of components. As interpretations, EIS should similarly be evaluated according to the form they take in the complex interplay between the technology, the user and the goals of the work to be done in the organisation.

Table 2: Objective and Subjective Objects of Evaluation

Objects of IT/IS Evaluation in Traditional IS - Objective	Objects of IT/IS Evaluation in EIS- Subjective
Improved product or service quality	Contextual fit
Improved competitive advantage	Value to user (or organisation)
Reduced Costs	Experiential benefits
Reduced risks	Interpretation of EIS
Better co-operation with partners	Time of use
Diversification	Situated use
Marketing	Informing, communication, and learning

Integral to the preceding issue is the question of who should do the evaluation. Though participative evaluation is suited for EIS, given the innovative and strategic nature of typical evolutionary IS and their penetration into the core of organisational functioning, one could argue that a high-level, broad perspective is needed to assess the system's implications for the business. Such a broad perspective can be brought into the evaluation only at a relatively high level in the organisational hierarchy. However, we would also argue that user involvement is also critical but mainly for ensuring acceptance and reducing resistance to change. This issue is discussed further in the section below.

Other issues to consider in any approach to evaluating EIS are: the involvement of all stakeholders, a focus on the wide range of benefits both tangible and intangible, subjective and objective, and whether it can accommodate change in the evaluation parameters. In particular, the subjective objects identified in Table 2 require further critical study. They could form the basis of an interpretive evaluation of EIS.

Table 3 below presents some existing evaluation techniques. A critique of these techniques can be levelled using the dimension of user requirements, system functionality, and adaptation shown in Table 1 above. Apart from prototyping, all the techniques in Table 3 do not allow for changing user requirements, changeable systems functionality and adaptation. Thus making most of the techniques unsuitable for evaluating EIS.

However some of the existing evaluation techniques shown in Table 3 below could be extended to incorporate the issues discussed above are worth mentioning. These techniques can be compared along the dimensions of complexity, communication, quantification, facilities and potential to extend it to interpretive IS evaluation.

Fisher (1995) compares traditional approaches to IT decision-making and shows that information economics focuses on change. Information economics' focus on change would be suitable, but it would need to be based on interpretative grounds. Experimental evaluation techniques like prototyping and simulation would be appropriate for the continuous aspects of EIS.

The Multi-Objective Multi-Criteria technique would be able to accommodate the interpretative aspects of evolutionary systems and facilitate the post-modernist notion of all perspectives being equally acceptable. Similarly, the Value Analysis technique would be suitable. However, it must be added that those techniques that initially seem appropriate may need to be further enhanced to address the unique requirements of evolutionary systems, and all would have to be re-rooted in interpretivism.

DISCUSSION AND FURTHER RESEARCH

There are EIS that are at present not evaluated appropriately and, possibly, there are IT/IS systems development proposals of an EIS kind that are turned down because the

Table 3: A Comparison of IS Evaluation Methods (Adapted from Giaglis, 1999)

Method	Complexity	Communication	Quantification	Facilities	Interpretive Potential
DCF Methods (NPV, ROI, IRR)	Easy to understand Large amounts of data required Focus only on cash flows	Easy to learn Easy to communicate	Precise Only monetary values	Some risk analysis (discount factor)	None
CBA, SESAME	Similar to DCF Intangibles taken into account	May involve controversy and discussion	Similar to DCF	Similar to DCF	None
ROM	Low data requirements	Difficult to learn, apply, and communicate	Precise (accounting data)	Targeted to MIS Suitable only for *ex post* evaluation	None
IE	Large amounts of data required Considerable expertise and resources required	Difficult to learn and apply	Precise measurement of tangibles, as well as ranking and rating of intangibles	Considerable risk analysis at all levels	None
MOMC methods, Value Analysis	Medium data requirements Focus on subjective measures of utility	Subjective and exploratory methods, involving discussion and controversy	Subjective, non-monetary measures	Stakeholder analysis	High Potential
Prototyping	Limited-scale system development required	Based on real data on system impacts	Precise	Congruent with IS development	High Potential
Simulation	Large amounts of data required	Expertise required in applying Easy to communicate	Precise (numerical) estimates	What-if analysis Sensitivity analysis Experimental control	Medium Potential

KEY: DCF = Discounted Cash Flow, NPV = Net Present Value, ROI = Return on Investment, IRR = Internal Rate of Return CBA = Cost-Benefit Analysis, ROM = Return on Management, IE = Information Economics, MOMC=Multi-Objective, Multi-Criteria

current evaluation techniques measures show them to be risky or non-beneficial. Such systems have been couched and presented here in the context of interpretivism, situatedness, and post-modernist thinking. In doing so, we have identified characteristics of modern economies and companies such as processes, uncertainty and change. It has been posited that IT/IS is used in this context and that it needs to be flexible, or in our terms, evolutionary. Examples of evolutionary systems development and EIS that are indicative of this trend have been provided.

Salient features of EIS have been identified and it has been shown how they may be used to construct an approach to evaluating them. It is not possible to set benchmarks for EIS. On the contrary, in the context of post-modernist thinking presented in this chapter, individuals and groups would be best suited to decide the utility or value they derive from evolutionary systems, and in the extreme case even the role of evaluating IT/IS may be questionable.

Farbey et al. (1995) state that IS evaluation can be ex ante, ex post or throughout the life of a system. It is posited that in EIS, evaluation should be done throughout the system's life. However, we emphasise that monitoring and control are critical, and that evaluation of evolutionary IS should include the management of benefits, risks and costs.

Our discussion has implications for research into IT/IS evaluation. As interpretation (utility and value, context and change, situation and action) is an important aspect of EIS, it is argued that interpretative notions of evaluation need to be researched further. Interpretative research into evolutionary systems cannot be a simple extension of the current research in qualitative techniques. Such research in based on the modernist, as opposed to post-modernists, view of society. Rather interpretative research would need to be based on the post-modernist view of society and consider phenomenology as a method of research (Husserl, 1970).

In particular, the hermeneutic method of understanding is appropriate for evaluating EIS. As Introna (1993) states, hermeneutic understanding is:

"Understanding that comes into being by active (in the situation) interpretation,

thus based on lived experience (*Erlebnis*) not on removed contemplation;

Always within a context and coloured by that context;

Part of the history and tradition of the person and the organisation;

The act of appropriate, thus genuinely making one's own what was initially alien."

Table 2 above listed situated use as a subjective object of evaluating EIS. Since EIS are deployed in an organisation their situated use consists of a social process. The evaluation of situated use in a social context is complex. Research into the evaluation of subjective objects is still in its infancy, but it would need to consider social anthropology, as exemplified by Suchman (1994).

There are practical implications for evaluation of EIS arising from the above discussion. In practical terms, how should EIS be evaluated? Another example of an EIS is the groupware package Lotus Notes. Its application in an organisation evolves over time as new uses for it arise, and its application is primarily in the social context of an organisation. Thus a subjective approach to evaluation is required. To evaluate such an IS it is necessary to develop evaluators trained in interpretative evaluation. They would require skills in analysing and, themselves, interpreting the value of IS to stakeholders and reporting their findings to key decision-makers. They may use subjective objects of evaluation as outlined above in Table 2.

Practitioners can use the outcomes of such an evaluation in various ways. The outcomes can be used to make users reflect on the utility of IS they use. This can be done by making the work responsibilities of users explicit and comparing them with the support provided by the IS being evaluated. Another way to use interpretative evaluation is to provide the qualitative data to management responsible for IT investment decisions. This data may be used in conjunction with quantitative data to make a rounded decision on the contribution of the IS in question to organisational objectives.

As IT/IS becomes ever more pervasive in organisations, and in society, we need to facilitate investments in them by expanding our perspective on evaluation. To assess the value and benefit of IT/IS decision-makers in companies need to change their attitude to investment decisions. Undertaking IT/IS evaluation is complex, but it becomes even more so when the target of the evaluation is a social process that involves the meanings and value that individuals, groups and organisations attach to their use of evolutionary information systems.

REFERENCES

Ali, S. M. and Zimmer, R. M. (1998), Emergence: A review of the foundations. *Systems Research and Info. Systems.* Vol. 8, pp 1-24.

Ballantine, J., Levy, M., and Powell, P. (1998), Evaluating information systems in small and medium-sized enterprises: Issues and evidence. *European Journal of Information Systems,* 7, 241-251.

Bjørn-Anderson, B. (1988), A Post-Modernistic Essay on Technology Assessment. In Børjn-Anderson, B. and Davies, G. B. (Eds.), *Information System Assessment: Issues and Challenges,* North Holland, Amsterdam.

Bowen, P. (1994), *Rapid application development: Concepts and principles.* Evolutionary Prototyping and Participative Systems Design. Unicom, London.

Ciborra, C.U.A. (1997). Theory of Information Systems Based on Improvisation. In Currie, W. L. and Galliers, B. (eds.), *Rethinking Management Information Systems.* Oxford University Press. pp 136-156.

Conradi, R., Fernström, C., and Fuggetta, (1994), Concepts for Evolving Software Processes. In Finklestein, A., Kramer, J., and Nuseibeh, B. (eds.) *Software Process Modelling and Technology.* Research Studies Press Ltd., Taunton, Somerset, U.K.

Davenport, T. (1993) *Process Innovation: Reengineering Work Through Information Technology.* Harvard Business School Press, Cambridge, Massachusetts.

Drucker, P. (1993), *Managing in Turbulent Times.* Butterworth Heinemann, Oxford.

Farbey, B., Land, F. F., and Targett, D. (1995), A taxonomy of information systems applications: The benefits' evaluation ladder. *European Journal of Information Systems,* 4, 41-50

Fisher, A. (1995), IE: A new paradigm for information management and technology decision-making. *Government Accounts Journal,* ISSN 0883-1483. 44(2), 30-34.

Fitzgerald, G. (1990), Achieving Flexible Information Systems: The Case for Improved Analysis. *Journal of Information Technology* 5, 5-11.

Giaglis, G. M. (1999), Dynamic Process Modelling for Business Engineering and Information Systems Evaluation. Unpublished thesis. Department of Information Systems and Computing, Brunel University, UK.

Hammer, M. (1990) Reengineering work: Don't automate, obliterate. *Harvard Business Review* July-August.

Hammer, M. and Champy, J. (1997) *Reengineering the corporation: A manifesto for business revolution.* Nicolas Brealey, London.

Handy, C. B. (1995), *The Age of Unreason.* Arrow Business Books, London.

ImachE (1999), Lean and Agile Manufacturing for the next Millennium, Seminar Papers for S602. *The Institute of Mechanical Engineers,* London.

Introna, L. D. (1993), Information: A hermeneutic perspective. In Whitley E (ed.) *The Proceedings of The First European Conference on Information Systems,* Henley on Thames, England. 20-30th March. 171-179. Operational Research Society, Birmingham, England.

Jacobson, I., Griss, M., and Jonsson, P., (1997), *Software Reuse.* ACM Press and Addison Wesley.

Johansson, H. J., Mchugh, P., Pendlebury, J. A. and Wheeler, W. A. (1993), *Business Process Reengineering.* Wiley, Chichester.

Land, F. F. (1982), Concepts and Perspectives - A Review. In *The Proceedings of the IFIP TC 8 Working Conference on Evolutionary Information Systems,* Budapest, Hungary, September (Hawgood J. Ed), North-Holland, Amsterdam.

Lehman, M. M. (1980), Programs, Life Cycles, and the Laws of Software Evolution. *Proceedings of the IEEE,* 68(9), September 1980.

Lehman, M. M. (1984), Program Evolution. *Information Processing and Management* 20 (1) 19-36.

Martin, J., (1992) *Rapid Application Development.* Prentice Hall, Englewood Cliffs.

Moreton, R. and Chester, M. (1997), *Transforming the Business.* MaGrawHill, Maidenhead. England.

Newman, W., Eldridge, M. and Lemming, M. G., (1991), Pepys: Generating autobiographies by tracking, *Proceedings E-CSCW 91* (Amsterdam, Kulwer).

Patel, N. V. (1999) The Spiral of Change Model for Coping with Changing and Ongoing Requirements. *Requirements Engineering.* 4:77-84.

Patel, N. V. and Irani Z., (1999) Evaluating Information Technology in Dynamic Environments: A Focus on Tailorable Information Systems. *Logistics and Information Management*, Special Issue: Investment Decision Making of Information Technology/Information Systems. 12 (1/2). 32-39. ISSN 0957-6053.

Paul, R. J. (1993), Why Users Cannot 'Get What They Want'. *ACM SIGOIS Bulletin* 14 (2), (December 1993). 8-12.

Pressman, R. S. (1997), *Software Engineering, A Practitioner's Approach*. McGraw-Hill. London.

Suchman, L. (1994), *Plans and Situated Action*. Cambridge University Press, Cambridge.

Symons, V. J. (1991), A review of information systems evaluation: content, context and process. *European Journal of Information Systems,* 1(3) 205-212.

Venkatraman, N. (1991) *IT-Induced Business Reconfiguration*. In Scott-Morton M S (ed.) The Corporation of the 1990s. Oxford University Press, Oxford. England.

W3C Consortium (1999), www.w3c.com.

Walsham, G. (1998) Interpretive Evaluation Design for Information Systems, in *Beyond the IT Productivity Paradox*, Willcocks L and Lester S eds. Wiley Chichester.

Walsham, G. (1995), Interpretative Case Studies in IS Research: Nature and Method. *European Journal of Information Systems* 4, 74-81.

Warboys, B. C. (1994), Reflections on the relationship between business process re-engineering and software process modelling. In *Proceedings ER '94., lecture notes in Computer Science*. Springer Verlag.

Warboys, B. C., Kawalek, P., Robertson, I., and Greenwood, M. (1999), *Business Information Systems, A process approach*. McGraw-Hill, London.

Wastell, D.G. (1996), The Fetish of Technique: Methodology as a Social Defence. *Information Systems Journal* 6 (1), 25-40.

Weiser, M. (1991), The Computer for the 21st Century, *Scientific American,* 265: 94-105.

Wilcocks, L. and Lester, S. (1999), In Search of Information Technology Productivity: Assessment Issues. In Wilcocks L and Lester S eds. *Beyond the IT Productivity Paradox*. Wiley, Chichester, 69-97.

Chapter IX

A Framework to Evaluate the Informatization Level

Soo Kyoung Lim
University of Wisconsin-Madison, U.S.A

INTRODUCTION

As information and communication technologies have rapidly developed in the 1990s, enormous changes have taken place everywhere. At work environment, these have been newer tools for increasing organizational productivity, and these are transforming organizations to the degree that *Taylorism* once did (Davenport, 1998). These trends have spread over various fields of society, and have over countries caused economical and cultural innovation and reformation. These phenomena can be summarized as *informatization*. Informatization is defined as "converting the main goods and energy of a social economy to information through the revolution of high data communication technology and utilizing information produced by gathering, processing and distributing data within the vast fields of the society" (National Computerization Agency [NCA], 1997).

Since The United States' NII project has been evaluated as one of the important success factors for economical growth, most countries have considered informatization as one of the most effective means for improving a nation's competitiveness. Similarly, many organizations have considered informatization as a strategy to improve quality of public service and productivity. They have tried to implement informatization and extensive investments are often budgeted and expanded to acquire information technology (IT).

An Information Strategy Plan (ISP) is needed at first to implement informatization of an organization. ISP usually includes business strategy, information technology strategy, project priorities, and an organization's structure strategy. Thus, when an ISP is set up, it describes whether the business or organization's strategic goals and objectives can be achieved through IT, in which field further IT investment will be needed, and whether efficient investment in IT will be made. In order to discuss these topics, the current organization's informatization level first must be known.

Moreover, since the middle of 1990, many countries have put emphasis on performance based management, in which the government has to set up investment plans according to its performance. For example, to budget IT, it is required to first evaluate its performance and results.

In this respect, evaluation of an organization's informatization level in order to review how much organization informatization it achieves is an important managerial concern.

Copyright © 2001, Idea Group Publishing.

However, this is not a simple problem because informatization includes many intangible factors such as the quality of information and an organization's culture.

In this chapter, framework and metrics are introduced to evaluate the organization's informatization level. This framework is designed to provide reasonable information by gathering and analyzing various IT metrics for determining whether organizations have made efficient and effective use of IT and have achieved the organizational strategic goals and objectives through IT. Therefore, the evaluation results can be used to improve the organization's informatization level.

The remainder of this paper is organized as follows: in the following section, some case studies and background information are presented. The next section introduces a framework, and then future trends are discussed in the next section. Finally, the summary and conclusion are presented.

BACKGROUND

Similar to other countries, Korea has been actively pursuing its vision and goals through informatization since the early 1990s, and will continue to do so. The Information Promotion Master plan was formulated following the Basic Act on Informatization on Promotion (BAIP) in 1996. According to this national master plan, every public organization such as the government, cities, agencies and so on, has established their Information Strategy Plan (ISP) and started to implement IT. The government has allowed a large budget for constructing infrastructure and implementing application software to improve quality of public service and productivity of government.

Recently, the government and public organizations have been interested in how their investments in IT have been made effective and efficient. Although the measurement of the performance of the government is more difficult than that of the business or private companies, where the investment strategy is more easily determined as the way to maximize its benefit, governments nevertheless need to evaluate their outcomes and use them to set a new strategy. They have placed more emphasis on evaluating their informatization level, and are interested in benchmarks to set up their baseline to improve their informatization and IT performance. Based on this circumstance, the development of evaluation methodology was required. To do this, some projects related to evaluation of IT were reviewed.

The first one is about benchmarking. For developing IT budgets and setting IT priorities, it might be helpful to check out how other companies are dealing with these topics. To that end, Dr. Rubin (2000) collects exclusive data which have coverage of key IT metrics and organizational trends, and publishes *The Worldwide Benchmark Report*. Schwarz (2000) stated that companies benchmark all kind of things, but one of the most valuable benchmarks can be the business results. They include measurements such as ROI, return on assets, IT spending per employee, etc. As IT becomes more important to companies, senior executives are demanding that their managers justify the expenditure, which requires measurement of IT productivity. Benchmarks of IT productivity involve analyzing revenue per employee as a function of IT spending per employee. Therefore, with various measures and metrics, benchmarks can alternate as an evaluator. Among Rubin's exclusive data, the following benchmarks are important: work profiles, work by maintenance type, life cycle distribution, language usage, support rates, new development productivity, defect rates, software process maturity, tool inventories, techniques inventories, staff profile, budget, and technology infrastructure/IT priorities (Rubin, 2000).

Since 1996, the Maxwell School of Citizenship & Public Affairs at Syracuse University has performed a project, called the Government Performance Project (GPP), which rates the management performance of local and state governments and selected federal agencies in the United States (Maxwell school at Syracuse University, 2000). The GPP evaluates the effectiveness of management systems and examines the role of leadership in government entities. The GPP examines government management in five areas: financial management, human resource management, information technology management, capital management, and managing for results. The following list shows information technology criteria, which evaluates city government management performance (Maxwell school at Syracuse University, 2000):

1. Government-wide and agency-level information technology systems provide information that adequately supports managers' needs and strategic goals.
2. Government's information technology systems form a coherent architecture.
3. Government conducts meaningful, multi-year information technology planning.
4. Information technology training is adequate.
5. Government can evaluate and validate the extent to which information technology system benefits justify investment.
6. Governments can produce the information technology systems they need in a timely manner.
7. Information technology systems support the government's ability to communicate with and provide services to its citizens.

These criteria can be summarized shortly in seven categories: information systems, architecture, IT planning, training, an evaluation process, procurement, and service to public.

The Progress and Freedom Foundation (PFF) studies the use of digital technologies by government and in the market place (The Progress and Freedom Foundation [PFF], 2000). In 1997, it published its first annual report on The Digital State. The report entailed a nationwide survey showing the extent to which state governments are utilizing digital technologies to streamline their operations and improve service. This report shows States making progress on implementing digital technologies. From 1997 to 1999, the study tracks state government IT uses in eight categories including education, business regulation, revenue and taxation, social services, and law enforcement. Measuring IT achievements against a series of benchmarks, the Digital State grades State efforts in these categories and also ranks their overall technological progress (Towns, 2000). Recently, a part of the *2000 Digital State Survey* was released, in which electronic commerce and taxation/revenue were included. Four key e-commerce areas are as follows (PFF, 2000):

1. the availability of download permit and licensing forms;
2. the ability of citizens and businesses to actually apply for licenses and permits electronically;
3. the availability of help or advice through a general on line mailbox; and
4. the availability of citizens and business to contact agency staff on line.

The five important survey areas for electronic taxation and revenue activities are as follows:

1. the availability of downloadable businesses and personal tax forms;
2. the ability of taxpayers to file tax returns on-line;
3. the ability of taxpayers to contact revenue department staff through a general electronic mailbox;
4. the ability of tax payers to contact specific revenue department staff members via e-mail; and
5. the percentage of tax records stored digitally rather than on paper.

A Framework to Evaluate the Informatization Level   147

Examining these areas included in each category, it can be found that these areas are composed of common features classified as information provided (i.e., quality of information), infrastructure (i.e., telecommunication, e-mail, database, etc.), accessibility to information (i.e., performance), and responsibility (i.e., user satisfaction).

Based on these projects, an evaluation framework for organization's informatization level was developed. In the next section, this framework is discussed in detail.

AN EVALUATION FRAMEWORK

In order to develop a framework that evaluates informatization, it is necessary to understand the concepts and procedures of informatization. From the definition of informatization described in section 1, the process of implementing informatization can be depicted as in Figure 1.

As shown in Figure 1, the informatization is implemented by using IT. That is, informatization is to implement IT by planning, designing and developing various IT resources, and then to help make more intelligent and qualitative society through the use of the information and knowledge produced from IT resources. In this view, the elements and procedure of informatization are considered as shown in Figure 2.

The flow chart in Figure 2 shows necessary components needed to achieve effective results through information technology. Elements composed of input and process are very important factors in making better outcomes. If they are well defined or well utilized, it can be said that it is highly possible to achieve effective and efficient IT performance. Therefore, all of these elements depicted in Figure 3 can be considered as subindexes composed of informatization level. Figure 3 represents a framework of the informatization level.

The IT strategy initiates the organization's informatization. However, if the higher executive (often decision-maker) does not agree with this concept, it is impossible to develop an effective IT strategy. Moreover, even if it is developed, without the approval or concern of the executive it is difficult to successfully have informatization in that organization. Therefore, to develop IT strategy, support for informatization from the executive is a very important factor. It is also important to note that IT policy, IT planning procedure and IT investment are necessary subindexes within the category of IT strategy.

In the category of IT resources, various elements related to the infrastructure, such as network, hardware, and security are included. Moreover, human resources, IT operation, IT facilities and IT training are impor-

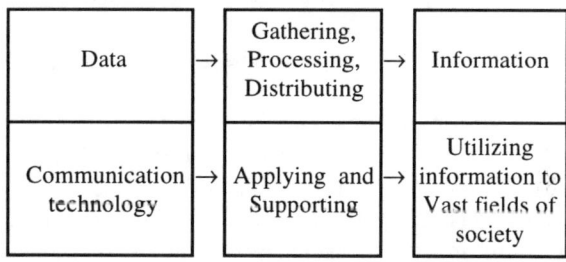

Figure 1: Implementing of informatization

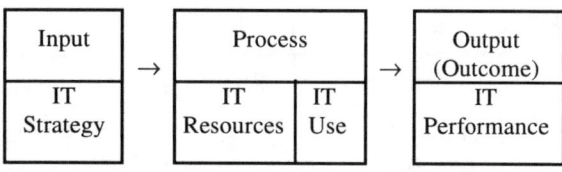

Figure 2: The elements and procedure of informatization

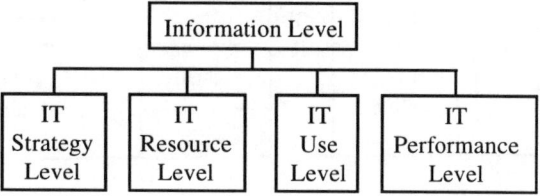

Figure 3: The sub-categories of the informatization level

tant sub-elements. This category can be classified into two groups; One group is concerned about infrastructure and the other about human IT resource.

The category of IT use is related to how many individuals in an organization utilize IT, for example, Internet, application systems, e-mail and so on. That is, this index can be measured by how many application systems are developed and how much those systems support individuals to improve their productivity and their IT capability.

IT performance is often defined as quality, customer satisfaction, and performance such as the delay-time of the project. However, it is relatively difficult to find appropriate metrics for this category compared to others. In GPP mentioned in an earlier section, they evaluate management system, not performance itself. The reasoning is as follows:

Most of the emphasis has been on *the measure of* performance and not on *capacity* to perform. Common sense tells us, however, that unless capacity is present, measurement of results can be a futile and dispiriting exercise. The GPP examines capacity by analyzing whether government can hire the right talent when it is needed, whether it has and can use the right information at the right time, and whether the systems that support both the visions and the strategies of leaders are present. These are *the platforms for performance*. Without them in place, high performance is not likely to occur. With them in place, both performance and its effective measurement become more likely. (Maxwell School, 1999)

This concept can be found in the PFF project, in which they use four categories including quality of information, infrastructure, accessibility to information (performance), and responsibility (user satisfaction).

In the framework, the IT capacity is considered a measure for IT performance. That is, the extent to which public services are provided through IT is considered as IT performance. For instance, through the home page of an organization, how much qualitative information provided, how fast government responds, and how often information is updated is shown. In addition, various financial or training supports for citizens are important sub-elements of this category. The following Figure shows the framework to evaluate the informatization level.

Various matrices have been used to measure each index. Although informatization has many intangible characteristics, in order to obtain objective evaluation results quantitative metrics can be used as possible in this framework. For example, the number of Local Area Network (LAN) ports and a network speed have been used as the network indexes. IT human resource index is measured by the ratio of IT resource to the total number of employees and their experiences.

Figure 4: The evaluation framework

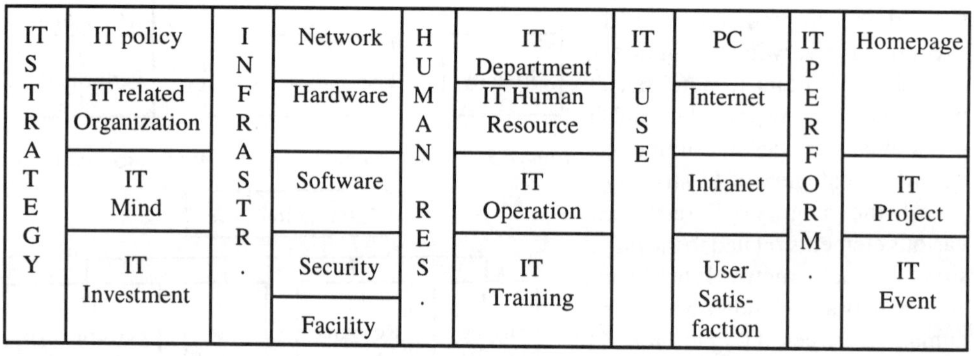

IT STRATEGY		INFRASTR.		HUMAN RES.		IT USE		IT PERFORM.	
	IT policy		Network		IT Department		PC		Homepage
	IT related Organization		Hardware		IT Human Resource		Internet		
	IT Mind		Software		IT Operation		Intranet		IT Project
	IT Investment		Security		IT Training		User Satisfaction		IT Event
			Facility						

Table 1: Indexes, Metrics and Weights

Level	Weight	Index		Weight	Metrics
IT Strategy	200	Policy		40	- ISP contents • organization's vision • Appropriateness of implementation Plan
		Organization Chart		20	- CIO • Authority, Skill - IT committee • Establishment, History
		Mind		100	- President's IT Mind - CIO's IT Mind
		Investment		40	- IT Budget, - Increase rates ('99/'98)
IT Resource	200	Infra	Network	30	# LAN port/ # total employee
			Hardware	40	# PC/# total employee(#TE) # PC connected to LAN/#TE # Printer/#TE
			Software	20	# Software/ #PC
			Security	20	#Virus vaccine SW/#PC Backup and Recovery
			Facility	10	Computer training center Conference facility
		Human	IT Department	10	# Division related
			Human resource	20	# Individuals qualified and certified
			IT Operation	30	# Guidelines Helpdesk Operation Y2K Problem
			IT Training	20	# Training/#TE # Training/#IT resource
IT use	200	PC		30	Mean PC use time/day PC use skill
		Internet		30	Mean internet use time/day #Email id/#TE
		Intranet		80	#Electronic document/#total doc. Intranet usage Information/data sharing Functions of intranet
		User satisfaction		60	Business process improvement IS satisfaction Motivation
IT Performances	400	Homepage		230	Contents, Information Interface • accessibility, design Public service • responsibility, transparency
		IT Project		70	# IT Project (SI Project) Amount of funds to IT company
		IT Event		100	# IT Training for citizen # various IT event

In addition, the sum of the weights assigned to each index is 100%. Through multiplying the score of each index by the weight, the total score can be obtained so that the organization's informatization level can be determined. Table 1 summarizes indexes, metrics, and weights.

Based on this framework, the NCA has performed a pilot project, which measured and evaluated the informatization level of 16 city governments in Korea in 1999. It has been recognized that the results could be varied according to which indexes are included in the framework and according to which weight is assigned to the indexes. However, the most important thing in this framework is that various meaningful metrics, which explain informatization, have been derived, and have been applied to a practical situation.

FUTURE TRENDS

The lesson learned from the pilot project using a framework described in the previous section is that a framework and metrics can be varied from evaluation objectives and evaluation subjects. Therefore, in the planning stage for evaluation the most important thing is to establish an appropriate objective and the coverage of evaluation. This framework was developed for evaluating an organization's informatization level, specifically public organizations. Moreover, the framework aims at suggesting the weaknesses and strengths of an organization's informatization so that improvements to reach a higher level can be made.

This framework has some limitations. First, in order to obtain objective evaluation results, the metrics used were mostly quantitative ones. The framework includes some qualitative metrics, but it was difficult to score with these metrics. Second, IT capacity was used to evaluate IT performance. There are various definitions for performance. For example Kraemer et al. (1995) defines performance as " metrics such as market share, customer satisfaction, defects rate, response time and service delivery to measure output, service and operation performance." In the framework described in the previous section, only IT capacity for public service was considered.

The third point is about the concept of informatization. Proposed framework is based on the process of informatization. Figure 5 shows another model focused on the classification with conceptual and physical elements of informatization.

A feature of Figure 5 is that the informatization level is classified into the IT capability and the IT resource levels. Compared to Figure 3, the IT capability level may include an IT strategy level, and an IT use/capacity level, and the IT resource level includes an IT infrastructure level and an IT human resource level. Therefore, IT capability has qualitative

Figure 5: Key elements of informatization

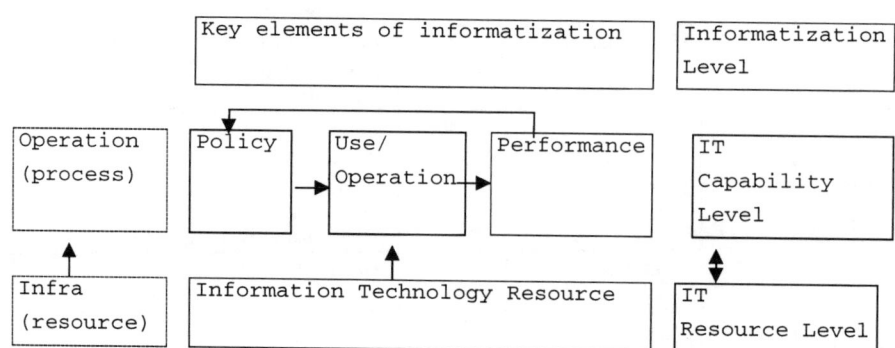

A Framework to Evaluate the Informatization Level 151

characteristics, while IT resource has quantitative ones. Quantitative metrics can be used as benchmarks so that improvement of an organization's IT level can be traced by this metric. With qualitative data of IT capability, some strategic information can be provided to improve an organization's informatization effectively and efficiently.

Dizard described a three-stage progression to the information age. Beginning with changes is already well in the basic information production and distribution industries. It leads to a greater range of services available for other industries and for government. With advanced information and communications resources, the third stage will result in a vastly expanded range of information facilities at the consumer level (Dizard, 1989).

Informatization can be considered as a stepwise process to implement information society. In this respect, the evaluation of the informatization level can be converted to determine the current informatization phase of an organization to give a direction for achieving its final goal. A similar concept has already been setup and used in the field of software development. The Software Engineering Institute (SEI) has developed the Capability Maturity Model (CMM) for determining capability maturity level of a software development organization (Paulk et al., 1993). Recently Kwak et al. (1999) has introduced the Project Process Maturity Model. Figure 6 shows a capability maturity model for organization's informatization.

In the initial stage, organizations focus on the building infrastructure. They acquire basic facilities such as network, hardware, system software, and so on. In the second stage, organizations try to develop application software such as management information systems and business application systems to improve productivity. In this stage, training for the members of organization is planned and performed.

The third stage is the integration stage dealing with networks and applications. An internal and external communication system is integrated to business applications, so that in this stage the organization's objective can be achieved through IT. In the forth stage, with an organization's data and information, a knowledge-base can be constructed. From this stage, an organization's culture is changed according to IT. In the fifth stage, information society of an organization can be realized.

This model is the only conceptualized model. Any essential element and evaluation criteria have not been studied yet. However, this model is meaningful, because it can follow social-trends and can give a direction for improvement of an organization's informatization. This model needs further research.

Figure 6: A informatization capability maturity model

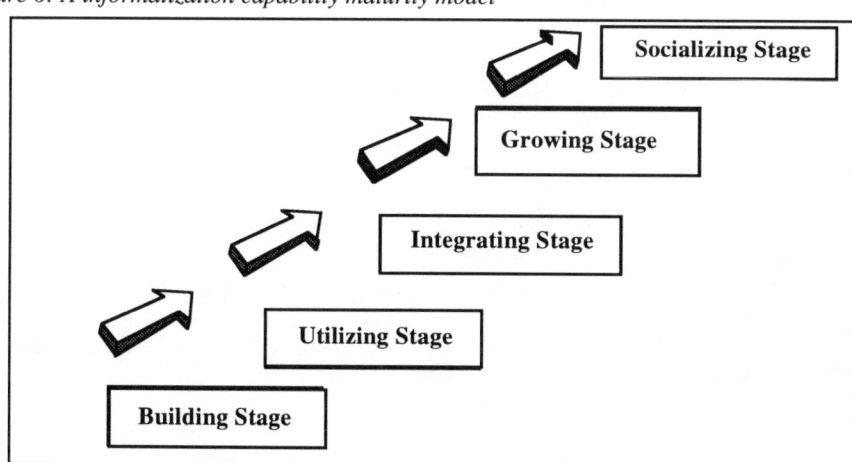

CONCLUSION

In this chapter, various evaluation model and framework have been reviewed. According to evaluation objectives and application fields, models and frameworks have been differently developed and used. However, some considerations have been taken similarly into these models. First, they have tried to evaluate performance of IT, such as customer satisfaction, management capacity, and quality of information. Second, they have released evaluation output as scoring. Therefore, evaluation outputs can be used as benchmarks. Third, in their frameworks, there were several common metrics such as policy (leadership), infrastructure, usability, etc.

This paper has provided various meaningful indexes, which can represent informatization. Moreover, an evaluation model with a different approach was introduced. In order to determine which framework and which metrics will be appropriate as an evaluation method, the evaluation objective and its coverage should be first determined.

In addition, by widely applying them for practical situations, it is expected that more significant metrics would be found and structured framework based on informatization theory would be suggested in further study.

REFERENCES

Davenport, T. H., & Short, J. E. (1998). The new industrial engineering: information technology and business process redesign. *IEEE Engineering Management Review*, Fall, 46-60.

Dizard, W. P. (1989). Chapter 1. The information age. *The coming information age (3rd. Eds.)*. London: Longman Group Ltd.

Hwang, B. et al. (1998). *Development of indexes for measurement informatization level of local autonomy*. Local Autonomy Informatization Foundation. Korea.

Kraemer, K. L. et al. (1995) *Performance Benchmarks for I/S in Cooperations*. Center for Research on Information Technology and Organization (CRITO) and CRC research and Advisory Service. University of California-Irvine. 1997.

Kwak, Y.H et al. (1999). The project management capability model and ROI. *Project Management & Technology, Summer.* PROMAT. Korea.

Maxwell school. (2000). Questions $ Answers About the GPP [On-line]. <http://www.maxwell.syr.edu/gpp/questions2000.htm> March.

Maxwell School at Syracuse University (2000). Project history [On-line]. Syracuse University. <http://www.maxwell.syr.edu/ gpp/history1.htm> March.

National Computerization Agency (1997). *National informatization white paper*. NCA : Korea

Oh, J. H. et al. (1998). *Development of indexes for measurement organization's informatization level*. NCA. Korea.

Paulk, M. et al. (1993). *Key practices for software capability maturity model (CMU/SEI-93-TR-024)*. Software engineering institute (SEI). Carnegie Mellon University.

PFF. Release Digital State Results (2000)[On-line]. <http://www.pff.org/pr/pr011100DigState2000.htm> March.

Progress & Freedom Foundation (PFF) (2000). Home Page. Mission Statement [On-line]. PFF. <http://www.pff.org/what_we_do.htm> March.

Rubin, H. The Worldwide Benchmark Report (2000)[On-line]. Cutter Information, Corp. <http://cutter.com/itgroup/ reports/ bench1999.html> February.

Schwartz, K. D. Benchmarking for dollars (2000) [On-line]. Datamation. <http://www.datamation.com/roi/02bench.html> February.

Towns, S. State Progress Toward Digital Nation (2000)[On-line]. PFF. <http://www.govtech.net/poblication...t/digitalstates/ digitalstates.shtm> March.

Part IV:

Evaluation of New Technologies

Chapter X

Using Cost Benefit Analysis for Enterprise Resource Planning Project Evaluation: A Case for Including Intangibles

Kenneth E. Murphy and Steven John Simon
Florida International University, USA

The goal of this chapter is to demonstrate how cost benefit analysis can be applied to large-scale ERP projects, and that these methods can incorporate the intangible benefits, e.g., user satisfaction. Detailed information on the business case utilized by a large computer manufacturer in their decision to implement the SAP system R/3 is presented. We illustrate how this organization utilized techniques to include intangibles in the implementation project's cost benefit analysis. The chapter concludes with a discussion on the state of valuing ERP projects and questions to be answered in the future.

INTRODUCTION

In 1998, expenditures for information technology (IT) accounted for more than 50% of corporations' annual capital investment in developed economies, and these outlays will average 5% of total corporate revenues by 2010 (Graeser, Willcocks, and Pisanias, 1998). Given the staggering amount of resources devoted to IT, $530 billion worldwide in 1995, one would expect managers to have a firm grasp of the anticipated contribution of their IT investments to the organization's profit margin. However, quantitative measurements of an IT project's expected return are not often used, primarily because they are unable to capture many of the qualitative and intangible benefits that are expected (Farbey, Land, and Targett, 1992). Still, managers must justify system investments, and hence "cost benefit analysis has assumed a pivotal position in the information systems revolution" (Sassone, 1988).

Information system project evaluation is challenging not because the projects cannot be justified but because they cannot be justified in terms which accountants and some senior managers are prepared to accept (Gunton, 1988). According to Mahmood and Szewczak (1999) the issue of measuring investments in IT is critical, these "measures may be quantitative in nature, but they must also be qualitative as well." The problem has grown as IS departments have advanced beyond implementing transaction processing systems with returns that are relatively easily to quantify to the implementation of management information, decision support and knowledge management systems. Systems in the latter category produce measurable benefits that are fuzzy at best, and defy conventional methods for quantifying the benefits. The failure of traditional measures to adequately capture the true value of the information technology systems was observed in the early days of MIS as an academic discipline (McRea, 1970). This measurement dilemma has grown worse as IT becomes part of the organization's nervous system or infrastructure and is a critical part of its structures and processes where all elements are integrated assessing returns on individual assets is impractical. Moreover, such integrated systems will also be extremely valuable as repositories to aid in strategic decision making. In addition to the factors listed above there is still a widespread lack of understanding of IT and information systems as a major capital asset (Willcocks and Lester, 1999).

In today's dynamic and competitive environment, senior managers are demanding figures that derive an IT project's return before the project is undertaken. Therefore, CIOs and their IT staffs are beginning to rely on both tangible and intangible measures to determine a system's contribution to an organization's bottom line. This procedure is a new endeavor for many IT staffs especially as they struggle to convert intangible measures such as user satisfaction to a tangible quantity suitable for inclusion in cost benefit calculations. This study examines Consolidated Computer Company's (CCC)[1] efforts to determine the contribution of a proposed enterprise resource planning (ERP) system they are seeking approval to implement.

The first part of the chapter presents a brief introduction to ERP systems, followed by a section where we define and discuss methods for evaluating IT investments and provide some detail on a family of techniques known as cost benefit analysis (CBA). We then discuss the wide range of intangibles that may enter into CBA calculations which leads to the presentation of a method for including intangibles in a CBA setting. The latter part of the chapter presents a case study in which Consolidated Computer Company seeks to use CBA to justify the implementation of an ERP system and demonstrates how an intangible factor, user satisfaction, can be included in the analysis.

ERP SYSTEMS[2]

Enterprise Resource Planning (ERP) is a term used to describe business software that is 1) multifunctional in scope, 2) integrated in nature, and 3) modular in structure. An ERP software solution is appropriate when an organization is seeking the benefits of business process integration and contemporary best practices in its information system, looking for a full range of functionality throughout its organization, and seeking to limit its implementation and ongoing support costs (Norris et al., 1998). Historically, the market for ERP solutions has been large multinational manufacturing companies (revenues over $1 billion) which operate in a discreet manufacturing environment. Today, however, the market is expanding to mid- ($250 million to $1 billion) and small- (under $250 million) companies across a wide range of industry sectors. The market is dominated by SAP-AG of Germany which holds over 70% of the software market with their R/3 system.

ERP systems have their origins in Manufacturing Resource Planning (MRP) software with installations traditionally in large scale manufacturing facilities. Although recently the trend is to extend installations to industry sectors including telecommunications, government, insurance, gas and oil, and high-tech manufacturing. Firms implementing the software generally seek process-oriented increases in productivity, up-to-the-minute access to timely information, and cost saving efficiencies. The systems are known for their process orientation rather than traditional functional orientation, and this may also enhance an organization's move toward breaking down departmental boundaries and thinking. The ultimate goal of many organizations implementing ERP packages is to carefully reengineer their processes, which in turn will hopefully benefit the bottom line.

Most ERP packages provide for a flexible organization structure with firms selecting which components they choose to use and how to implement them. Since the system is fully configurable, the organization turns on only those portions of the package it requires. The systems are designed with an open architecture facilitating expansion for future modules and allowing bolt-on applications from their approved vendors. Currently, the packages are being enhanced with tools such as data warehousing, advanced planning and scheduling optimizers, executive decision support systems, and tools to enable electronic commerce.

The cost of implementing an ERP system varies. Approximate costs for implementations in smaller companies are $10 million, and the projects take an average of 23 months with the total cost of ownership estimated at $15.6 million (Meta Group, 1999). In large companies, e.g., Fortune 1000, implementations can exceed several hundred million dollars and may take five years to accomplish. Yet, despite the considerable investment in time, capital, and resources the return on system investment is not clear. A survey of 63 companies with ERP systems discovered an average negative value of $1.5 million when quantifiable cost savings and revenue gains were balanced against spending on hardware, software, consulting, and support (Meta Group, 1999). Additionally, anecdotal evidence suggests that ERP failures have contributed to the bankruptcy of companies, e.g., Fox-Meyer Drugs (Bulkely, 1996). Yet despite negative reports such as the Meta Group's, ERP implementations are proceeding at a record pace with an abundance of success stories such as Cara Airport Services that saved 7% on production costs after their implementation of J. D. Edwards' application (Stedman, 1999).

Given the growing expenditures required to implement large scale systems such as ERP and infrastructure projects, senior management is demanding that chief information officers (CIOs) provide metrics to measure system contribution to the organization. CIOs are being counted on to deliver cost savings while adding competitive value as opposed to merely deploying systems. These managers are required to justify ERP investments through metrics while demonstrating that projects are satisfying business objectives. The use of metrics allows the CIO to show what business problems are being solved while providing evidence of the project's profitability.

The process of quantifying the value of IT projects becomes much more difficult as the scope and magnitude of projects grow. ERP system implementations generally require large capital investments and, because of their integrated nature, possess a wide and complex scope. Early business information systems were transaction processing systems (TPSs) designed to replace workers who performed repetitive tasks, e.g. payroll clerks. The determination of the costs and the benefits for these systems was relatively easy. The salary of workers to be replaced was compared against the cost of the system and hence the project's value was estimated. As systems became more complex and began to support other types of activities, e.g., decision making, the ability to quantify their payback became more difficult. It was clear that more and better information improved decision making, but it was very challenging to quantify the value of a better decision. Better decisions represent one form of intangible benefit derived from the IT system.

This process has become ever more complicated with the advent of large-scale projects such as ERP systems and other infrastructure technologies. In the case of infrastructure projects the benefits in part found in the services that they support and enhance. For instance, upgrading the telecommunication infrastructure could be measured in increased bandwidth or even how much faster files/information is transferred. While the improvements in these metrics are tangible, it is challenging to link their relationship to the monetary benefits resulting from increased market share as a result of a customer's repeat visits to an electronic commerce site. One of the ERP system's greatest benefits is its ability to integrate, standardize and provide real time visibility to an organization's data. This benefit has been attributed to increasing productivity and improving management decision making. Yet, given the task of quantifying this benefit, many organizations state that it is beyond their ability to measure and hence list it as an intangible.

EVALUATION TECHNIQUES FOR TECHNOLOGY INVESTMENTS

In this section we briefly present a variety of techniques for measurement of technology investments and argue why a family of methodologies known as cost benefit analysis (CBA) is often the methodology utilized in practice. To determine the benefits of IT projects Wehrs (1999) differentiates between ex ante and ex post evaluation. In ex ante evaluation the focus is on justifying the IT investment before it is made, and in ex post evaluation the goal is to justify costs that have been incurred so as to guide future IT expenditures. In this chapter we are taking the ex ante view of the IT investment, that is, our goal is to focus on the investment decision and not the justification of costs already incurred from an IT project. Wehrs partitions ex ante analysis into three major sets of techniques: decision theory, user information satisfaction, and CBA.

Decision theoretic approaches include information economics (Parker and Benson, 1988) and multi-criteria approaches (Keeny and Raiffa, 1976; Land, 1976). In this class of approaches the decision-maker that attempts to maximize the organization's utility or value function by taking the preferred action over the set of choices for the firm. Information economics sets out to rank or measure the financial impact of changes brought about by implementing the new information system. The multi-criteria approach measures the value of an IT project, perhaps in terms other than economic, and allows for appraisal of the relative value of different project outcomes. Both approaches allow for the advent of uncertainty and for the fact that different stakeholders may have different views on the benefits of the project. Implementation of this class of models is complex requiring the specification of outcomes for the firm, the organizational utility or value derived from each outcome, and the potential actions available to the organization. In information economics one must also define the transition matrix describing likelihood of the moving from one outcome state to another given the action taken. While both approaches have deep theoretical roots, they have been criticized by Kleijnen (1980a; 1980b) and Treacy (1981) for the extent of knowledge required by the decision maker to estimate model parameters and equations, the assumed rationality of the decision maker and the relative simplicity of this class of models in general.

The literature on user information satisfaction includes two major schools, user satisfaction and system use (Melone, 1990). Ives, Olson, and Baroudi (1983) introduce the concept of User Information Systems (UIS) as a method to evaluate IT investments by measuring the extent to which users believe the information system available to them meets their information requirements. UIS is accepted as a surrogate measure for changes in organizational effectiveness-the real goal of information system implementation. However, authors have criticized the UIS measure because of its lack of theoretical basis and because of a lack of empirical work validating the relationship between subjective assessment outcomes and economic performance (Wehrs, 1999). According to Melone (1990), the literature on UIS indicates that, on its own, UIS cannot be a surrogate for effectiveness of an information system. Swanson (1988), on the other hand, argues that the user attitudes are measured because information systems exist for the purpose of serving client interest, and hence, individual assessments of information systems are held to matter.

Cost Benefit Analysis (CBA) has been widely utilized to compare the costs and benefits of all types of capital investment projects (Prest and Turvey, 1965). It seems to be the family of techniques most often utilized in calculating the economic value of IT projects (Farbey, Land, and Targett, 1992; Bacon, 1992). In all CBA approaches the future stream of economic benefits and costs is estimated and the value of each project option is calculated. One major benefit of the family of CBA approaches to managers is that the results are relatively easy to interpret, while the greatest challenge involves the adequate measurement of project costs and benefits (Brynjolfsson, 1993). Detailed lists of the set of methodologies that fall under the CBA umbrella can be found in King and Schrems (1978), Sassone and Schaffer (1978) and Sassone (1988). Because of the challenges associated with providing economic value for the costs and benefits in IT projects, Dos Santos (1991) has proposed a more complex IT project valuation models based on the idea that subsequent project investments are optional. The subsequent decisions can be made when additional information concerning project success has been revealed.

It is generally accepted that all CBA activities require estimates of costs and benefits in future time periods. These estimates should account for the relevant costs at each stage of the project or system's life cycle. Most authors (Sassone and Schaffer, 1978; Hares and Royle, 1994) make strong arguments for the inclusion of discount factors in CBA analysis. Additionally it is the consensus that proper implementation of CBA includes the use of marginal versus average value analysis and that careful sensitivity analysis should always be part of the process. The challenge of effective assessment of costs and benefits is at the heart of any CBA activity, and Sassone (1988) lists seven generic methods for accomplishing this task. Following the assessment of costs and benefits, one of a number of outcome measures is calculated. The most common measures include net present value (NPV), internal rate of return (IRR) and payback period.

The formula for computing net present value, NPV (Sassone & Schaffer, 1978) is

$$NPV = \sum_{i=0}^{T} \frac{B_i - C_i}{(1+d)^i}$$

In this formula B_i and C_i are the values of the benefit and cost for the ith period in the future and d is the discount factor. In this formula the NPV is calculated from the current period (time 0) until period T. A greater value for NPV is assumed to be an indication of a more desirable project. One principal criticism of the use of NPV as a criterion to judge a project is the choice of the discount factor, d. However, sensitivity analysis performed on this parameter will allow for examination of how the decision factor may affect the decision. In many cases, management will fix a value of d, known as the hurdle-rate, for which the project must a have a positive NPV.

A second criterion that is often used is payback period. The payback period is simply the earliest period in which the project's cost is recovered. In using this criterion, it is assumed that the project with the earliest payback period is the best. However, this may not be reasonable if a competing project has large anticipated benefits further in the future. A third method for evaluating projects is the internal rate of return, IRR, i.e, the annual rate at which project is estimated to pay off. The quantity IRR is found by solving the following equation:

$$C_0 - B_0 = \sum_{i=1}^{T} \frac{B_i - C_i}{(1 + IRR)^i}$$

If using IRR to compare project alternatives, one chooses the project with the largest value for IRR.

As mentioned previously CBA has been criticized on a number of fronts. Keen (1975) was the first to note that the many of the costs and benefits of information systems are challenging to measure. This challenge is heightened as the use of information systems moves from transactional towards strategic. Keeny and Raiffa (1975) and others have criticized CBA on the basis that it does not include a method of coping with uncertainty that is usually present in information systems projects, e.g., many IT projects may be subject to user acceptance uncertainties. These criticisms make CBA challenging to utilize, however other authors have argued that this not evidence that CBA should be abandoned. Kaplan (1986) argues that one should choose a comparator, like missed strategic opportunity or declining cash flows to compare information systems projects. Dos Santos (1991) argues that since IT investments are made over time, some decisions can be foregone until more information about project success becomes available. In either case, a project may then appear to be more valuable than it otherwise would have been had these factors not been accounted for.

Practically speaking, when financial analysis is called for, CBA is often utilized to analyze IT investment. Using CBA, Boehm (1993) found that the investment in software technology at the Department of Defense paid back $5 for every $1 invested. The Glomark Group (1996) with over 300 client companies uses an ROI approach to assess the value of IT investments. Still, the issue of quantifiability seems to provide the major excuse for many organizations not to use CBA at all. Hogue and Watson (1983) found that in the case of DSS systems that a great majority (83%) of the organizations investigated did not bother to even try and quantify the benefits either in an ex post or ex ante setting. Farbey, Land and Targett (1992) found that only 4 of 16 firms surveyed used any kind of quantitative analysis in estimating the value of IT projects. Bacon (1992) found that approximately 52% of companies used CBA approaches, but it was only applied to 56% of the projects within those organizations.

Sassone and Schaffer (1978) explicitly admit that costs and benefits of IT projects lie along a "spectrum of quantifiability" which makes the use of this procedure challenging. Hares & Royle (1994) agree and argue for the use of a broader methodology, "investment appraisal" which extends CBA to include discounting, the effects of over projects, project risk and flexibility and the inclusion of intangible assets. In the next section, we define and discuss classes of intangible assets that organizations may want to include in their CBA activities.

INTANGIBLES AND TECHNOLOGY EVALUATION

The new International Accounting Standard (IAS) 38 defines an intangible as an identifiable nonmonetary asset without physical substance held for use in the production or supply of goods or services, for rental to others, or for administrative purposes. In many areas, investment results in economically valuable, legally recognized intangible assets, including copyrights (*Titanic* and Windows2000), patents (Viagra), changes in processes for making existing goods, and other assets such as brand names and trademarks. If companies fail to include intangible assets or their marketplace results, then corporate profits are vastly understated (price/earnings ratios are overstated) which in turn impacts national income, savings, and investment.[3] Intangibles can result in any combination of (1) a higher price for a premium product or service, (2) more sales from existing product or service, (3) cost savings from an existing or new product or service, and (4) new business/ new sales from a new product/service (Hares and Royle, 1994).

Not every business goal or benefit from a project can be quantitatively measured. A recent Ernst & Young (1999) study found that information not quantified on a company's balance sheet is increasingly becoming an important criterion for potential investors. Nonfinancial criteria, such as quality and credibility of management, market share, quality of investor relations, and customer satisfaction, accounted for 35% of the investor's decision. Davidow (1996) indicates that in the information age four-fifths of a firm's assets are intangible and that double-entry bookkeeping and measures of return on investment which do not consider intangibles are understating corporate value and profitability. Lev (1997) also indicates that conventional accounting performs poorly with internally generated intangibles, e.g., R&D, brand names, and talent, which are considered the engines of modern economic growth. Hares and Royle (1994) suggest that none of the methods for CBA are able to show how to measure and value the intangible benefits in financial terms.

Historically, the different treatment of tangibles and intangibles can be traced to the distinction between goods and services. As far back as Adam Smith, goods were material and could be stored while services were immaterial and transitory. This transitory nature meant services could not be counted as assets, but goods could. Logically, then, things counted as investment must be tangible. This led to a definition of wealth as "material objects owned by human beings." Therefore, what is material is tangible, and can constitute wealth which underlies the national income accounting conventions used to determine asset value, profit, saving, and investment. This logic fails to consider that more investment in today's economy are intangible, and these investments yield higher profits which equal greater output and savings. One estimate suggests that an adjustment for R&D alone would raise U.S. GDP roughly 1.5%. Extending this argument to project evaluation, the payback period would be reduced, and the return to the business would become proportionally greater.

Annie Brooking suggests that IT cannot be measured in isolation. In her book, *Intellectual Capital*, she indicates that there is a shift in the makeup of the net value of a company. In 1977, 1% of the net value of a UK company was based on intangible assets. In 1986 the make up shifted to 44% and it is growing rapidly. She decomposes intangibles into four areas: market assets, items which yield market power, e.g. brand names; intellectual property - copyrights; human-centered assets like knowledge; and infrastructure assets. IT falls into the last category. It is not the value of computers and software in the business, but their impact on the business' performance. Brooking examines Barclay's bank, whose computer and software assets equal approximately £100million. If those assets suddenly disappeared, the bank would not open, so clearly the value of the assets is much greater than the cost of the assets themselves. The difference is their intangible benefit or worth. The knowledge and expertise of an information technology department is an intangible asset in and of itself, but so is the way IT applies that knowledge to make other departments function more smoothly (Schwartz, 2000).

Computer systems are increasingly being developed for what are at first sight non-price factors and hence intangible. Hares and Royle (1994) indicate there are two main intangible benefits in IT investment. The first is internal improvement or infrastructure investment and the second related to customers. The latter, customer viewed intangible benefits are overwhelmingly those that the customer sees now and wants in the future - particularly related to customer service and user satisfaction (See Figure 1). These qualities posse an intrinsic value potentially greater than the immediate and calculable financial returns. They categorize intangible benefits as ongoing and future oriented.

The ongoing intangible benefits are those concerned with internal improvement of company operations or output performance. These are perhaps the most tangible of intangible benefits and can include 1) changes in production processes, 2) methods of management operations, and 3) changes to production value and process chains with resulting benefits as increased output or lower production costs. The second group of intangible benefits is more difficult to measure with their effectiveness being decided by external forces. This grouping involves services to customers that increase the utility (value added) of products or services. The benefits are converted into retained sales/customers, increased sales, customer satisfaction, and increased prices and include 1) quality of product or service as a market differentiator, 2) improved delivery of a product or service, and 3) improved service provided with products and services.

The next two groups of intangibles relate to future benefits and include the ability to identify new business opportunities leading to competitive advantage. The first of these benefits embodies spotting market trends. If new trends can be ascertained then a business is able to convert products or services to gain new sales and market position. Another example of this intangible benefit is the development of process through which to conduct business operations. This method provides the opportunity to cut prices and gain market dominance as was the case with Dell Computers. The final group of intangible benefits is the ability to adapt to change. As with the identification of market trends the benefits derived include adapting products and services to market trends and the modification of production processes. This ability is critical for firms in rapidly changing industries and can potentially converted into increased sales and higher margins.

IT projects deliver intangible benefits that cannot be quantified using mathematical equations like NPV, such as better information access, improved workflow, and increased customer satisfaction (Emigh, 1999), which are listed among the key attributes of ERP systems. One key function of IS departments has been the support of high quality decision making. This has

Figure 1: What are intangibles?

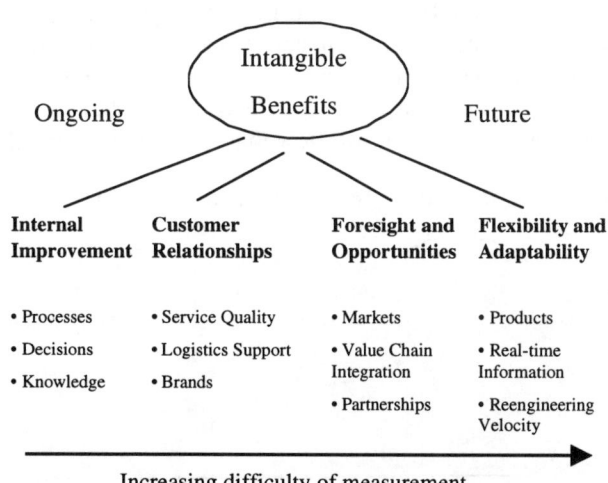

been difficult to quantify, especially at the higher levels of the organization where results are deferred. In the IS literature, Davis (1976), Emery (1971), and Keen and Scott-Morgan (1978) point out the importance of intangible benefits. Litecky (1981) indicates that despite the perceived importance of intangibles, there has been little if any guidance in the quantification of derived benefits. He proposes several assumptions as a precondition to quantifying benefits. First, tangible costs and benefits are relatively easy to estimate whereas intangible benefits are quite difficult to estimate. Second, tangible costs are ordinarily much greater than tangible benefits, and intangible costs are insignificant.

Parker and Benson (1988) explain that in order for enterprises to gain competitive advantage, the way IT is financially justified must change. Cost-benefit analysis is not adequate for evaluation of IT applications, except when dealing with cost-avoidance issues. If CBA is to be expanded, additional measures such as the perceived value to the business, increased customer satisfaction, or the utility of IT in supporting decision making must be considered (Katz, 1993). Clark (1992) found little guidance on IT's contribution to corporate profits in the literature but found reliability of service, technical performance, and business plan support all items difficult to accurately quantify. Other studies found varying measures of IT assessment including productivity (increases of), user utility, impact on value chain, business alignment (Wilson, 1988), system quality, information quality, use, and user satisfaction (DeLone and McLean, 1992). Accampo (1989) contends that CBA can be hard to apply to activities where information is the key commodity. Given that many of the measures found in the IS literature and listed above to evaluate system success are intangible, traditional methods of project evaluation fall short. This problem becomes even more difficult when analysis encompasses changes to business processes and information flows which impact productivity and decision support.

A survey conducted by Ernst & Young (1988) found that 60% of all UK companies concerned with the manufacturing of automobile components made no attempt to quantify the intangible benefits gained from the use of CAD/CAM systems, that only 20% quantified the benefits in physical terms, and only 20% quantified the intangible benefits in monetary terms. To accomplish the task of incorporating the intangible benefits into the financial analysis one must create multiattribute justification techniques which permit the inclusion of both monetary and nonmonetary factors in the analysis (Badiru, 1990). This method can lead to a single financial model, but requires a technique that bridges the gap between the intangible and tangible factors. Illustrated in Figure 2, this quantification technique (Hares and Royle 1994) applies a set of steps to express intangibles in monetary terms. The steps include 1) identify benefits, 2) make the benefits measurable, 3) predict the results in physical terms, and 4) evaluate the cash flow. As will be explained, this technique strives to convert the intangible benefit into cash flow that can be incorporated into CBA.

Figure 2. The Quantification Technique "Bridging the Gap"

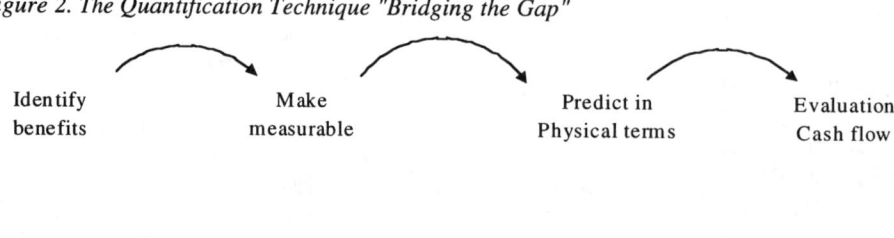

The first step of quantification is the *identification* of the intangible benefits. Two useful sources of information to assist identification include 1) critical success factors (CSFs) and 2) a checklist of intangibles. Many CSFs include quantifiable items. For instance, improved customer service, a CSF, could be measured by a stated reduction in the number of customer complaints. A checklist of intangibles is generally easier to create but perhaps harder to quantify. The IS literature suggests a number of factors for evaluating system success including customer satisfaction, product/service quality and reliability, speed of service, improved service, and reduction of errors. All of these intangibles can be converted into monetary terms through the ability to 1) maintain and increase sales, 2) increase prices, 3) reduce costs, and 4) create new business.

The second step is to make the intangible benefits *measurable*. This consists of re-expressing the benefits described above in more measurable terms. The third step is to *predict the benefit in physical terms*. This is generally the most difficult of the quantification steps with multiple methods to convert measures into actual numbers. The first method is the market survey. Market surveys are most attractive because 1) the perceptions of the company and customer can be aligned and an agreement of monetary equivalence can be agreed upon and 2) surveys can include a forward-looking component which could lead to proactive actions which potentially increase the value of the project. The second method is management estimates, which is usually used when surveys are not possible. Senior management, whose operations are supported by the project but who are removed from project responsibility generates these estimates. The problem with management estimates is that they are based on past evidence and therefore reactive. The third method is comparative case study of a similar business. The advantage of this method is that the firm gains from the lessons learned from the past exercise. The disadvantage is that past projects are conducted in a different business environment, and the method uses backward looking methodology. The final step in the quantification technique is the *evaluation in cash flow terms*. This is a simple mathematical process with the volumes from the previous steps related to the monetary value of the benefit. It is at this point that the technique can be merged with CBA.

THE BUSINESS CASE FOR ERP AT CONSOLIDATED COMPUTER COMPANY

Consolidated Computer Company (CCC) is a global software and hardware manufacturing, distribution and consulting organization with a long and very successful history delivering and implementing a wide range of business solutions. In the early 1990s CCC faced very significant challenges that included high operating costs, a bloated workforce and many redundancies across the globe in manufacturing, research, and design. To revive its legacy, CCC set forth a number of strategic imperatives that included firm wide cost reduction, reduction of product development and deployment cycle, marketing as a single global organization and streamlining the relationship between CCC and its customers. One part of implementing the plan for accomplishing these objectives was to put in place the integrated supply chain in its Personal Computer Division (PCD) which would include procurement, production and fulfillment.

In examining the costs of operating the company, among other items CCC's corporate management found that IT expenditures were excessive. Like many other multinational organizations CCC had, over the years, implemented and now operated hundreds of nonintegrated information systems to support their business throughout the world. Management felt that reduction of a number of legacy systems could be achieved through the use of an integrated software solution, i.e., an ERP system. To accomplish their goal of reducing costs and integrating the supply chain, corporate managers established a relationship with a major ERP solution provider. While corporate management moved ahead with the new initiatives, the operating problems at PCD were only getting worse.

In 1993, the management of PCD had authorized significant expenditures on new parts in anticipation of strong sales. Customer dissatisfaction with product stock outs had been on the rise for some time, and using strategic procurement PCD management hoped to prevent this from reoccurring. When sales of certain models did not reach expectations, the outdated procurement and inventory systems were unable to stop the flow of parts from vendors increasing inventory levels sharply. Across the globe each production facility was operating its own applications and systems for supporting supply chain processes, which made it impossible to coordinate purchasing and inventory management activities. This crisis, which resulted in heavy operating and bottom line losses, made a massive restructuring of PCD division an imperative. For the personal computer division the reengineering would entail removing many redundant elements and replacing fragmented information systems with integrated solutions.

At the beginning of 1997, managers in charge of reengineering the supply chain in PCD were directed to justify the large expenditure that would be required to implement the ERP system. To cut costs in operations PCD's management believed that decreasing work in progress and increasing inventory turns in the production facility was necessary. One part of achieving these goals included the implementation of a just-in-time procurement system. Improvements to the order fulfillment and customer management processes would also be required to improve declining customer satisfaction indices. Furthermore, it was clear that the large number information systems used in operations did not offer the functionality required to attain these goals. To justify the substantial investment, the CBA methodology was utilized to build the business case for implementing the ERP solution.

The scope of the PCD systems implementation project was to bring the ERP solution to three major production facilities across the globe. In their analysis PCD management used NPV, IRR and payback period to assess the project's return on investment. The business case was built assuming a 10-year time horizon utilized a 20% hurdle rate. To calculate benefits, very conservative revenue growth and profit margin assumptions were made and no benefit was assumed to commence until one year following the implementation at each site. Productivity savings of between 5 and 20% were assumed for the production and order fulfillment processes. The new system would also enable a onetime inventory reduction of 10% that was assumed to occur one year after implementation. Cost savings resulting from the reduction in number and complexity of systems in operation were also included. Major capital expenditures were assumed to occur over three years and the high-end of development and deployment cost estimates were used for each site. The cost benefit estimates resulting from the above assumptions are shown in the Table 1.

Table 1: Cost Benefit Analysis (Tangibles Only) NPV($ millions)/IRR

Productivity	Inventory	IT Operations	Implementation Cost	Total*	IRR*
18.8	49.1	23.4	(73.4)	28.1	39.20%

*Additional factors are included in these figures

INCLUDING INTANGIBLES IN THE BUSINESS CASE

PCD management utilized calculations with several intangible items in building the business case including customer satisfaction. In this example, only customer satisfaction will be used simplifying the illustration. Customer satisfaction was selected since CCC's management indicated that of all the intangible items it had the greatest potential impact on project evaluation.

Long before PCD embarked on their ERP evaluation project the company's managers knew there was serious problem with customer satisfaction. Their annual surveys of customers and suppliers indicated that levels of satisfaction were down 21% and 15%, respectively, with customers indicating significantly better relations with PCD's competition. Realizing that sagging satisfaction would soon translate into smaller market share and falling profit margins, senior management ranked satisfaction improvement as a key goal during system evaluation. Using a method very similar to the quantification technique of Hares and Royle (1994), the project managers identified improved customer/user satisfaction as a key system deliverable.

Satisfaction as a deliverable was critical to customers and suppliers, and existed as a metric within PCD with data collected on an annual basis. To convert satisfaction from an intangible to a measurable factor, PCD's IS department compiled a list of customer reported system deficiencies from the last satisfaction survey. Upon completion of the deficiencies list, IS managers examined each item's performance in the current system and its expected result on the proposed ERP system (see Figure 3 for a sample of the items). This procedure established a baseline from which the managers could project the level of satisfaction improvement once the proposed system was in place. From the proposed improvements, managers throughout PCD, not just members of the project team and IS department, consulted customers and projected that once deployed the proposed ERP system could improve customer satisfaction by 5% initially and approximately 2% per year thereafter (assuming system performance expectations were met).

The next major task in this analysis was to predict the economic value of an increase in customer satisfaction to CCC. This step was also accomplished through management interviews and surveys with CCC's key customers and suppliers.

The managers undertaking this project were those most familiar with customers, particularly those in the sales and marketing organizations. The results of several hundred interviews suggested that for each 5% improvement in customer satisfaction, CCC could expect a 1% gain in market share. The results, while nonscientific, indicated that market share increases would result in significant benefits for the company. The final step of this process was to evaluate the potential cash flow resulting from market share gains.

Figure 3. Sample items for Customer Satisfaction Survey

Item	Current system	Proposed ERP system
Enter pricing data	5-80 days	5 minutes
Committed ship date	1 day	Realtime
Schedule orders	Overnight	Realtime
Credit check	15-20 minutes	Realtime
Enter order	30 minutes	5 minutes
Inquiry response	15-20 minutes	Realtime
Ship and build	Overnight	Realtime

Table 2: Cost Benefit Analysis (Tangibles and Intangibles) NPV($ millions)/IRR

Productivity	Inventory	IT Operations	User Satisfaction	Implementation Cost	Total*	IRR*
18.8	49.1	23.4	228.7	(73.4)	228.9	124.00%

*Additional factors are included in these figures

Based on the cash flow resulting from the customer market share improvements the CBA was redone and the results appear in Table 2 below.

DISCUSSION AND CONCLUSION

In the face of significant threats to market share and shareholder value, the Consolidated Computer Company (CCC) embarked on a major organizational restructuring. As part of that effort, the personal computer division (PCD) was asked to justify in economic terms a large investment in an ERP solution. In the first cost benefit analysis (Table 1), PCD management used estimates of productivity savings, inventory savings and savings resulting from more effective information systems operations to calculate the net present value (NPV) and internal rate of return (IRR) for the project. The NPV of productivity improvements and inventory reductions resulting from new system implementation were estimated at $18.8 million and $49.1 million respectively over the 10-year time horizon. The NPV for IT operations cost savings was estimated to be $23.4 million. The total NPV for the project was estimated to $28.1 million and all NPV calculations assumed a 20% hurdle rate. Table 1 shows that internal rate of return for this project was estimated to be 39.2%. Using these estimates the ERP implementation project appears to meet the hurdle rate criteria set by CCC management.

In a subsequent CBA many of the same cost and benefit figures were utilized, however, the intangible item of customer satisfaction was also included. The assumption made by PCD management was that implementing the ERP system would have a significant and positive impact on customer satisfaction and hence improve market share. This view had been justified by data gathered through customer interviews. The NPV resulting from increasing the customer satisfaction was estimated to be $228.7 million. This resulted in the new total NPV of the project when customer satisfaction was included in the project to be valued at $228.9 million with an IRR was 124%. Upon including the intangible of customer satisfaction the ERP implementation project was significantly more beneficial to improving organization bottom line.

CCC's managers reported that at the inception of evaluation, the company's senior management sought an economic analysis oriented towards immediate bottom-line returns. The evaluation team reported some of their concerns that potentially could skew short-term oriented analysis. Their concerns included: 1) long implementation lead times; 2) large initial investments for hardware and software; 3) increasing frequency of changes in technology; and 4) shortages of skilled technically knowledgeable personnel and knowledge workers. The evaluation team also raised a number of issues that influenced decisions that were beyond the range of quantification but impacted the project's acceptance. A sample of these issues include: 1) IS departments limited control of technology budgets, 2) dynamic and complex technology cycle, 3) socioeconomic and socio-technical factors, and 4) organization politics. As a result of the successful evaluation and awareness of the related issues listed above, CCC's senior managers were better prepared to evaluate the project, able to reorganize organizational priorities, and view the system's contribution holistically. Therefore, the decision to implement the ERP system was not based completely on numerical evaluation of tangibles and quantified intangibles, but on more substantial issues such as market position, growth potential, benefits to customers, product effectiveness, and of course, profitability.

As a result of the evaluation and analysis, CCC's IS department was given the approval to implement the Enterprise Resource Planning (ERP) package. The implementation of the new system, including replacement of legacy hardware and software, was conducted on time and on budget. During its first year of operation the ERP system contributed over $225 million in savings and productivity improvements. The results of the CBA and intangible analysis were convincing factors that led CCC to the decision to implement the system. Without the numerical results senior management would not have agreed to move the project forward regardless of anticipated benefits. Even with the analysis of tangible factors and the quantification of intangibles, all benefits of a large scale IS project are not included in the overall equation.

The first and most important criteria when undertaking IS project evaluation is whether the system contributes to the strategic objectives of the organization. This initial criteria is more complex than "does the system help the organization achieve its goal" or make the organization better at what it does. One benefit often cited upon installation of an ERP system is the improvement in information quality, access, and use. Benchmarking Partners (1999), a research and technology advisory firm, reported 88% of respondents in the banking and finance industries cited information quality and accessibility as a key benefit of their implementation. Janus Capital Management found that their finance group spent 65% of their time tracking down data so that analysis could be conducted. Their implementation cut storage costs, improved efficiency, and decreased time to provide clients with information. In addition to accessibility of information, quality and standardization of information rank high on the list of strategic benefits. Most ERP systems replace a number of legacy systems, each with their own databases and data formats. The ability for management throughout the organization to understand data while having it in a standard format, located in a single database, improves the quality of the information in turn leading to better decisions.

Movement to enterprise integration and process orientation was also cited as a key strategic benefit. This benefit is manifest in the breaking down of organization barriers, previously represented in departmental structures, and replacing them with integrated optimized processes. Standardization of an organization's technology platform is another benefit area. Similar to the situation for business processes, many organizations have found themselves with widely disparate IT platforms, systems, and data standards throughout their organizations. As a matter of fact, the Y2K problem prompted many organizations to replace old legacy systems. While there is no guarantee that replacing legacy systems is cheaper in the long run (there is actually some debate regarding total cost of ownership with ERP systems), in many situations new systems are easier and cheaper to maintain than their predecessors.

Most of the benefits discussed above can be grouped under the category of keeping the organization competitive. Some of the costs and benefits, e.g., system operations and maintenance, are easily measured. Others, such as better access to information, information quality, and improved decision making are harder, if not impossible, to quantify. The managers at CCC made a conscious decision to measure some factors and omit others due to time and fiscal constraints. They also made all their assumptions based on the worst case scenario. All estimates and numerical justification were made using the most conservative estimates. This allowed them to report confidently to senior management knowing their results reflected the most realistic findings and expectations.

The future will bring about an even greater need to focus on intangibles when justifying systems implementation projects. The projected economic value of process and systems integration efforts will need to be measured not only on internal projects, but also across suppliers, customers and other partners in the value chain as business to business electronic commerce becomes more prevalent. The value of information that will be available in data warehouses, from knowledge management and executive information systems will be another challenge for managers performing ex ante systems implementation analysis. A cohesive set of methodologies and techniques to measure the value of these intangibles will be required if management is to succeed in better estimation of project value.

Information systems' escalating expense and growing importance to organizations have made the justification of projects increasingly critical. This study demonstrated that traditional cost benefit analysis could be applied to large-scale information systems projects such as infrastructure and enterprise resource planning. Extending the traditional methodology, the study illustrated how intangible measures can be used to augment CBA analysis and include what was once believed not considered measurable. This improved analysis provided CCC's managers with a more accurate and realistic look of the returns expected as a result of undertaking the ERP implementation.

REFERENCES

Accampo, P. (1989). Justifying Network Costs. *CIO, 2(5)*, 54-57.

Bacon, C.J. (1992). The Use of Decision Criteria in Selecting Information Systems/Technology Investments. *MIS Quarterly.* 16(3), 335-354.

Badiru, A.B. (1990). A Management Guide to Automation Cost Justification. *Industrial Engineering, 22(2)*, 26-30.

Barchan, M. (1998). Beyond the Balance Sheet: Measuring Intangible Assets. *Chief Executive, 139*, 66-68.

Barua, A., Kriebel, C., & Mukhopadhyay, T. (1989). MIS and Information Economics: Augmenting Rich Description with Analytical Rigor in Information Systems Design. *Proceedings of the Tenth International Conference on Information Systems*, 327-339.

Beenstock, S. (1998). The Calculation IT Can't Make. *Management Today.* (June).

Boehm, B.W. (1993). Economic Analysis of Software Technology Investments. In T.R. Gulledge & W.P. Hutzler (eds.) *Analytical Methods in Software Engineering Economics.* Berlin:Springer-Verlag.

Brooking, A. (1996). *Intellectual Capital.* London: Thomson.

Bulkeley, W.M. (1996, November 18). A Cautionary Network Tale: Fox-Meyer's High-Tech Gamble. *Wall Street Journal Interactive Edition.*

Clark, T.D. (1992). Corporate Systems Management: An Overview and Research Perspective. *Communications of the ACM, 35(2)*, 61-75.

Davidow, W. (1996, April 8). Why Profits Don't Matter: *Until We Measure Intangible Assets. Forbes, 157(7)*, 24.

Davis, G.B. (1976). *Management Information Systems.* New York:McGraw-Hill.

DeLone, W.H. & McLean, E. (1992). Information System Success: The Quest for the Dependent Variable. *Information Systems Research, 3(1)*, 60-95.

Dos Santos, B.L. (1991). Justifying Investments in New Information Technologies. *Journal of Management Information Systems*, 7(4), 71-90.

Emery, J.C. (1971). Cost/Benefit Analysis of Information Systems. *The Society for Management Information Systems Report #1*, 41-47.

Emigh, J. (1999, July 26). Net Present Value. *ComputerWorld, 33(30)*, 52-53.

Ernst & Young (1988). *The Use of CAD/CAM Systems in the UK Automotive Components Industry.*

Farbey, B., Land, F., & Targett, D. (1992). Evaluating IT Investments. *Journal of Information Technology*, 7(2), 109-122.

Glomark Group, Inc. (1996). *The Glomark ROI Approach.*
Graeser, V., Willcocks, L. & Pisanias, N. (1998). *Developing the IT Scorecard: A Study of Evaluation Practices and Integrated Performance Measurement.* London: Business Intelligence.
Gunton, T. (1988). *End User Focus.* New York: Prentice-Hall.
Hares, J. & Royle D. (1994). *Measuring the Value of Information Technology.* Chichester: Wiley.
Hogue, J., & Watson, H. (1983). Management's Role in the Approval and Administration of Decision Support Systems. *MIS Quarterly,* 7(2), 15-26.
Ives, B., Olson, M., & Baroudi, J. (1983). The Measure of User Information Satisfaction. *Communications of the ACM,* 26, 785-793.
Kaplan, R. (1986). Must CIM be Justified by Faith Alone? *Harvard Business Review.* 64(2), 87-95.
Katz, A.I. (1993). Measuring Technology's Business Value: Organizations Seek to Prove IT Benefits. *Information Systems Management,* Winter, 33-39.
Keen, P.G.W. (1975). Computer Based Decision Aids: The Evaluation Problem. *Sloan Management Review,* 16(3), 17-29.
Keen, P. & Scott-Morton, M. (1978). *Decision Support Systems: An Organizational Perspective.* Readings:Addison-Wesley.
Keeny, R. & Raiffa, H. (1976). *Decisions with Multiple Objectives,* New York: John Wiley & Sons.
Kleijnen, J. (1980). Bayesian Information Economics: An Evaluation. *Interfaces.* 10(3), 93-97.
Kleijnen, J. (1980). *Computers and Profits.* Reading, MA: Addison-Wesley.
Kroll, K. (1999). Calculating Knowledge Assets. *Industry Week, 248(13),* 20.
Lev, B. (1997, April 7). The Old Rules No Longer Apply. *Forbes,* 34-36.
Litecky, C.R. (1981). Intangibles in Cost/Benefit Analysis. *Journal of Systems management, 32(2),* 15-17.
Mahmood, M.A. & Szewczak, E.J. (Eds.) (1999). *Measuring Information Technology Payoff: Contemporary Approaches.* Hershey, Pa: Idea Group.
Markus, M.L. & Tanis, C. (Forthcoming). The Enterprise Systems Experience - From Adoption to Success. Zmud, R.W. (Ed). *Framing the Domains of IT Research: Glimpsing the Future Through the Past.* Cincinnati: Pinnaflex.
McRea, T.W. (1970). The Evaluation of Investment in Computers. *Abacus,* 6(2), 20-32.
Melone, N.P. & Wharton, T.J. (1984). Strategies for MIS Project Selection. *Journal of Systems Management.* 35, 26-33.
Melone, N.P., (1990). A Theoretical Assessment of the User-Satisfaction Construct in Information Systems Research. *Management Science,* 36(1), 76-91.
Meta Group (1999). *Extract of META Group Survey: ERM Solutions and Their Value.* Meta Group.
Nakamura, L. (1999). Intangibles: What put the New in the New Economy? *Business Review - Federal Reserve Bank of Philadelphia,* 3-16.
Norris G., Wright, I., Hurley, J.R., Dunleavy, J., & Gibson, A. (1998) *SAP An Executives Comprehensive Guide.* New York: John Wiley.
Parker, M. & Benson, R. (1988). *Information Economics: Linking Business Performance to Information Technology.* London: Prentice Hall.
Parker, C. & Soukseun, D. (1998). IAS 38:How Tangible is the Intangible Standard. *Australian CPA,* 68(11), 32-33.
Sassone, P.G. & Schaffer, W.A. (1978). *Cost Benefit Analysis.* New York: Academic Press.
Sassone, P.G. (1988). A Survey of Cost-Benefit Methodologies for Information Systems. *Project Appraisal,* 3(2), 73-84.
Schwartz, M. (2000, February 28). Intangible Assets. *ComputerWorld.*
Stedman, C. (1999, June 14). Airline Food Vendor Seeks 7% Savings on Production. *ComputerWorld.*
Swanson, E. (1988). Business Value as Justificatory Argument. In ICIT Research Study Team#2, *Measuring Business Value of Information Technologies.* Washington, D.C.: ICIT Press.
Tam, K.Y. (1992). Capital Budgeting in Information Systems Development. *Information & Management, 23,* 345-357.
Treacy, M. (1981). Toward a Behaviorally Ground Theory of Information Value. *Proceedings of Second International Conference on Information Systems,* 247-257.

Wehrs, W. (1999). A Road Map for IS/IT Evaluation. In *Measuring Information Technology Investment Payoff: Contemporary Approaches* (Mahmood and Szewczak eds.), Hershey, PA: Idea Group.

Wilcocks, L.P. & Lester, S. (Eds.) (1999). *Beyond the IT Productivity Paradox*. Chichester: Wiley.

Wilson, D.D. (1988). *Assessing IT Performance: What the Experts Say*. MIT Working Paper. Cambridge, MA.

ENDNOTES

[1] The name is fictional, the company and case study are real.
[2] Additional information on ERP and SAP's R/3 can be found at www.sap.com or in one of many current business publications.
[3] For instance, Titanic sold $1 billion in theater tickets and Viagra sales exceeded $700 million in its first month.

Chapter XI

Evaluating the Management of Enterprise Systems with the Balanced Scorecard

Michael Rosemann
Queensland University of Technology, Australia

The management of Enterprise Systems (ES) software consists of two main tasks: the implementation and the use, stabilisation and change of this comprehensive software. The Balanced Scorecard, a framework originally developed in order to structure the performance measurement for an enterprise or a department, can also be used for the evaluation of ES software. Adapting the Balanced Scorecard and adding a new fifth project perspective allows the comprehensive evaluation of Enterprise Systems and represents an alternative IT evaluation approach. It supports the time consuming implementation of enterprise systems as well and the benefits realization stage. Furthermore, the application of the Balanced Scorecard for IT evaluation represents a novel application area for this strategic management concept.

INTRODUCTION

Enterprise Systems (ES) (synonyms are Enterprise Resource Planning (ERP), Enterprise-wide Systems, Integrated Vendor Software, Integrated Standard Software Packages, and Enterprise Application Systems) can be defined as customizable, standard application software which includes integrated business solutions for the core processes (*e.g..*, production planning and control, procurement) and the main administrative functions (*e. g.,*. accounting, human resource management) of an enterprise (Rosemann 1999). In order to configure and use ES software efficiently, several components, like implementation tools (procedure models, reference information models, configuration guidelines, project management software), workflow functionality, tools for the development of add-on solutions and system administration, and office suites are usually embedded. ES software, which also includes integrated solutions for the management of transactions with business partners, especially supply chain management and customer relationship management, is called extended ES software. Currently, the main ES vendors are SAP AG, BAAN, J. D. Edwards, Oracle and PeopleSoft.

The GartnerGroup 1999) forecasts that the ES market will be greater than $20 billion by 2002 (with a probability of 80%). More than 50% of this will be ES service revenue, while the total ES license revenue will cover approximately $9 billion. They estimate that more than 90 percent of Fortune 500 enterprises have purchased a module or a set of modules from an ES vendor. 50 percent have made a commitment to one vendor, while only less than 20 percent went actually live. They also estimate that the SME market is the main customer group, as more than 50% of these enterprises still haven't selected a next-generation ES. For 2000 (2001, 2002) the GartnerGroup anticipates a market growth of 22% (25%, 28%). ES software accounts for more than half of the software licenses and maintenance revenues. In Western Europe, the top-tier ES vendors account for 64 percent of ES market revenue (see AMR Research Inc., Boston 1998, in Electronic Buyer). These figures show that ERP-initiatives are among the biggest investments enterprises are currently conducting. In addition to the huge initial investment, which often is beyond $5 million US, the necessary ongoing costs for system maintenance, stabilisation and upgrades are enormous. Thus, enterprise systems represent a *long-term financial commitment,* and elaborated forms of evaluating this investment are required.

Furthermore, it is usually reasonable easy to collect the related costs, but far more difficult to estimate the benefits and opportunity costs related to ES. Classical indicators like ROI are not appropriate as they do take not the qualitative benefits of ES (service provision to the business, investment into IT infrastructure) into account. Therefore, an evaluation of ES has to *cover more than just financial indicators.*

Because of its comprehensive functionality, ES software is very complex. As an example, the following indicators demonstrate the complexity of SAP R/3. This enterprise system includes more than 20 industry specific solutions and covers the areas of material management, production planning, sales and distribution, human resource management, financial accounting, asset management, and cost controlling. Separated applications support among others customer relationship management, supply chain management, or knowledge management. Current state of the art technologies like workflow management, data warehousing and data mining, Internet-interfaces or Internet portals (mySAP.com) are parts of the SAP product family. Thus, ES software is not only regarding the necessary investment, but also its *comprehensiveness and complexity* in application, that demands sophisticated evaluation to gain transparency. As core business operating systems ES are of relevance for most IT applications.

This chapter suggests using a modified version of the Balanced Scorecard for the evaluation of enterprise systems. It will be discussed how the perspectives of the Balanced Scorecard can be used for an evaluation of software that goes far beyond financial figures. The next section introduces briefly the concept of the Balanced Scorecard and motivates its application in the area of IT evaluation in addition to the typical use for departments or other organisational units. Structured in the two main tasks of ES management, the next two sections discuss the application of the Balanced Scorecard for the ES implementation and the operational ES use. The chapter concludes with an overview of future trends in this area.

BACKGROUND: THE BALANCED SCORECARD

The management of ES software can be subdivided into two main stages of implementing ES software and its operational use. The Balanced Scorecard can be applied for the evaluation of both tasks (Beeckman, 1999; Brogli, 1999; Reo, 1999; Walton, 1999; van der Zee, 1999). The Balanced Scorecard is a framework which aims to structure the relevant key indicators for performance management (Kaplan and Norton, 1992; Kaplan and Norton, 1993;

Kaplan and Norton, 1996) into four interrelated perspectives. For a more comprehensive introduction to the Balanced Scorecard see the relevant chapters in this book. Besides the traditional financial measures, the Balanced Scorecard accounts for a wider range of effects (Martinsons et al. 1999) as it also consists of indicators for the performance of the internal processes, the customer relationship management, and the innovation and learning activities. Each of the four perspectives consists of various key indicators. The different indicators are linked across the four perspectives according to cause-effect relationships. The assumption of the presented concept is that the Balanced Scorecard addresses exactly the two main tasks of ES management.

First, the Balanced Scorecard allows building up a comprehensive evaluation of an ES implementation. The process of implementing enterprise systems can take up to 24 months and represents besides complex organizational, and IT challenges the migration from the business strategy into an ES strategy, and finally, in the product specific approach for the selected ES software (Figure 1). This can be seen as an analogy to the approach of Balanced Scorecards to transfer visions into strategies and operations. In this application, the Balanced Scorecard reports about one major process, the implementation project. This form of IT evaluation is of interest for ES project managers as well as for the involved external consultants and the ES provider, who might consolidate these data for internal purposes. These different stakeholders perceive the effectiveness and the efficiency of enterprise systems differently and require individual, but consistent indicators.

Second, the Balanced Scorecard may be useful for the continuous evaluation of the use, stabilisation and change of the ES software. In this role the responsible manager for the ES software or certain ES modules would be interested in these data. The Balanced Scorecard could become the main evaluation tool that reports about the ongoing costs, as well as about criteria like user acceptance or process performance.

It has to be stressed that using the Balanced Scorecard for IT evaluation is not a typical Balanced Scorecard application. More often the Balanced Scorecard evaluates the performance of an enterprise or a department. A main application area in this context is the IT department. Designated IT Balanced Scorecards (Beeckman, 1999; Walton, 1999) aim to measure the performance within an IT department as well as the links of the IT department to other business departments. Wright et al. (1999) present a comprehensive Balanced Scorecard analysis of Compaq Computer Corporation. In one of their Balanced Scorecards the ES software SAP R/3 is a part of the innovation and learning perspective (Wright et al. 1999, p. 33).

However, it seems to be reasonable to apply the entire Balanced Scorecard not only for organisational units, but also for comprehensive software packages. The motivation is that from a more generic level no major differences exist between corporate and organizational Balanced Scorecards and IT Evaluation Scorecards. With a focus on financial measures this is comparable to the differences between a cost center and a cost object perspective. Though elaborated reports usually exist for both, the maintenance of the cost object perspective requires more efforts, as the number of cost objects is usually higher than the number of cost cen-

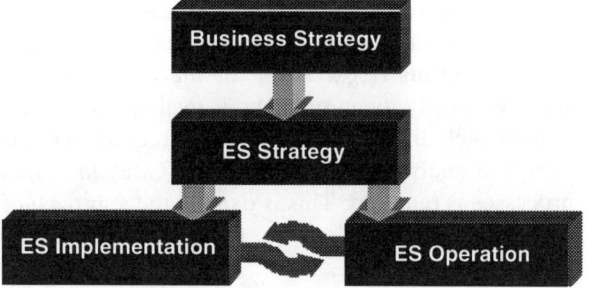

Figure 1: Business strategy and ES management

ters, their life span is shorter and the corresponding organisational responsibilities often harder to identify.

Consequently, in order to make an IT evaluation based on the Balanced Scorecard cost-effective, it is indispensable that only important, huge and complex IT investments with well defined project managers, support for entire processes and a well defined user community are analysed in this way. Enterprise systems are an ideal candidate from this perspective.

MEASURING THE PERFORMANCE OF IMPLEMENTING ES SOFTWARE
The Need for a Strategic Approach to Controlling the Implementation of ES Software

The enterprise systems initiative is for many companies an IT investment that radically redesigns the entire IT landscape. Nevertheless, it is amazing that the success of an ES implementation project is, at least by project managers, often reduced to two facts - the ES system is configured and running, and the entire project is (more or less) on time and within budget. This narrows the success of the ES investment down to very obvious selected indicators only. Besides the reduction to these types of indicators, the following observations (Densley 1999) stress the need for better ES evaluation:

- 78% of all ES projects are over budget;
- they took on average 2.5 times longer than intended; and
- they delivered only 30% of the promised benefit.

The lack of a more elaborated analysis of the success of ES can also be found in current publications. They focus on the discussion of procedure models for ES implementation (Kirchmer, 1998), critical success factors in ES projects (Sumner, 1998), issues related to the project organization (Mahrer, 1999), the explanation of individual projects (Clemons, 1998) or product-individual implementation methodologies (Brand 1999, Slooten and Yab, 1999) like ValueSAP.

Regarding the extent of an ES project, a comprehensive set of key performance indicators is necessary that does not only look backwards on financial key performance indicators. Consequently, during the process of implementing ES software, project managers, implementation partners and involved end-users should be informed whether

- the ES software is aligned to the organization and its business processes,
- the selected ES solution supports the needs of the system users (the internal customers),
- the development and the adaptation of the system to changing parameters are guaranteed.

In addition to this extension of a reporting system for the ES implementation with perspectives beyond the financial consequences, it is necessary to *integrate* these perspectives. An ES project - the adaptation of a standard software package to the individual needs - usually includes many compromises. In any case, an analysis, which only focuses on financial results (*e.g.,*. costs for the development of add-on solutions), would not be sufficient. Only the simultaneous evaluation of financial, customer and process-oriented criteria including the integration of requirements of future developments guarantees a complete analysis, especially in the cases in which a modification of the standard ES processes is required. This is exactly the approach of the Balanced Scorecard.

Adaptation of the Standard Balanced Scorecard to ES Software Implementation

The implementation of ES software is not a typical domain for the Balanced Scorecard approach because only one process – the implementation process – is evaluated. Nevertheless, it seems worthwhile to adapt the perspectives within the Balanced Scorecard for this purpose. Two reasons motivate the use of the Balance Scorecard for controlling and evaluating an ES implementation.

First, the Balanced Scorecard highlights the four perspectives discussed above. In addition to these classical perspectives (financial/cost, customer, internal processes, and innovation and learning), it is recommended to add at least for the evaluation of the implementation phase, but not for the ES usage (see section 3), a fifth perspective - the *project perspective*. Balanced Scorecards are typically designed to monitor business processes. Using the Balanced Scorecard for a project like the introduction of ES software is slightly different from this purpose as it focuses in most cases on only one process - the implementation. The individual project requirements (identification of the critical path, definition of milestones, evaluating the efficiency of the project organization) are covered by this perspective, which represents all the project management tasks. Such an evaluation of the project performance is not only of interest for an enterprise that is actually implementing the ES solution, but also for the involved external consulting partner and the ES provider. They can design benchmark cases and deliver reference project plans within the project perspective. In the next section selected examples for the "classical" four perspectives of the Balanced Scorecard as well as for the added project perspective will be discussed.

Second, one main objective of the Balanced Scorecard is the consistent transformation of visions into strategies, objectives and measures. The implementation of ES software can be interpreted as an example of this general objective. *Visions* describe the general motivation for the selection of ES software (*e.g.,*. use of only one integrated information system). *Strategies* like the selection of the relevant ES modules, the configuration of the companies' system organizational structure in the ES software, the design of the project plan, or the project organization represent the framework for the definition of objectives. The main *objective* of the implementation process is an economically efficient customization of the ES software that follows strategic goals and is conform to these objectives. As a consequence, the use of the Balanced Scorecard within an ES project leads to a documentation of the expected benefits (e.g. increased user satisfaction, decreased processing time, less process variants) of the ES system, which otherwise does not necessarily exist. The Balanced Scorecard is an addition to available ES-specific implementation tools (like SAP's ValueSAP including Accelerated SAP - ASAP, Brand 1999) of high importance as it serves as a guideline for the evaluation of the entire ES project. By contrast, current implementation tools mainly focus on the project perspective and the step-by-step completion of all relevant configuration tasks.

Definition of Selected Specific Measures Concerning the ES Implementation

The four main perspectives of a Balanced Scorecard, plus the project perspective, form a structured reporting system for the entire implementation process. Like all indicators discussed within the Balanced Scorecard, certain requirements exist concerning the measures (Hoffecker and Goldenberg 1994):

- It is important that all measures can be controlled. The personnel in charge should be

able to influence the indicators. A weakness concerning the ES software itself (e.g., insufficient support of certain customer needs even with a tricky customizing solution) is outside the scope of these measures.
- The key indicators should be easy to quantify. An economically efficient use of the Balanced Scorecard demands that data required to be collected is readily available. A good example for this is the business process performance measurement system developed by the IDS Scheer AG (Hagemeyer, 1999). It consolidates selected SAP transactions, designs a process model based on these data and derives automatically process indicators like the total processing time.
- The involved project members have to understand the measures. The indicators have to have a certain conceptual simplicity to guarantee that all project members (coming from different business departments, IT, quality management, external consultants) are able to interpret the measures correctly and in the same way.
- Finally, the measures must be relevant, reliable and as precise as possible.

Currently, only the *financial perspective* and the project perspective are (weakly) supported within the process of ES implementation (Figure 2). Every ES project has a detailed budget, which is divided into different cost elements like consulting, software or hardware costs. In regular project meetings a comparison of the actual project costs with the project budget takes place. However, more sophisticated measures like *total-cost-of-ownership* (TCO) are often not determined. The TCO would, among others, enable the identification of (sub-) modules where over-customizing took place - an intensive configuration of the ES software which does not represent the desired simpler solution. A planned TCO can limit the customizing activities and avoid the project being driven by the software (complexity) rather than by the business needs.

Though every ES project has an underlying project plan, the *project perspective* is often not integrated with other relevant activities. Thus, the project progress has to be documented in the configuration tool of the ES as well as in a separated project management tool. The project perspective includes typical project controlling measures, like processing time along the critical path or remaining time to the next milestone. It is also an example for the need to balance different measures. A comprehensive requirements analysis may lead to a delay concerning the first milestone (project perspective) and increase the budget for this activity (financial perspective), but in total it may reduce the project costs as it accelerates the selection and customizing tasks.

Though a business process oriented ES implementation (Kirchmer 1998) becomes more important, the implementation of ES software in most cases still follows a functional module-driven approach (e.g., materials management, financial accounting). Consequently, it is difficult to customize a single, isolated business process and to report from a process-oriented point of view. Within the *internal process perspective* it is possible to maintain measures, like the processing time before and after the ES implementation or the coverage of individual requirements for a process, by the selected ES software.

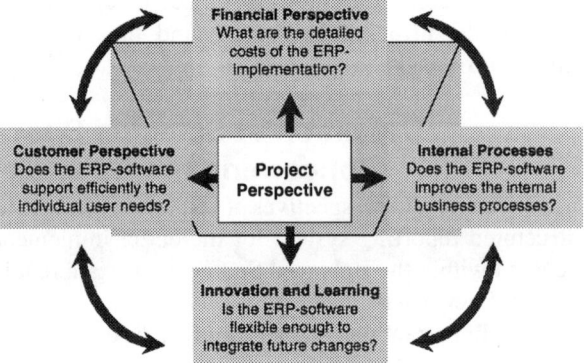

Figure 2: The ES implementation balanced scorecard

Evaluating the Management of Enterprise Systems

The other two perspectives: customer; and innovation and learning are almost neglected in every ES project. One of the most challenging tasks is the development of key performance indicators for the *customer perspective* on the ES implementation process. This perspective can be differentiated between internal and external customers. It should be possible to link certain tasks, like the selection of the relevant business processes and the definition of process related objectives to certain

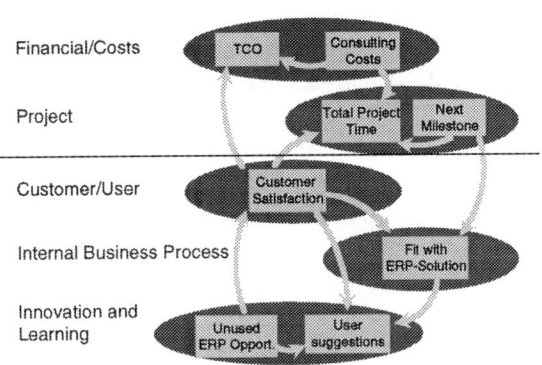

Figure 3: Selected cause-effect-relationships in the ES

customers. This avoids "over-customization" as well as, from a customer point of view, inappropriate resource allocation within the ES project.

For the *innovation and learning perspective* it is indispensable to anticipate the expected development of the enterprise and to evaluate the correlation between the customizing tasks and different scenarios. Thus, this perspective has to include alternative values for different conceivable development paths to support a flexible system implementation. Potential measures are the number of alternative process paths, the number of parameters representing unused customizing potential, and also the number of documents describing the customizing decisions (as an indicator for the quality of capturing knowledge).

Besides the design of adequate indicators for every perspective, the linkage of these indicators in cause-effect relationships is important and an innovative aspect of the Balanced Scorecard. Figure 3 includes some examples of these interrelations.

After successful implementation, the project perspective is relevant only in two cases. First, the experiences can serve as a kind of benchmark if further similar implementation projects take place (*e.g.,*. in a multinational company or for ES implementation partners). Second, minor implementation projects will be necessary with every new ES release. Both cases are of less importance for controlling the actual use of the ES software, which is discussed in the next section. Consequently, this section focuses only on the traditional four perspectives of the Balanced Scorecard.

MEASURING THE BUSINESS PERFORMANCE: CONTROLLING ES SOFTWARE

The Need for a Strategic Approach to Controlling the Use of ES Software

In evaluating the performance of IT, two categories of measures are traditionally used, the first being the financial, the second the technical view (Fabris, 1996). A financial evaluation is primarily concerned with deviations of the actual from the budgeted costs. Important cost categories include license and maintenance fees for software, leasing payments for hardware, network and communication costs as well as payroll costs. Financial evaluations have a number of flaws:

- Most of the costs "controlled" have been decided on in the past and cannot easily be revised today. Therefore, the impact of such information is rather limited.
- Another difficulty lies with the valuation of intangible goods connected with an ES

information system such as training, support, or self-developed software.

Evaluation of the technical ES performance focuses on criteria such as MIPS achieved. Those system properties are easily measured but bear a rather loose relationship to the underlying business and ES strategy. Faced with these drawbacks of both traditional perspectives of evaluation, the Balanced Scorecard extends the control of ES software use.

Adaptation of the Standard Balanced Scorecard to Evaluate the Use of Enterprise Systems

For the purpose of using the Balanced Scorecard to evaluate the use of ES software, it is necessary to adjust the four standard perspectives of the Balanced Scorecard to the specific object of an ES system.

The financial and customer perspectives are essentially the same in relation to the ES information system: the company figures both as shareholder and customer of the IS department in charge of running the ES system. Therefore, there needs to be a differentiation between the input side – looking at the system's cost or financial perspective – and its entire output - represented by the customer, internal process, and the innovation and learning perspectives. That said, the following questions relate to the four views (see Figure 4).

Definition of Specific Measures Concerning an ES Software

In defining specific measures for each of the four perspectives presented above, two basic approaches can be used, a top-down and a bottom-up approach. Each has specific merits, which points to using a combination of the two approaches.

The top-down approach works by deducing specific measures from those primary aims the ES strategy was devised to achieve. This is to ensure that those goals are being pursued and no resources wasted. The bottom-up approach, on the other side, aims at bottlenecks of the ES system, which can hinder an effective and efficient use of the ES system.

Financial Perspective

Financially, an ES system represents a capital investment that entails expenses as well as revenues. The latter, however, are not easily quantifiable in an objective way (Gattiker and Goodhue, 2000). This is true even in cases where transfer prices are being used for the IT department's services. In as much as the system is but one option among many competing investment possibilities, a financial follow-up is nevertheless required and can usefully take on the form of a gap analysis concentrating on the actual expenses versus those expenses budgeted. Information about cost evolution will in most cases serve as a lagging indicator for feedback purposes with a view towards evaluating the quality of past decision-making. In some cases, it may have feed-forward value; negative deviations of actual training costs versus budgeted costs may indicate that the

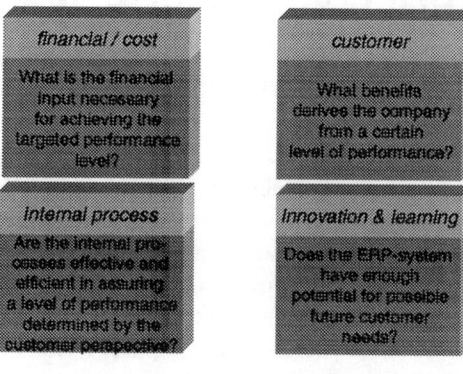

Figure 4: The ES operation balanced scorecard

system's functions are not efficiently used by staff members. By contrast, a continuous increase in external consulting expenses may point to deficiencies in the internal training staff's competence (Figure 5).

Figure 5: The financial perspective

Goal	Measure				
		Cost center	Maintenance		Training	
	Cost category		budget	actual	budget	actual
Compliance with budget	Hardware	absolute change				
	Software	absolute change				
	Consulting	absolute change				

Customer Perspective

With respect to an ES system, the customer(s) of the system have to be determined first. On one side, and directly dealing with the system, are employees using the system. More indirectly concerned are external business partners like suppliers, subcontractors, and customers in the strict sense of the word. For our purposes, concentrating on internal users seems more adequate, since the system's effects on external partners are rather remote and indirect. Following first the top-down approach in choosing measures for the customer perspective, an ES system's ability to functionally cover the entire range of the company's business processes figures preeminently among useful measures. There are two aspects of coverage to be differentiated:

- One important measure is the share of types of business processes covered by the system, as opposed to the complete range of process types. An example of this in respect to an ES system is the retailing sector with business process types like classical retailing, third party orders, settlement, promotion and customer service.
- Second, the share of total transaction volume handled by the system versus transactions performed outside of it needs to be considered.

Both measures are useful to collect whenever the respective number of transactions varies between the process types. Data needed for calculating them can be derived from system analysis. This information can be added by periodic user evaluations of the ES system's performance and utility to them. This source can also be used for compiling time series on both measures and thereby allowing the identification of trends.

Following the bottom-up approach, measures should be designed so as to allow easy identification of bottlenecks connected with the system. In the wholesaling area, useful criteria might be the proportion of business transactions not finished on schedule or the proportion of phone orders cancelled due to noncompetitive system response time (Figure 6).

Internal Process Perspective

The internal process perspective focuses on the internal conditions for satisfying the customer expectations as presented above. These conditions can be grouped into processes needed for operating the system on the one hand and those for improving and enhancing its capabilities on the other hand.

As to the day-to-day operation of an ES system, essential measures for evaluating its internal processes are the number and type of trends in user complaints. Analysis of these measures should lead to a ranking of system defects by disutility to users, and these are tackled accordingly.

Figure 6: The customer perspective

Goal	Measure
Coverage of business processes	% of covered process types
	% of covered business transactions
	% of covered transactions valued good or fair
Reduction of bottlenecks	% of transaction not finished on schedule
	% of cancelled telephone order processes due to non-competitive system response time

Further important criteria relate to system bottlenecks that inhibit the permanent and effective handling of business processes. Prominent among these criteria figures the technical system availability. The 24 hour/7 days/12 months availability is increasingly judged necessary for remaining competitive, as more and more companies have global activities spanning several time zones or offer e-commerce not restricted to rigid opening hours. Other such bottleneck measures include response time, transaction volume, and their respective evolution over time, which are early indicators of the need for capacity augmentation. Distinction should be made between OLTP and query transactions because of differences in capacity requirement (Figure 7).

Figure 7: The internal process perspective - operational view

Goal	Measure
Reduction of operational problems	# of problems with customer order processing % of problems with customer order processing # of problems with warehouse processes # of problems with standard reports # of problems with reports on demand
Availability of the ERP-system	average system availability average downtime maximum downtime
Avoidance of operational bottlenecks	average response time in order processing average response time in order processing in the peak time average # of OLTP-transactions maximum # of OLTP-transactions

The system development process aims at eliminating defects as well as improving the system's present capabilities and introducing new functions. A prerequisite for maintaining and enhancing an IT system is the use of the latest releases. Therefore, attention should be given to the time elapsed between the introduction of a new release into the market and the productive start at the company. Another indicator to be surveyed is the number of releases not (yet) introduced into the enterprise. In order to evaluate the effectiveness of the enhancement process, standardised indices can be employed concerning the actual time needed for development as compared to schedule, as well as an index to measure the quality of the developed software.

Bottlenecks in system development are mainly caused by employee shortage, since a heavy workload has adverse effects on development quality. A leading indicator of quantitative resource availability is, therefore, the average number of hours worked per employee, supplemented by the time spent on sick leave. The qualitative bottleneck aspects can be followed by analysing the extent to which vital ES system knowhow is spread among the IT staff (Figure 8).

Innovation and Learning Perspective

The innovation and learning perspective is dedicated to an examination of the company's ability to effectively make use of the ES system's functions as well as to enhance and improve it. Since that ability depends on the knowhow of personnel, employee-centred measures covering both users and IT staff are called for. A useful indicator is the level of training courses, measured by the amount of time or expenses spent. Specifically for system developers, their type of formal qualification can additionally be surveyed.

Another important measure in this field is dependence on external consultants. As a rule, a standard ES system is implemented with the help of external consultants. However, the company desires a quick

Figure 8: The internal process perspective - development view

Goal	Measure
Actuality of the system	average time to upgrade the system release levels behind the actual level
Improvement in system development	punctuality index of system delivery quality index
Avoidance of developer-bottlenecks	average workload per developer rate of sick leave per developer % of modules covered by more than 2 developers

transfer of system knowhow to its staff in order to reduce its need for highly paid consultants and to strengthen its ability to cope with problems flexibly on its own. To measure the success of such a transfer, the evolution over time of the number of consultant days spent within the firm should be followed.

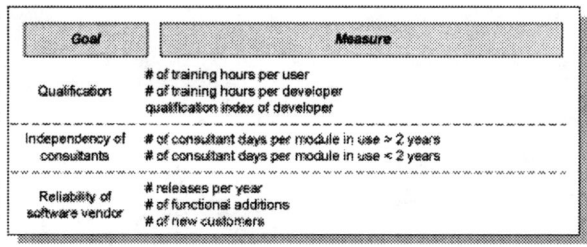

Figure 9: The innovation and learning perspective

An eminent success factor for successfully using an ES system is the possibility to fall back on the system provider for support and maintenance. The probability of long-term survival of the provider can be estimated by taking into account licenses sold, the number of releases, or the number of functional enhancements provided (Figure 9).

FUTURE TRENDS

Regarding the development of the Balanced Scorecard, the following trends can be observed:

- Besides the structure in the four perspectives, various approaches exist to add industry-specific content to the Balanced Scorecard. These Balanced Scorecards include specific reference key performance indicators and thus accelerate the implementation time. ES software targets often industries. Parts of these industry-specific Balanced Scorecards may be reused in Balanced Scorecards for the evaluation of ES industry solutions.
- These approaches are especially of interest for ES providers integrating functionality of the Balanced Scorecard into their solutions. The integration of Balanced Scorecards into enterprise systems facilitates the integration of transactional data. The ES serves in this case as the core repository for the Balanced Scorecard.
- In addition to the design of industry-specific Balanced Scorecards and their integration in comprehensive ES solutions, various approaches exist to further increase the transparency over the results.

An ERP-related example for an approach to visualize the results of multidimensional executive information systems is SAP's Management Cockpit, which was developed together with N.E.T. Research. As a part of the SAP controlling submodule Strategic Enterprise Management (SAP SEM), the Management Cockpit includes, similar to the classical Balanced Scorecard, four dimensions, so called walls.

- The black wall displays the primary success factors and financial indicators,
- the red wall shows the market performance,
- the blue wall presents the performance of the internal business processes and employees, and
- the white wall reveals the status of strategic projects.

More impressive than these four dimensions is the physical realization in form of the Management cockpit room (see Figure 10). Various screens and different graphics using colour coding represent the four critical dimensions. A "flight deck" enables the drill down to more detailed data within the data warehouse. This entire arrangement is similar to control panels found in airplanes or automobiles and facilitates the decision-making process between the managers. The idea of the management cockpit can be easily adapted for the proposed evaluation of enterprise systems. It would just be required that the object of the Balanced Scorecard (the entire enterprise, a certain department or the enterprise system in use) could be selected.

Figure 10: SAP's management cockpit

CONCLUSION

Just as vital as the selection of measures for each of the Balanced Scorecard perspectives itself is the planning of targets for every criterion. Here, general planning principles have to be applied, i.e., being demanding in order to motivate the personnel responsible while still being sufficiently realistic to ensure that the target can be achieved. Last, in order to have not only an effective ES, but also an efficient one, the gathering, processing, and formatting of information should be automated as much as possible. Traditionally the process of controlling an investment like an ES system ends with the delivery of the system to the users. Approaches that are related to the productive phase of the ES system focus mainly on the financial effects in terms of costs. Based on the Balanced Scorecard we also want to develop a set of interrelated measures that give a more complete insight into the performance of the ES system beyond financial key performance indicators.

In the *customer perspective* the effects on customers, both internal as well as external of the ES system, are measured. Based on those measures one can learn whether the system meets the expectations and goals that were intended with the implementation and the configuration of the system. This is a means to evaluate the fit of the ES system in relation to the underlying business strategy. Within the *internal process perspective,* the process of information processing in particular is measured. Relevant measures have to be derived on the one hand from the business strategy, like response-time, and on the other hand from typical bottlenecks, like downtime. Information concerning this perspective, as well as concerning the other perspectives, is mainly based on nonfinancial measures. With the *innovation and learning perspective* the aspect of intangible investments and continuous improvement is integrated into the reporting system. With regard to this perspective, we want to ensure that investment in system improvements - the configuration of innovative business processes like e-commerce, or training for users and developers - is not only regarded in terms of costs but also in terms of intangible investments that yield not only short-term economic profit but long-term economic health. Currently, we are working on an empirical study to identify the main key performance indicators for every perspective. Every indicator will be classified concerning the availability of the underlying data. The final objective is to design a reference Balanced Scorecard for the evaluation of enterprise systems. As far as possible the Balanced Scorecard functionality within ES software will be used to realize an IT-based solution.

REFERENCES

Beeckman, D. (1999). IT Balanced Scorecards: Filling the Missing Link between IT-Effectiveness and IT-Efficiency. *Proceedings of the Symposium on IT Balanced Scorecard*, Antwerp, Belgium, 15-16 March.

Brand, H. (1999). *SAP R/3 Implementation with ASAP: The Official SAP Guide*. Bk&Cd Rom edition.

Brogli, M. (1999). Using the Balanced Scorecard in the IT of a Financial Services Company. *Proceedings of the Symposium on IT Balanced Scorecard,* Antwerp, Belgium, 15-16 March.

Carlson, W., and McNurlin, B. (1992). Basic Principles For Measuring IT Value, *I/S Analyzer,* 10.

Clemens, Chr. (1998). Successful Implementation of an Enterprise System: A Case Study. *Proceedings of the 4th Americas Conference on Information Systems,* Baltimore, Maryland, USA, August 14-16.

Curran, Th., and Keller, G. (1998). *SAP R/3 Business Blueprint*. Prentice Hall PTR: Upper Saddle River, New Jersey.

Densley, B. (1999). The magnificent 7. Getting the Biggest Bang from the ERP Buck. *Proceedings of the First International Workshop on Enterprise Management and Resource Planning: Methods, Tools and Architectures.* J. Eder, N. Maiden, and M. Missikoff (Eds.). Venice, 25-27 November, 59-65.

Fabris, P. (1996). Measures of Distiction. *CIO*, November 15, 68-73.

GartnerGroup (1999). Enterprise Resource Planning Vendors: The Going Gets Tough. *Proceedings of the GartnerGroup Symposium/Itxpo 99*. 19-22 October, Brisbane, Australia.

Gattiker, Th. F., and Goodhue, D. L. (2000). Understanding the Plant Level Costs and Benefits of ERP: Will the Ugly Duckling Always Turn Into a Swan? *Proccedings of the 33rd Hawaii International Conference on System Sciences.* Maui, January.

Hagemeyer, J. (1999). *Process Performace Measurement. Concept and Implementation*. IDS Scheer AG, Saarbruecken.

Hoffecker, J., and Goldenberg, C. (1994). Using the Balanced Scorecard to Develop Companywide Performance Measures. *Journal of Cost Management*, 8, Fall, 5-17.

Kaplan, R. S., and Norton, D. P. (1993). Putting the Balanced Scorecard to work. *Harvard Business Review,* 71, September-October, 134-142.

Kaplan, R. S., and Norton, D. P. (1992). The Balanced Scorecard – Measures that Drive Business Perfomance. *Harvard Business Review*, 70, January-February, 71-79.

Kaplan, R. S., and Norton, D. P. (1996). *The Balanced Scorecard: Translating Strategy into Action*. Harvard Business School Press.

Kaplan, R. S. (1984). The evolution of management accounting. *The Accounting Review, Vol. LIX*, 3, 390-419.

Kirchmer, M. (1998). *Business Process oriented Implementation of Standard Software*. Springer-Verlag, Berlin, Heidelberg.

Mahrer, H. (1999). SAP R/3 Implementation at the ETH Zurich. A Higher Education Management Success Story? *Proceedings of the 5th Americas Conference on Information Systems*, Milwaukee, Wisconsin, USA, August 13-15.

Martinsons, M., Davison, R., and Dennis K.C. Tse (1999). The Balanced Scorecard: A Foundation for Strategic Management Information Systems, *Decision Support Systems,* Vol. 25, 1, 71-81.

Reo, D. A. (1999). The Balanced IT Scorecard for Software Intensive Organizations: Benefits and Lessons Learnt Through Industry Applications. *Proceedings of the Symposium on IT Balanced Scorecard*, Antwerp, Belgium, February.

Rosemann, M. (1999). ERP-Software - Consequences and Characteristics. *Proceedings of the 7th European Conference on Information Systems - ECIS '99.* Vol. III, Copenhaven, Denmark, June, 1038-1043.

Rosemann, M., and Wiese, J. (1999). Measuring the Performance of ERP Software - a Balanced Scorecard Approach. *Proceedings of the 10th Australasian Conference on Information Systems (ACIS).* B. Hope, and P. Yoong (Eds.), Wellington, 1-3 December, 773-784.

SAP America (1999). Management Cockpit. West Chester Pike.

Slooten, K. v., and Yap, L. (1999). Implementing ERP Information Systems using SAP. *Proceedings of the 5th Americas Conference on Information Systems,* Milwaukee, Wisconsin, August 13-15.

Sumner, M. (1998). Critical Success Factors in Enterprise Wide Information Management Systems Projects. *Proceedings of the 5th Americas Conference on Information Systems,* Milwaukee, Wisconsin, August 13-15.

van der Zee, J. T. M. (1999). Alignment is not Enough: Integrating Business & IT Management with the Balanced Scorecard. *Proceedings oft the Symposium on IT Balanced Scorecard,* Antwerp, Belgium, February.

van Es, R. (Ed.). (1998). *Dynamic Enterprise Innovation,* 3rd ed., Ede.

Vitale, M. R., Mavrinac, S. C. and Hauser, M. (1994). New Process/Financial Scorecard: A Strategic Performance Measurement System. *Planning Review,* July/August.

Walton, W. (1999). The IT Balanced Scorecard: Linking IT Performance to Business Value. *Proceedings of the Symposium on IT Balanced Scorecard,* Antwerp, Belgium, February.

Willyerd, K.A. (1997). Balancing Your Evaluation Act, *Training,* March, 52-58.

Wright, W. F., Smith, R., Jesser, R., and Stupeck, M. (1999). Information Technology, Process Reengineering and Performance Measurement: *A Balanced Scorecard Analysis of Compaq Computer Corporation, Communications of the Association for Information Systems,* 1.

Chapter XII

A Balanced Analytic Approach to Strategic Electronic Commerce Decisions: A Framework of the Evaluation Method

Mahesh S. Raisinghani
University of Dallas, USA

Choice, not chance, determines human destiny (*God's Lil' Instruction Book II*)

This chapter presents a comprehensive model for optimal electronic commerce strategy and extends the relatively novel Analytic Network Process (ANP) approach to solving quantitative and qualitative complex decisions in electronic commerce strategy. A systematic framework for the identification, classification and evaluation of electronic commerce (e-commerce) strategy using the Internet as an information, communication, distribution, or transaction channel that is interdependent with generic business strategies is proposed. The proposed methodology could help researchers and practitioners understand the relation between the benefits organizations seek from an information technology and the strategies they attempt to accomplish with the technology. As companies all over the world come out of restructuring, downsizing and business process re-engineering, many are realizing that in order to achieve a competitive edge they must formulate and implement strategies based on innovation and development of e-commerce. This chapter identifies and analyzes the methodology for synergistic integration of business and Internet domain strategies.

INTRODUCTION

A recent study by *Computer Economics Report* estimates that about three-fourths of information systems investments, ranging from data centers to websites, offer no calculable business value. At most companies, the money spent on servers, information systems (IS) salaries, and systems maintenance is not linked to specific projects and the deliverables are not always tied to the end user. Although there is little agreement on which of the new capital budgeting approaches works best, Chief Information Officers (CIOs) use some combination

of the following approaches to determine optimum spending levels:
- Activity-based costing: This most commonly used approach determines the cost of providing a service to a particular business unit, with the goal of calculating how much should be charged for the service.
- Performance-based budgeting: This approach relies on a rigid system of measures, where percentages of expenditure are tied to performance goals, in order to evaluate the effectiveness of technology.
- Benchmarking: IS budgets are based on what other companies are doing or to the return on investment (ROI) of the firm. However, companies recognize that they must have more tangible goals linked directly to the business.
- Economic value-added: This quantitative concept tries to measure benefit by assessing productivity gains. It is the cash-adjusted operating profit minus the cost of capital used to produce earnings. The approach tries to identify the contribution of a particular business initiative (e.g., increased sales from a new e-commerce application offered on the WWW) and compares it to capital costs. If there is a positive return on capital, then the initiative has succeeded.
- Options analysis: This approach is better for evaluating major technology initiatives such as migrating to an intranet- based on a client-server architecture. The real-options theory of evaluating and analyzing IT investments considers factors such as systems upgrade from Novell Netware to Microsoft Windows NT, or timing of a project launch. The idea is to evaluate corporate goals according to a variety of scenarios, since IT investments cannot always be traced to a tangible business benefit. The value of options is not considered by traditional accounting measures.
- Cumulative anecdotal evidence and "gut instinct", not just accounting measures.

According to Erik Brynjolfsson, an associate professor of Information Systems at MIT's Sloan School of Management, there is a need for new metrics that go beyond the traditional industrial-age measures that focus on cost analysis and savings, due to the difficulty of measuring the true economic benefits of IT and determining the accurate accounting of IT returns. For example, top management would like to know how to determine e-commerce technology's contribution to areas such as competitive differentiation and how to measure the advantage that their company has over the competition in its e-commerce applications. The primary question for the decision maker is; *In order to assess the business value of information technology for e-commerce, how can the qualitative and quantitative factors of a strategic decision be evaluated before investing in a particular e-commerce technology?*

The key focus of this chapter is to address the measurement of the linkage between business and information technology objectives for e-commerce applications using the analytic network process (ANP). This chapter discusses the measures of success for e-commerce and ways to improve on the business processes in the physical world by adopting e-commerce. The ANP is described as an integrative methodology for decision structuring and decision analysis of the various e-commerce technology/business model alternatives. Managers can benefit from this holistic approach to formulating the optimal e-commerce strategy based on the interdependencies between the Internet-level strategy domain and the business-level strategy domain.

BACKGROUND

Showing how investment in IT has been put to productive and profitable use is one of the biggest challenges faced by the IT manager. Research on IT investment refers to the phenomena of the IT Black Hole or the Productivity Paradox where large sums invested in IT seem to disappear with no return on investment and/or disproportionate returns to productivity. The "Balanced Scorecard" is an effective and increasingly popular way to set IT objectives and track success.

The Balanced Scorecard (BSC) methodology combines traditional financial measures such as return on investment (ROI) with supplemental measures such as gauging innovation and customer satisfaction. Objectives and measures are set from four perspectives:
- Financial perspective (traditional measures of profitability, revenue, and sales growth);
- Customer perspective (customer retention, customer satisfaction, and market research);
- Internal business processes perspective (processes instituted to meet or exceed customer expectation);
- Learning and innovation perspective (how the organization and its people grow and change to meet new challenges).

Renaissance Worldwide, an IT consulting firm that makes extensive use of the BSC methodology, suggests the following considerations for a good balanced scorecard:
- Cause and effect relationships: every one of your measures should be part of a chain of cause and effect relationships that represent the strategy.
- Performance drivers: A good BSC should have a mix of "lag indicators" (traditional measures like market share and customer retention) and "lead indicators" (drivers of performance that tend to relate to what is different about the strategy).
- Linked to financials: In a world of rapid change it is easy to become preoccupied with goals such as quality, customer satisfaction, and innovation. These goals must be translated into measures that are ultimately linked to financial indicators.

Since this is a comprehensive solution that requires participation from top to bottom, it begins with building a consensus on strategic vision, setting goals, and then deciding the best indicators for tracking progress toward those goals. The following checklist could be used to for deciding if you are ready to tackle a BSC.

1. Is there a strategic vision? A scorecard doesn't provide you with a vision. If there is no vision, there is no way to apply measures.
2. Does the scorecard have executive buy-in? Without buy-in on strategy and objectives you can't set measures that will have an impact.
3. Do your initiatives and measurements tie into your strategy? If initiatives and measurements don't link to strategy, you've expended resources on activities that won't contribute to your success.
4. Does your compensation program link to strategy? Measurement motivates, but if there are no rewards for improving measurements there is no motivation to pursue those activities.
5. Do you have a real purpose for implementing a scorecard?
6. Are you communicating the scorecards' importance at every level?
7. Are you revisiting measures to confirm their continued relevance?

The balance between strategy and technology is achieved when an organization not only uses technology to differentiate its products and/or services but also provides a high level of integrated service to the customers (Rumelt, 1974; Kalakota and Whinston, 1997; Lederer et al., 1997). The alignment between technology and the organizational processes requires a redefinition of the processes affected by electronic commerce. This could be achieved through the use of technology to redesign the process in order to reduce the cost and minimize errors associated with the process, while increasing the service level.

In the internal organizational functioning dimension, e-commerce has facilitated an information-based organizational model as a dramatic shift from the traditional, hierarchical command-and-control organization. This techno-organizational structure, illustrated in Figure 1, involves changes in work-group structures, information flows, and managerial responsibilities.

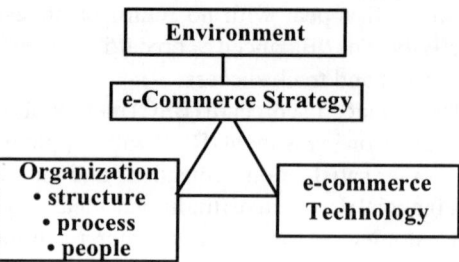

Figure 1: E-Commerce task-technology fit

The issues of alignment that are relevant to e-commerce systems are:
- between strategy and technology
- between technology and the organizational processes
- between technology and people

From an ontological perspective, diffusion of e-commerce strategy usually proceeds through a series of phases. E-commerce replaces the traditional inventory-based model with an information-based model and the basis of competition shifts from companies with strong distribution systems to those with strong information systems. A systematic framework for identification and classification of e-commerce strategy using the Internet as an information, communication, distribution, or transaction channel is illustrated in Figure 2 (Angehrn and Meyer, 1997), which correlates closely with the evolutionary diffusion of e-commerce innovations in organizations.

A virtual information space (VIS) presents new channels for economic agents to display and access company, product, and services-related information (e.g., marketing and advertising). A virtual communication space (VCS) presence includes strategies aimed

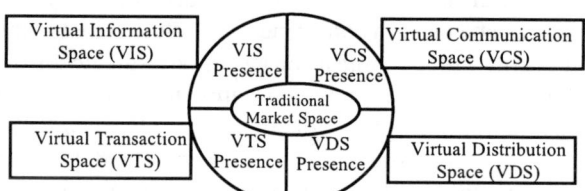

Figure 2: Four domains of Internet expansion

at monitoring and influencing business-related communications between economic agents operating on the Internet (e.g., negotiations between potential and existing customers, partners, government agencies, and competitors). A virtual distribution space (VDS) presence provides new channels for economic agents to distribute products and services (e.g., digital goods and content, software, and teleconsulting services). Lastly, a virtual transaction space (VTS) presence reflects strategies for economic agents to initiate and execute business-to-business or business-to-customer transactions such as orders and payments. These Internet domain strategies are now integrated with the business level strategy using the Analytic Network Process (ANP).

THE ANALYTIC NETWORK PROCESS MODEL

The ANP is a generalization of the Analytic Hierarchy Process (AHP), which is a decision methodology that uses a set of axioms to develop a hierarchy of attribute values, based on relative values obtained from pair-wise comparisons of attributes. AHP is "a theory of measurement concerned with deriving dominance priorities from paired comparisons of homogeneous elements with respect to a common criterion or attribute" (Saaty, 1980; 1994). AHP imitates the natural tendency of humans to organize decision criteria in a hierarchical form starting with general criteria and moving to more specific detailed criteria. In contrast

A Balanced Analytic Approach

ANP, a generalization of AHP, uses a feedback approach that replaces hierarchies with networks. In both AHP and ANP, judgments are brought together in an organized manner to derive priorities. However, the advantage of ANP over AHP for multi-criteria decision making is the allowance of interdependency between levels. The work on systems with feedback is extended to show how to study inner and outer dependence with feedback. Inner dependence is the interdependence within a component combined with feedback between components, and outer dependence is the dependence between components that allows for feedback circuits (Saaty and Takizawa, 1986).

The ANP/BSC methodology discussed in this section can be used by members of the top management team and any other stakeholders (i.e., strategic partners, key customers, and so forth) that the members of the top management team think should be involved in the decision making process. However, it is critical for the users of this methodology to be able to focus on the business reasons for the investments in e-commerce solutions. The following five steps illustrate the ANP/BSC methodology (Saaty, 1996; Meade, 1997):

Step 1: Model Construction and Problem Structuring—The relevant criteria and alternatives are structured in the form of a hierarchy, where the higher the level, the more strategic the decision. The topmost elements are decomposed into subcomponents and attributes. The conceptualization of the four phases (i.e., VIS, VCS, VDS, VTS) of e-commerce can be used by the managers as a communication tool to better understand and promote consensus regarding the appropriate role of e-commerce in their firms. In the context of e-commerce strategy, the only interdependence or feedback occurs between the four domains of Internet expansion and generic business strategies as illustrated in figure 3. This is the framework for the ANP that incorporates the business-level and the Internet-

Figure 3: Strategic link between e-commerce and business strategy

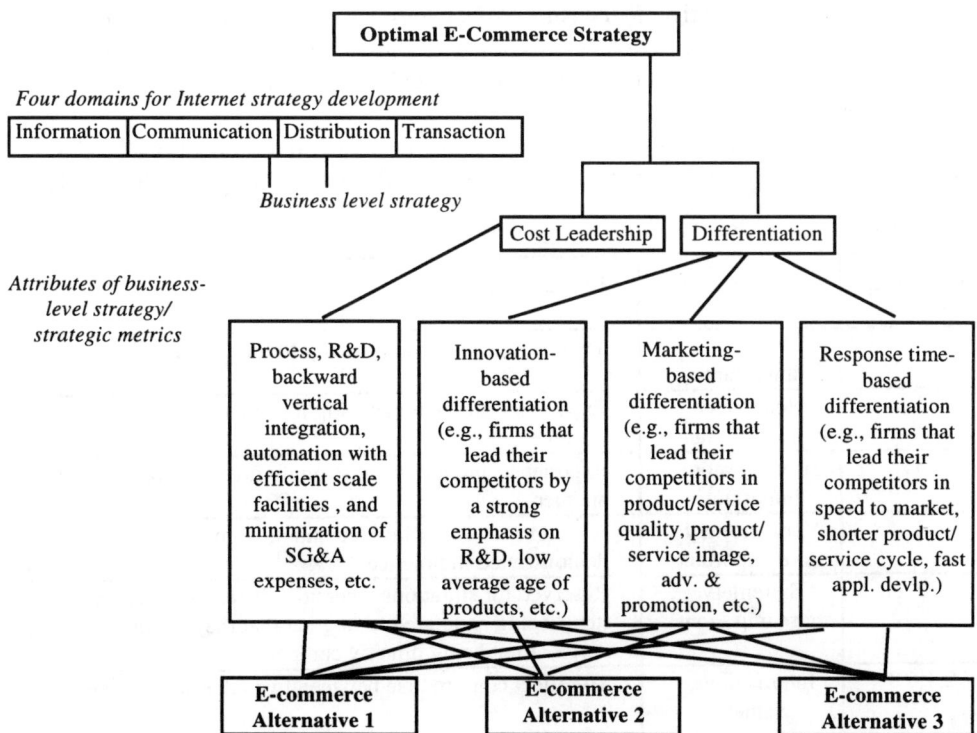

level strategies as an adaptation of the balanced scorecard methodology. It provides a measure of the consensus in an organization and facilitates the evaluation of viable alternatives. ANP/BSC provides a method to evaluate the e-commerce strategy alternatives with an interdependent link to the business strategy that incorporates the balanced scorecard perspective.

Step 2: Pairwise comparisons matrices of interdependent component levels—In the second step, a decision maker sequentially compares two components at a time with respect to an upper level control criterion. For example, the relative importance of cost leadership strategy and differentiation strategy is determined with respect to the optimal e-commerce strategy. A pairwise comparison matrix is required for each of the four domains of Internet expansion to calculate the impacts of each of the two generic business level strategies compared.

In many cases, AHP and ANP use a verbal rather than a linear quantitative scale for comparisons, which are then interpreted as a ratio scale. A nine-point ratio measurement scale developed by Saaty (1988b) is used to make the comparisons (see Table 1).

AHP/ANP assumes that the decision maker must make comparisons of importance between all possible pairs of attributes, using a verbal scale (from the most important to the least important) for each variant. The decision maker also makes similar comparisons for all pairs of subcriteria for each criterion. The information obtained in this process is used to calculate the scores for subcriteria, with respect to each criterion.

When AHP/ANP is used to make choices, the alternatives being considered are compared with respect to the subcriteria/criteria included in the lowest level of the hierarchy, and the global weights are determined for each of the alternatives within each subcriterion. The global weights summed over the subcriteria, are then used to determine the relative ranking of the alternatives. The alternative with the highest global weight sum is the most desirable alternative. In focusing on decisions, tactical as well as strategic, trade-offs among multiple, competing objectives, a basic approach based on preference/utility theory and including subjective probabilities, are utilized (Raisinghani and Schkade, 1997a; 1997b; Raisinghani, 1997b; Kim, 1997).

Just as in AHP, when scoring is conducted for a pair in ANP, a reciprocal value is automatically assigned to the reverse comparison within the matrix. In order to assign

Table 1. AHP/ANP Scale

Numerical value	Definition	Explanation
1	Equally important	Two factors contribute equally to the objective
3	Moderately more important	Experience and judgment slightly favor one factor over the other
5	Strongly more important	Experience and judgment strongly favor one factor over another
7	Very strongly more important	A factor is strongly favored and its dominance is demonstrated in practice
9	Extremely more important	Reserved for situations where the difference between the items being compared is so great that they are on the verge of not being directly comparable
2,4,6,8	Intermediate values	To reflect compromise between two adjacent judgments

meaningful values to these pairwise comparisons, use of additional strategic group decision making tools, such as the Delphi approach or scenario planning, are recommended.

Finally, the local priority vector w (defined as the eVector) is computed as the unique solution to:

$$Aw = \lambda_{max} w$$

where λ_{max} is the largest eigenvalue of A. Next, in order to aggregate over normalized columns, each element in a column is divided by the sum of the column elements and then, after summing the elements in each row of the resultant matrix, divided by the n elements in the row. Mathematically, the equation is as follows:

$$w_i = \frac{\sum_{i=1}^{I} \{a_{ij} / \sum_{j=1}^{J} a_{ij}\}}{J}$$

where:
w_i = the weighted priority for component i
J = index number of columns (components)
I = index number of rows (components)

Table 2 illustrates the business strategy pairwise comparison matrix within the Virtual Information Space (VIS). The weighted priorities for this matrix are shown in the column labeled eVector which stands for eigenvectors in Table 2.

A similar table is constructed for VCS, VDS, VTS and the weighted priorities for each of the four strategy relationship matrices are combined to create a matrix A shown in Table 3.

Next, the procedure detailed above is repeated and the eigenvector of local priority weights for the relative impact of the Internet expansion strategy, given a business strategy,

Table 2: Business Strategy Pairwise Comparison Matrix

Virtual Information Space (VIS)	Cost-Leadership	Differentiation	eVector
Cost Leadership	1	3	0.75
Differentiation	0.333	1	0.25

Table 3: The A Matrix Formed from Eigenvectors

A MATRIX	Virtual Information Space (VIS)	Virtual Communication Space (VCS)	Virtual Distribution Space (VDS)	Virtual Transaction Space (VTS)
Cost Leadership	0.75	0.23	0.45	0.37
Differentiation	0.25	0.77	0.55	0.63

are calculated. These are shown in a 2 X 4 Matrix B in Table 4. The formation of matrices A and B is required for each level of the hierarchy considered in relation to the parent/control node (i.e., innovation-based/ marketing-based/response-time based differentiation considered with respect to differentiation strategy as illustrated in Figure 1).

Step 3: Consistency Ratio Calculation—Saaty (1988b) argued that perfect consistency is difficult to attain in practice. The level of inconsistency in a set of AHP responses is called the *inconsistency ratio*. The inconsistency ratio (CI) is calculated as follows:

$$CI = (\lambda_{max} - N) / (N-1)$$

Table 4. The B Matrix formed from eigenvectors

B MATRIX	Cost Leadership	Differentiation
Virtual Information Space (VIS)	0.342	0.137
Virtual Communication Space (VCS)	0.143	0.422
Virtual Distribution Space (VDT)	0.245	0.376
Virtual Transaction Space (VTS)	0.270	0.065

where:

λ_{max} is the largest eigenvalue of an NxN pair-wise comparison matrix, and
N is the number of items being compared

The issues of transitivity and magnitude are encapsulated by this inconsistency ratio in the AHP/ANP. For example, if a decision maker prefers factor A over factor B by a multiple of three, and factor B is preferred to factor C by a multiple of four, then factor A should be preferred to factor C by a multiple of twelve.

Although Saaty (1988a; 1983) suggested that the maximum allowable inconsistency ratio should be in the neighborhood of 0.10., general agreement regarding the acceptable level of inconsistency in AHP survey research has not been reached by AHP scholars. On one hand, Harper (1988) argued that an inconsistency ratio of higher than 1.0 would be highly suspect and that models with an inconsistency ratio of less than 1.0 are used without modifications in many applications. On the other hand, recent accounting research, by Apostolou and Hassell (1993a; 1993b), has shown that average decision weights of AHP models calculated across subjects are not sensitive to the inclusion of subjects with inconsistency ratios up to and including 0.20.

The AHP/ANP can also be used to predict choices by rating absolute measurement of alternatives against standards established for each criterion, rather than a comparison of alternatives against each other (Saaty, 1988b). Standards are established by inserting a set of nodes, referred to as *intensities*, under each criterion or subcriterion. The use of standard intensities for the evaluation of alternatives also avoids the problem of rank reversal that sometimes occurs in AHP, when additional alternatives are added to the decision (Saaty, 1987). Priorities are established for the intensities, using paired comparisons of the intensities under each criterion or subcriterion.

Step 4: Supermatrix formation—The supermatrix comprises submatrices which are composed of a set of relationships between two levels in the graphical model. It allows for a resolution of the effects of interdependence that exist between the elements of the system by representing: a) independence from succeeding components; b) interdependence among components; and c) interdependence between levels of components. Figure 4 illustrates the network model and the matrix model of the supermatrix relationships. Since interdependencies of a level of components on itself (represented by matrices C and D in figure 4) are not considered to be significant, they are assigned a value of zero.

The 7 x 7 supermatrix M shown in table 5 is formed by combining the two compiled matrices A and B (i.e., Tables 3 and 4).

In order to reach convergence of the interdependent relationships between the two levels being compared it is necessary to raise the supermatrix to the power 2k+1 where k is an arbitrary large number. These "long-term" stable weighted values should be used for further analysis.

Figure 4: Network and matrix model of supermatrix relationships

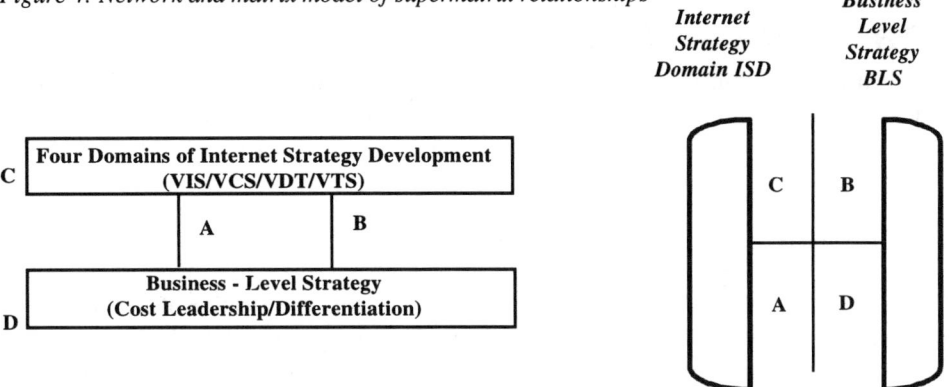

Table 5: Supermatrix M Compiled from Matrices A and B

SUPER MATRIX	VIS	VCS	VDS	VTS	Cost-Leadership	Differentiation
VIS	0	0	0	0	0.342	0.137
VCS	0	0	0	0	0.143	0.422
VDS	0	0	0	0	0.245	0.376
VTS	0	0	0	0	0.270	0.065
Cost Leadership	0.75	0.23	0.45	0.37	0	0
Differentiation	0.25	0.77	0.55	0.63	0	0

Step 5: Selection of the Best Alternative—The "desirability index" for an alternative i (D_i) determines the best alternative. It is computed as follows:

$$D_i = \sum_{j=1}^{J} \sum_{k=1}^{K_j} P_j A_{kj} S_{ikj}$$

where:

P_j is the relative importance weight of benefit j,
A_{kj} is the relative importance weight for attribute k of principle j,
S_{ikj} is the relative impact of alternative I on attribute k of principle j,
K_j is the index set of attributes for principle j, and
J is the index set of principles.

The decision maker/e-commerce strategist selects the alternative with the largest desirability index among the various alternatives for e-commerce such as buying carrier provided Virtual Private Network (VPN) (which is based on standard Internet Protocol (IP) but transports data over secure tunnels that are functionally similar to conventional phone lines) services that include varying degrees of management, build end-to-end VPNs using hardware and software from a host of vendors, leased lines, private data networks for Electronic Data Interchange (EDI), or e-commerce on the Internet/Intranet/Extranet with dependence and feedback to the desired level of implementation determined by the four domains of Internet strategy (i.e., VIS, VCS, VDS, and VTS) discussed previously. Organizations of all sizes and from all industries can weigh the pros and cons of linking with partners and remote sites while reducing telecommunication costs.

DISCUSSION AND RECOMMENDATIONS FOR MANAGEMENT

With tactical information technology investments, a straightforward return on investment (ROI) calculation is often the best way to evaluate and ultimately "green light" a project. These clear ROI decisions are one thing, but what can be done with possible investments that have clear business benefits yet are difficult or impossible to quantify?

Investments like e-business cannot be ignored simply because a clear ROI cannot be calculated and does not fit within an organization's model for allocating capital to projects. Consider these two types of e-business investments:

- Efficiency oriented: E-business initiatives can include the creation of efficiencies like reducing processing costs and improving the supply chain. For these projects that create cost reductions through an investment, an ROI calculation provides a relatively clear method of evaluation.
- Customer oriented: conversely, customer oriented projects that provide improved customer service (e.g. 24 hour support) or improve customer satisfaction (e.g. access to information) are extremely difficult to quantify, making any ROI calculation rather sketchy. This is not to say that these projects deliver less value; no one would question the value of repeat customers, but actually measuring the impact any specific initiative has on customer retention is a practical impossibility.

The bottom line is to perform an ROI calculation whenever possible. Although it is not advisable to shortchange the planning and goal setting portion of a project or the creation of a full business case, it is critical to recognize that some IT initiatives that improve customer relations are difficult to quantify with a meaningful ROI calculation. It should be made clear to management that initiatives designed to deliver efficiencies can be adequately measured with hard ROI calculations, while customer-focused initiatives must be decided using a softer approach.

If the decision maker cannot differentiate between these various types of projects and make a clear case for evaluation outside of the traditional capital allocation process, the decision maker may have trouble getting non efficiency-oriented projects approved, which could prove fatal in the long run (35/roi.htm" http://www.iweek.com/735/roi.htm, 6/1/99).

Since it is critical to align a company's IT department's goals and objectives with that company's business strategy, Craig Goldman, former CIO at Chase Manhattan Bank, lays out the following strategy on how one can come up with a "simple, logical, repeatable, and proven process" to help one align business and IT goals. It not only provides one with techniques to change attitudes and ensure buy in, but also allows "both IT and business executives to visualize the IT portfolio as a whole and see the degree to which IT is aligned to meet the corporate mission." (http://www.cio.com/archive/011599_expert_content.html, 1/26/99).

1. "Identify differences of opinion" – conduct surveys of the different business units and execs involved in the decision-making process. Have them identify and rank what each of them sees as the major factors (such as minimizing operating costs or improving service quality) that influence your company's ability to achieve its overall business objectives. Follow up by showing the different units and execs how their thinking and priorities differ from one another in order to promote mutual understanding.
2. "Prioritize business drivers" – once the major factors (drivers) have been identified, have the group compare them to your corporation's actual vision and assign them a rank. All drivers should be ranked according to their business value, not how they play

into an individual project. This will help in deciding later on which projects will be considered first for resource allocation. It's a difficult process, but critical if consensus is to be reached.
3. "Link vision, process and projects" – next, compare basic business processes and proposed IT projects one by one against the ranked drivers. If improving service quality is a high-ranking driver, then 24-hour customer support will be a high-ranking process. In turn, any IT project that supports achieving 24-hour customer support will also rank highly. Again, always ensure that decisions are being made with corporate vision and business success in mind.
4. "Incorporate resource constraints" – constraints such as cost and availability of human resources can keep any project from getting off the ground. Identify all constraints and apply them to the new ranked list of IT projects. This process will let you come up with a final "portfolio" of projects that maximizes available financial and human resources while optimizing each project's strategic value.

Finally, the ANP methodology can be used by the decision maker in conjunction with the balanced scorecard technique for a holistic, comprehensive and integrated approach to optimal decision. As Thomas L. Saaty aptly puts it, "Our lives are the sum of our decisions—whether in business or in personal spheres."

LIMITATIONS OF THE ANP/BSC METHODOLOGY

Neither revenues nor development costs are clear indicators of the business value of an application of e-commerce technology. Since most e-commerce applications are relatively new, reliable measures of return on investment, internal rate of return, net present value, and other financial measures are not available. Revenue generation estimates are subjective projections for applications that have either not yet earned a profit, or if they have, are subject to dramatic changes. Development costs of e-commerce applications are difficult to quantify, since these application costs are often hidden in other budgets. With costs, value and applications being so variable, comparing the relative merits of projects can become extremely difficult.

With respect to basic cultural differences in the strategic evaluation of e-commerce technology, internal validity of this methodology may be affected because there is no control over independent variables (i.e., the alternative/s evaluated) and the interaction between the subjects (i.e., top management team members). In addition, the strength and range of variables studied are limited due to the need for reasonably fast and easily understood communication in the interview and the time constraints faced by the members of the top management team members.

An additional threat to internal validity concerns the lack of cross references between written business and information technology plans in this methodology, due to the lack of written long term plans for e-commerce applications and the impracticality of obtaining written short term plans. Besides, asking people to comment on their written plans in a survey instrument/interview might overly challenge their ability to answer correctly, since many people interviewed may not have looked at the written plans in the last few months. Finally, the lack of respondent anonymity may cause the responses to be biased by the methodological artifact that respondents are known to the researcher.

CONCLUSION

In the final analysis, it is the notion of business value which ultimately justifies any innovation. This is implicit in the ANP/BSC methodology proposed in this chapter. The conceptualization of the four phases (i.e., VIS, VCS, VDS, VTS) of e-commerce can be used by the managers as a communication tool to better understand and promote consensus regarding the appropriate role of e-commerce in their organizations. Managers can use an e-commerce strategy for better overall business strategy formulation and implementation, thus leading to reduced planning problems and increased organizational impact. This internal focus contributes to more efficient business practices and leads to more successful websites. In contrast, the "build it and they will come" mindset, without careful consideration to service delivery and appropriateness of channel, characterizes the unsuccessful websites of organizations that do not incorporate capabilities and competencies into their electronic commerce strategy.

In order to take a strategic approach to using e-commerce, the entire organization needs to be re-engineered with e-commerce as a major business objective. This re-engineering includes developing a new business plan, a technology strategy and a total business process redesign. Efficiency and success of a website is measured by its ability to seamlessly integrate with back-office systems that handle order entry, confirmation and fulfillment, warehousing and inventory management, catalog management, financial accounting, reporting, and customer profiling. As Bryan Plug, president and CEO of Pandesic, the Intel-SAP joint venture, told the audience at Networld+Interop 1997 trade show, "Doing business on the Web is not as hard as doing business in a traditional way...it's harder."(www.pandesic.com). It is harder mainly because of the reasons a company gets into e-commerce and the extent to which its website is seamlessly integrated with its back-office support systems. The alignment between technology and the organizational processes using the ANP/balanced scorecard methodology requires a redefinition of the processes affected by electronic commerce. Overall a commitment to integrating e-commerce with the overall business strategy can provide unique competitive advantage to an organization.

REFERENCES

Angehrn, A. and Meyer, J. (1997). Developing mature Internet strategies—Insights from the banking sector. *Information Systems Management*, Summer, 37-43.

Apostolou, B. and Hassell, J. (1993a). An empirical examination of the sensitivity of the analytic hierarchy process to departures from recommended consistency ratios. *Mathematical and Computer Modeling*, 17: 163-170.

Apostolou, B. and Hassell, J. (1993b). An overview of the analytic hierarchy process and its use in accounting. *Journal of Accounting Literature*, 12: 1-27.

Bonafield, C. H. (1997). Cerf envisions what's in store for the Internet. *Network Computing*, August 15, 28-32.

Cross, K. (2000). The Ultimate Enablers: Business Partners, *Business 2.0*, February, 139-140.

Davydov, M. (1999). Who Knows, Who Cares? *Intelligent Enterprise*, 21 December, 60-61.

Harper, R. M., Jr. (1988). AHP judgment models of EDP auditors' evaluations of internal control for local area networks. *Journal of Information Systems*, Fall: 67-85.

Kalakota, R. and Whinston, A. (1997). *Electronic Commerce—A Manager's Guide*. Addison-Wesley Longman, Inc.

Kaplan, R.S. and Norton, D.P. (1992). The Balanced Scorecard - Measures That Drive Performance. Harvard Business Review, January-February, 71-79.

Kim, J. (1997). *A survey study on the relative importance of intranet functions: Using the analytic hierarchy process*, International Decision Sciences Institute Meeting, July, Sydney, Australia.

Lederer, A. L., Mirchandani, D. A., and Sims, K. (1997). The link between information strategy and e-commerce. *Journal of Organizational Computing and Electronic Commerce*, 7(1), 17-34.

Meade, L. (1997). *A methodology for the formulation of agile critical business processes*. Ph.D. dissertation, University of Texas at Arlington.

Pitkow, J. and Kehoe, C. (1996). *GVU's sixth WWW user survey*, Georgia Tech Research Corporation, December 11. http://www.cc.gatech.edu/gvu/user_surveys/survey-10-1996/

Raisinghani, M. S. and Schkade, L. L. (1997a). *Strategic evaluation of e-commerce technologies*. World Multiconference on Systemics, Cybernetics and Informatics, July, Caracas, Venezuela.

Raisinghani, M. S. and Schkade, L. L. (1997b). *An analytic tool for strategic evaluation of advanced networking technologies*. 1997 International Decision Sciences Institute Meeting, July, Sydney, Australia.

Raisinghani, M. S. (1997a). *Internet-Intranet empowered global organizations: Planning and strategic implications*. Association of Information Systems (AIS) Annual Meeting, Global Information Technology Track, August, Indianapolis, Indiana.

Raisinghani, M. S. (1997b). *Strategic evaluation of e-commerce technologies: An application of the analytic hierarchy process*. Association of Information Systems (AIS) Annual Meeting, Research-in-progress Minitrack, August, Indianapolis, Indiana.

Reich, B. H. and Benbasat, I. (1996). Measuring the linkage between business and information technology objectives. *MIS Quarterly*, March, 55-77.

Reich, B. H. (1993). *Investigating the linkage between business and information technology objectives: A multiple case study in the insurance industry*. Unpublished Ph.D. dissertation, University of British Columbia, Vancouver, B.C.

Rumelt, R. P. (1974). *Strategy, structure, and economic performance*. Cambridge, Massachusetts, Harvard Graduate School of Business Administration.

Saaty, T. L. (1980). *The analytical hierarchy process: planning, priority setting, resource allocation*. McGraw Hill, New York.

Saaty, T. L. (1987). Concepts, theory and techniques: Rank generation, preservation, and reversal in the Analytic Hierarchy Process. *Decision Sciences*, 18: 157-177.

Saaty, T. L. (1988a). *Decision Making for Leaders: The Analytical Hierarchy Process for Decision in a Complex World*. Pittsburgh: RSW Publications.

Saaty, T. L. (1998b). *Multicriteria Decision Making: The Analytical Hierarchy Process*. Pittsburgh: RWS Publications.

Saaty, T. L. (1990). An Exposition of the AHP in Reply to the Paper 'Remarks on the Analytic Hierarchy Process,' *Management Science*, Vol. 36, 259-268.

Saaty, T. L. (1994a). Highlights and Critical Points in the Theory and Application of the Analytic Hierarchy Process, *European Journal of Operational Research*, 74: 426-447.

Saaty, T. L. (1994b). How to Make a Decision: The Analytic Hierarchy Process, *Interfaces*, vol. 24, no.6, 19-43.

Saaty, T. L. (1995). *Decision Making for Leaders: The Analytic Hierarchy Process for Decisions in a Complex World*, Third Edition, RWS Publications, Pittsburgh, Pennsylvania.

Sawhney, M. and Davis, J. (2000). How IT Works, *Business 2.0*, February, 112-140.

Varney, S. E. and McCarthy, V. (1996). E-Commerce: Wired for profits. *Datamation*, October, Vol. 42, No. 16, 43-50.

Violino, B. (1997). ROI: The intangible benefits of technology are emerging as the most important of all. *Information Week*, June 30, 36-44.

Part V:

IT Evaluation through the Balanced Scorecard

Chapter XIII

Information Technology Governance through the Balanced Scorecard

Wim Van Grembergen
University of Antwerp (UFSIA) and University of Leuven (KUL)

Ronald Saull
Great-West Life, London Life, Investors Group

The balanced scorecard (BSC) initially developed by Kaplan and Norton, is a performance management system that enables businesses to drive strategies based on measurement and follow-up. In recent years, the BSC has been applied to information technology (IT). The IT BSC is becoming a popular tool with its concepts widely supported and dispersed by international consultant groups such as GartnerGroup, Renaissance Systems, Nolan Norton Institute, and others. Purcciafelli et al. (1999) predict that "by 2003, 60 percent of large enterprises and 30 percent of midsize enterprises will adopt a balanced set of metrics to guide business-oriented IT decisions (0.7 probability)." In this chapter, a generic IT BSC is proposed and its relationship with the business balanced scorecard (BU BSC) is established. It is shown how a cascade of balanced scorecards can support the IT governance process and its related business/IT alignment process. Further, the development and implementation of an IT BSC is discussed and an IT BSC Maturity Model is introduced. The chapter concludes with the findings of a real-life case.

INTRODUCTION

Kaplan and Norton (1992; 1993; 1996a; 1996b) have introduced the balanced scorecard at an enterprise level. Their fundamental premise is that the evaluation of a firm should not be restricted to a traditional financial evaluation but should be supplemented with measures concerning customer satisfaction, internal processes and the ability to innovate. Results achieved within these additional perspective areas should assure future financial results and drive the organization towards its strategic goals while keeping all four perspectives in balance. For each of the four perspectives they propose a three layered structure: 1. Mission (e.g., to become the customers' most preferred supplier); 2. Objectives (e.g., to provide the

Copyright © 2001, Idea Group Publishing.

customers with new products); 3. Measures (e.g., percentage of turnover generated by new products). The balanced scorecard can be applied to the IT function and its processes as Gold (1992; 1994) and Willcocks (1995) have conceptually described and has been further developed by Van Grembergen and Van Bruggen (1997), Van Grembergen and Timmerman (1998) and Van Grembergen (2000).

IT BALANCED SCORECARD

In Figure 1, a generic IT balanced scorecard is shown. The *User Orientation* perspective represents the user evaluation of IT. The *Operational Excellence* perspective represents the IT processes employed to develop and deliver the applications. The *Future Orientation* perspective represents the human and technology resources needed by IT to deliver its services over time. The *Business Contribution* perspective captures the business value created from the IT investments.

Each of these perspectives has to be translated into corresponding metrics and measures that assess the current situation. These assessments need to be repeated periodically and aligned with preestablished goals and benchmarks. Essential components of the IT BSC are the cause-and-effect relationships between measures. It enables the connections between the measures to be clarified in order to determine two key types of measures: outcome measures and performance drivers. A well developed IT scorecard contains a good mix of these two types of measures. Outcome measures such as programmers' productivity (e.g., number of function points per person per month) without performance drivers such as IT staff education (e.g., number of educational days per person per year) do not communicate how the outcomes are to be achieved. And performance drivers without outcome measures may lead to significant investment without a measurement indicating whether the chosen

Figure 1 Generic IT balanced scorecard

USER ORIENTATION	BUSINESS CONTRIBUTION
How do users view the IT department?	How does management view the IT department?
Mission	**Mission**
To be the preferred supplier of information systems.	To obtain a reasonable business contribution from IT investments.
Objectives	**Objectives**
• Preferred supplier of applications	• Control of IT expenses
• Preferred supplier of operations vs proposer of best solution, from whatever source	• Business value of IT projects
• Partnership with users	• Provision of new business capabilities
• User satisfaction	
OPERATIONAL EXCELLENCE	**FUTURE ORIENTATION**
How effective and efficient are the IT processes?	How well is IT positioned to meet future needs?
Mission	**Mission**
To deliver effective and efficient IT applications and services.	To develop opportunities to answer future challenges.
Objectives	**Objectives**
• Efficient and effective developments	• Training and education of IT staff
• Efficient and effective operations	• Expertise of IT staff
	• Research into emerging technologies
	• Age of application portfolio

Figure 2 Cause-and-effect relationships

IF IT employee's expertise is improved	(future orientation)
THEN this may result in a better quality of developed systems	(operational excellence)
THEN this may meet better user expectations	(user orientation)
THEN this may enhance the support of business processes	(business contribution)

strategy is effective. These cause-and-effect relationships have to be defined throughout the whole scorecard (Figure 2): more and better education of IT staff (future perspective) is an enabler (performance driver) for a better quality of developed systems (operational excellence perspective) that in turn is an enabler for an increased user satisfaction (user perspective) that eventually must lead to a higher business value of IT (business contribution perspective).

LINKING THE IT BSC WITH THE BU BSC

The proposed standard IT BSC links with business through the business contribution perspective. The relationship between IT and business can be more explicitly expressed through a cascade or waterfall of balanced scorecards (Van der Zee, 1999; Van Grembergen, 2000). In Figure 3, the relationship between IT scorecards and the business scorecard is illustrated. The IT Development BSC and the IT Operational BSC both are enablers of the IT Strategic BSC that in turn is the enabler of the Business BSC. This cascade of scorecards becomes a linked set of measures that will be instrumental in aligning IT and business strategy and will help to determine how business value is created through information technology.

BSC CASCADE AS A SUPPORTIVE MECHANISM FOR IT GOVERNANCE

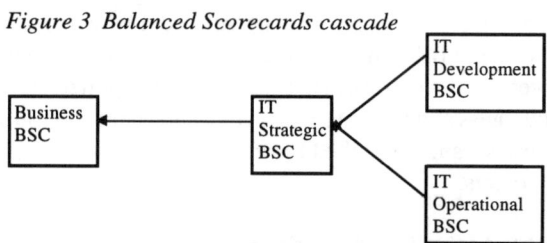

Figure 3 Balanced Scorecards cascade

The balanced scorecard approach can be instrumental in the IT governance process. A formal definition of IT governance is "the organizational capacity to control the formulation and implementation of IT strategy and guide to proper direction for the purpose of achieving competitive advantages for the corporation" (Ministry of International Trade and Industry, 1999). IT governance is part of the broader concept of corporate governance (enterprise governance). In their survey of corporate governance Shleifer and Vishny (1997) give the following definition: "corporate governance deals with the ways in which suppliers of finance assure themselves of getting return on their investment." They translate this definition into concrete questions: "How do the suppliers of finance get managers to return some of the profits to them?" "How do they (suppliers of finance) make sure that managers do not steal the capital they supply or invest in bad projects?" "How do suppliers of finance control managers"?

The same questions can be worded for IT governance (Van Grembergen, 2000):
1. How does top management get their CIO and their IT organization to return some business value to them?
2. How does top management make sure that their CIO and their IT organization do not steal the capital they supply or invest in bad projects?
3. How does senior management control their CIO and their IT organization?

These three crucial IT governance questions are respectively about effectiveness, efficiency and control of IT. The effectiveness question refers to the business/IT alignment process.

IT governance has to provide mechanisms for IT councils, business alignment, and implementation processes (Broadbent, 1998). The IT balanced scorecard with its linkage to the business balanced scorecard is such a mechanism. The cascade of scorecards can provide answers to the three fundamental IT governance questions and at the same time can support the business/IT alignment process. The business BSC identifies the strategies of the organization and within the IT balanced scorecards it is shown how these strategies will be enabled through information technology (question 1 and 2). In this way, these scorecards may uncover major problems: e.g. it may be possible that the Board of Directors of a retail bank decides to introduce web banking and that its IT organization is not at all acquainted with this technology as delineated by its IT balanced scorecard. IT governance also means that control mechanisms are to be provided to senior management (question 3). The IT scorecard provides the board with crucial measures on IT expenses, user satisfaction, efficiency of development and operations and expertise of the IT staff. This certainly avoids the potential of IT reporting to the board being restricted to technical matters such as the selection of a new telecommunications network. Instead, IT reporting can begin to detect inhibitors to new business strategies for decision and action by the board.

Figure 4 applies the concept of the cascade of scorecards to a generic retail bank. The cascade of the scorecards of the bank example fuses business and IT and in this way supports the IT governance process. The scorecards of Figure 4 illustrate that IT is fully involved in the new business processes of the bank. The business BSC shows a marketing strategy of reaching more and new customers through alternative distribution channels. The IT governance process and its related IT/business alignment process is shown in the IT Strategic BSC and the IT Development BSC: the website technology is chosen and a rapid website development approach is to be applied. The different balanced scorecards drive the business and IT strategies on measurement and follow-up. In this way, there is assurance (or no assurance) that the IT organization returns some business value and does not invest in bad projects.

CONSTRUCTION AND IMPLEMENTATION OF AN IT BSC

We identify five phases when building and implementing an IT BSC:
1. Presentation of the concept of the balanced scorecard technique to senior management and IT management.
2. Establishing a project team.
3. Data-gathering phase where information is collected regarding the corporate and IT strategy, the business/IT alignment processes, the IT governance processes, the (traditional) IT metrics already in use for performance measurement.
4. Development of the company-specific balanced scorecards inspired on a "standard-

Figure 4. Fictitious bank example of a cascade of scorecards

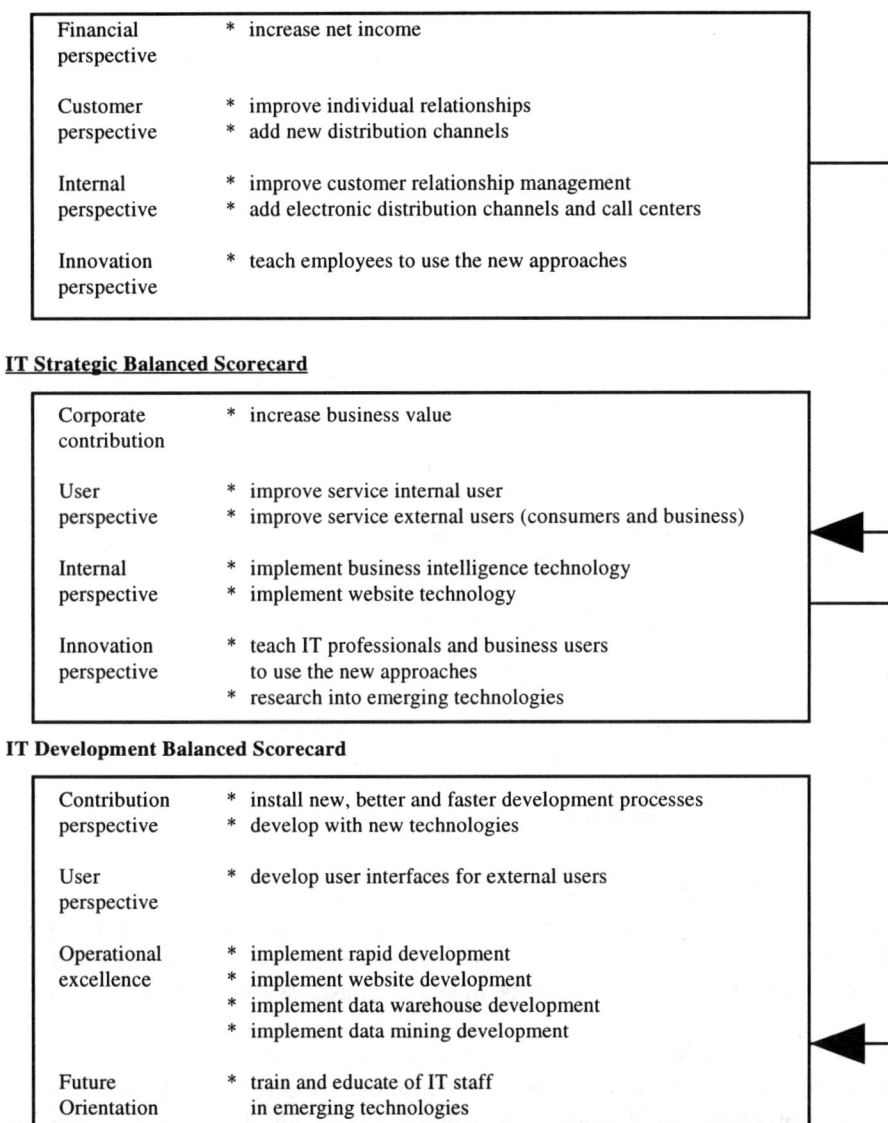

ized" model as presented in this chapter and based on the Kaplan and Norton (1996b) principles.

5. Implementation of the balanced scorecard mechanisms and improving it to an agreed upon maturity level (cf. infra).

Following Kaplan and Norton three principles have to be complied with in order to develop a scorecard that is more than a set of isolated and eventually conflicting strategies and measures. As we already explained, (1) cause-and-effect relationships have to be built-in and (2) sufficient performance drivers have to be included. Moreover, a balanced scorecard must (3) retain strong emphasis on financial outcomes: "A failure to convert improved operational performance into improved financial performance should send executives back to the drawing board to rethink the company's strategy or its implementation plans" (Kaplan and Norton, 1996b).

Further, the developers of a balanced scorecard must keep continuously in mind that measurements are not enough and that they must be used and acted upon by management. The balanced scorecard is not only a measurement system but is in essence a strategic management system. The steps to implement effectively a scorecard as a strategic management system are (Kaplan and Norton, 1996b):
- clarify and translate vision and strategy :
  - align IT and corporate strategy
  - clarify cause-and-effect relationships
  - build-in sufficient performance drivers in order to visualize how IT strategy will be achieved
  - build-in sufficient outcome measures in order to monitor whether the strategy is successful
  - link to financial objectives in order to visualize how IT strategy is improving the company's financial performance
- link strategy to resource allocation:
  - stretch long term targets
  - define long term targets that are realistic
  - take strategic initiatives to achieve the stretch targets
  - define short term milestones for the IT BSC
  - link priority settings for IT investment projects to the IT BSC
- link strategy to team and individual goals:
  - communicate the IT BSC to the employees
  - link individual objectives of the IT professionals to the IT BSC
  - link incentive system to the IT BSC measures
- organize strategic feedback
  - act upon the measurement results

MATURITY MODEL FOR THE IT BSC

Recently, many organizations started establishing an IT BSC and question if they are doing the right thing in an optimal way. Another crucial question is how far one should go with the scorecard and whether it should be linked to a business balanced scorecard from the start.

As we indicated in the introduction of this chapter, the concepts of an IT BSC at this point are relatively new for most business organizations. A question may be whether a developed IT BSC can be qualified as a good or best practices case. To answer this question we will, hereafter, develop a maturity model for the IT balanced scorecard which is an instrument for management to match their company's scorecard practice against this model.

The proposed Maturity Model (MM) for the IT balanced scorecard is based on Software Engineering Institute's Capability Maturity Model CMM (Paulk et al, 1993). Our IT BSC Maturity Model highlights five maturity levels with the following characteristics (Figure 5):

IT BSC IN PRACTICE

In Saull (2000) and Van Grembergen and Saull (2000) the development and implementation of an IT BSC within a Canadian financial group is described. The financial group under review is a merger of three companies: The Great-West Life Assurance Company, London Life, and Investors Group. Their IT divisions were merged in November 1997, and this was also the start of the construction and implementation of the IT BSC.

The CIO of the merged IT department begun focusing on the scorecard with the objective to ensure that the new IT organization was fairly evaluated. In his own words "through the balanced scorecard I would know what was important to the business, and I would not fall victim to the early termination syndrome. Or at least I would have a better chance

Figure 5. Maturity levels for the IT balanced scorecard

Level 1 **Initial**
There is evidence that the organization has recognized there is a need for *a measurement system* for its information technology division. There are ad hoc approaches to measure IT with respect to the two main IT processes, i.e. operations and systems development. This measurement process is often an individual effort in response to specific issues.

Level 2 **Repeatable**
Management is aware of the concept of *the IT balanced scorecard* and has communicated its intent to define appropriate measures. Measures are collected and presented to management in a scorecard. Linkages between outcome measures and performance drivers are generally defined but are not yet precise, documented or integrated into strategic and operational planning processes. Processes for scorecard training and review are informal and there is no compliance process in place.

Level 3 **Defined**
Management has standardized, documented and communicated the IT BSC through formal training. The scorecard process has been structured and linked to business planning cycle. The need for compliance has been communicated but compliance is inconsistent. Management understands and accepts the need to integrate the IT BSC within the alignment process of business and IT. Efforts are underway to change the alignment process accordingly.

Level 4 **Managed**
The IT BSC is fully integrated into the strategic and operational planning and review systems of the business and IT. Linkages between outcome measures and performance drivers are systematically reviewed and revised based upon the analysis of results. There is a full understanding of the issues at all levels of the organization which is supported by formal training. Long term stretch targets and priorities for IT investment projects are set and linked to the IT scorecard. A business scorecard and a cascade of IT scorecards are in place and are communicated to all employees. Individual objectives of IT employees are connected with the scorecards and incentive systems are linked to the IT BSC measures. The compliance process is well established and levels of compliance are high.

Level 5 **Optimized**
The IT BSC is fully aligned with the business strategic management framework and vision is frequently reviewed, updated and improved. Internal and external experts are engaged to ensure industry best practices are developed and adopted. The measurements and results are part of management reporting and are systematically acted upon by senior and IT management. Monitoring, self-assessment and communication are pervasive within the organization and there is optimal use of technology to support measurement, analysis, communication and training.

of survival." The stakes were raised considerably with the merger of the three IT groups because now the new IT division had exposures on multiple fronts with stakeholders who were concerned about the perceived loss of control over their vital IT services. This prompted an executive request for a formal measure of factors to measure IT's success. The response of the merged IT division was to formalize the criteria into an IT scorecard.

Figure 6. IT Balanced Scorecard of the Canadian financial group

USER ORIENTATION	BUSINESS ORIENTATION
How should IT appear to the internal customers (users and division managers)?	How should IT appear to the Executive Committees and Boards in order to be considered a significant contributor to company success?
Mission To be the supplier of choice for all information services, either directly or indirectly through supplier partnership	**Mission** To enable and contribute to the achievement of business strategies through the effective application of information technologies and methods
Objectives • Customer satisfaction • IT/business partnership • Application delivery performance • Service level performance	**Objectives** • Strategic contribution • Synergy achievement • Business value of IT projects • Management of IT investments
Typical corresponding measures • Score on annual customer survey • Index of business involvement in developing new applications • Delivered within time • Weighed % of application and operations services meeting service level targets	**Typical corresponding measures** • Completion of strategic initiatives • Completion of single systems solutions • Evaluation based on traditional financial measures (ROI) or Information Economics • Actual versus budgeted expenses
OPERATIONAL EXCELLENCE	FUTURE ORIENTATION
At which services and processes must IT excel to satisfy the stakeholders and customers?	How will IT develop the ability to change and improve in order to better achieve the IT and company's vision?
Mission To deliver timely and effectively IT services at targeted levels and costs	**Mission** To develop the internal capabilities to learn & innovate and to exploit future opportunities
Objectives • Process excellence • Responsiveness • Backlog management and aging • Security and safety	**Objectives** • Service capability improvement • Staff management effectiveness • Enterprise architecture evolution • Emerging technologies
Typical corresponding measures • Process maturity rating • Systems time to market • Number of days of effort of budgeted work in backlog status • Absence of major, unrecoverable failures or security breaches	**Typical corresponding measures** • Internal process improvement • % of staff with completed professional development plans • Systems adherence to Enterprise Architecture Plan and technology standards • % of IT budget allocated to research for new technologies

The current situation is that the IT scorecard effort is not yet approached as a formal project and as a result, progress has been somewhat limited. It is the intention of IT management to increase the formality of the project in 2000 as they progress toward their target state for the IT organization. To date, the sponsor of the IT BSC has been the CIO and its project leader has been the Director of Management Services who within the IT department is responsible for ensuring IT is managed as a business.

It was recognized that building an IT BSC was only meaningful under two conditions which required (1) clearly articulated business strategy, and (2) the new IT division moving from a commodity service provider to a strategic partner. The new constructed IT department is viewed as a strategic partner and its vision, strategy, measures of success and value are created jointly by IT and business. These issues go to the heart of the relationship between IT and the business and will be reflected in the IT balanced scorecard as is illustrated in Figure 6.

Within the Business Contribution, the main measurement challenges are with the areas of Strategic Contribution and the Business Value of IT Projects. In the strategic area, although the measure is focused on the successful completion of strategic initiatives, it is recognized that the perception of IT success or value added is highly dependent on the specifics of each initiative. It is now accepted that one must negotiate appropriate measures for each initiative with its corporate sponsor. For the business value of IT projects, the measures flow from the nature of the business case prepared for each project. Those focused on cost reduction use traditional financial measures such a ROI. Those based on service improvements will be measured on attainment of higher service level targets, and those based on enabling the achievement of corporate strategy will be based on factors similar to strategic initiatives i.e. negotiated measures which demonstrate achievement of intended benefits.

Within the User Orientation, the main measurement challenges are the softer relationship areas, i.e., customer satisfaction and IT/business partnership. In the customer satisfaction area, the IT BSC of the merged IT organization is relying on annual interviews with key business managers. It is the intent to obtain external assistance in developing and implementing a systematic survey process which will provide better insights into customer (user) perceptions of the IT services. In the IT/business partnership area, currently regular steering committee meetings between the IT account team and their business partner are organized. However, IT management would like to improve the understanding of the degree of mutual influence in key areas. In particular, the degree of influence IT staff have in the key investment decisions concerning the use of information technology in business and the degree of business involvement in developing new applications. It is believed these areas are critical in the development of the strategic relationship and ensure the success of both the business and IT in obtaining value for IT investments.

In relation to the Operational Excellence perspective (specifically for the process excellence objective), significant use will be made of external benchmarking to assist in developing and tuning the IT organizational structure and IT processes. Over the past several years the services of a consultant have been used to benchmark the efficiency and effectiveness in three service areas: data center, the client server environment and application delivery. It is the intent to continue to conduct such comparisons to top performing companies around the world, every two years or so. The services of another consulting firm are engaged to conduct an operations process maturity assessment as a step in planning to adopt the ITIL operational process models (see e.g. http://www.proactive-sv.com.au/core.htm). ITIL (Information Technology Infrastructure Library) contains a comprehensive description of the processes involved in managing IT infrastructures and is considered as a best practice and a standard of quality for IT service management. It was introduced in the late 1980s and has become the de facto IT standard in Europe. It is also planned to conduct an applications process maturity assessment using the SEI (Software Engineering Institute) Capability Maturity Model (see e.g. Paulk et al., 1993). It is the initial objective to reach Level 3 (Defined) in all process areas. Level 3 (highest level is 5) means that for the development activities: "The software process for both management and engineering activities is documented, standardized, and integrated into a standard software process for the organization. All projects use an approved, tailored version of the organization's standard software process for developing and maintaining software."

Within Future Orientation, staff management effectiveness is an important objective as there is a great concern for developing, retaining and deploying IT personnel because of the contemporary shortage of these professionals. Typical measures in this area are: the percentage of staff with completed professional development plans and percent of work done by contractors.

Although a lot of ground is covered, there is still a long way ahead to put in place an effective overall scorecard in the IT division under review. The status to-date of the IT BSC within the Canadian financial group is that the IT management team has discussed the IT BSC and accepted its value and necessity. Each individual manager has developed a measurement and reporting framework based on the IT BSC. However, these individual frameworks have not yet been consolidated into an overall reporting structure for executive management and confirmed with the major stakeholders.

According to the IT BSC Maturity Model (Figure 5) as presented in this chapter, the case company is at the Repeatable stage (Level 2). The challenge is to reach stage 4, the Managed level within two to three years. It is understood that major milestones in this further development will be:
- output measures and performance drives have to be systematically identified;
- the cause-and-effect relationships between these two measures have to be established;
- the IT scorecard has to be linked with the business scorecard;
- long term targets have to be defined;
- individual and group objectives of IT employees have to be linked to the IT BSC;
- scorecards have to support the IT governance process and the IT/business alignment process in a more direct way;
- the scorecards have to be integrated in the strategic and operational management short and processes.

Based on the Maturity Model of Figure 5, these actions for the two subsequent years should put the financial group close to Level 4. It is the belief of the CIO that to date these plans are realistic but "this desired time-line is probably quite optimistic and it may well take twice as long to accomplish these changes." However, the most important aspect is that all stakeholders in the process are engaged by he end of 2000 and that progress is made each subsequent year.

LESSONS LEARNED

The following lessons can be attributed to the IT BSC project of the Canadian financial group:

1. Start simultaneously constructing a business and IT scorecard

The IT BSC within the case company was started within the IT organization primarily with the objective to ensure that IT is fairly evaluated by the business. This is a rather defensive approach and focuses merely on the internal IT processes. Although it is clearly recognized within the case company that a more explicit linkage with the business (with a business balanced scorecard) has to be developed and supported, the question still remains whether it is more appropriate (a) to start with a business balanced scorecard followed by the subsequent creation of the corresponding IT scorecards or (b) to develop both scorecards simultaneously? It is now our conclusion that it is probably more ideal to start simultaneously with both scorecards which requires both IT and senior management to discuss the opportunities of information technologies which supports the IT/business alignment and IT governance process.

2. Consider the scorecard technique as a supportive mechanism for IT/business alignment and IT governance

Recurring issues in IT practice and IT academic publications focus on how to align IT and business and how to control IT. It is our strong belief that a cascade of business and IT balanced scorecards may support both processes. However, as is shown in this case study, the balanced scorecard is only a technique that can only be successful if the business and IT work together and act upon the measurements of the scorecards. The balanced scorecard approach will only have results when other mechanisms such as a well functioning Board and IT Steering Committee are in place.

3. Consider the construction and implementation of an IT balanced scorecard as an evolutionary project

Constructing an IT balanced scorecard is not a one week project. It requires considerable time and other major resources. Moreover, it is a project that is to be matured over time and that is characterized by different stages as is illustrated by the IT BSC Maturity Model introduced in this chapter. This iterative approach is confirmed by this case. The described IT BSC began at a lower level with actions currently in place to reach a higher level where a more explicit connection exists between outcome measures and performance drivers, and where an explicit linkage is established with business requirements.

4. Provide a formal project organization

Good project management is a critical success factor for effective construction and implementation of an IT balanced scorecard. IT management of the case company confronted with the question of how the IT BSC project was organized, had to admit that there was no real formal organization in place and that this delayed the progress of its implementation. Currently, the sponsor of the IT BSC is the CIO, and the project leader is the IT division. A formal project team to support the further development will be established with other key IT people along with key business representatives.

5. Provide best IT practices

Introducing an IT balanced scorecard in an IT environment with poor management and IT practices is too large a challenge. The implementation of the IT BSC within the case company was certainly supported by practices already in place such as ROI-evaluation of IT projects, the existence of IT steering committees, service level agreement practices, etc. If it is decided to implement the Information Economics approach to score and evaluate projects and to integrate this method within the IT BSC, this will take considerable time and is to be seen as a separate project.

CONCLUSION

In this chapter, the balanced scorecard technique has been applied to the IT division. A generic IT BSC is proposed with four perspectives: the User orientation perspective representing the user evaluation of IT; the Operational Excellence perspective representing the IT processes employed to develop and deliver the applications; the Future orientation perspective representing the human and technology resources needed by IT to deliver its services over time; the Business Contribution perspective capturing the business value created from the IT investments.

It is demonstrated that the IT BSC should be linked with the Business Balanced Scorecard and that this cascade of scorecards is a mechanism to support the IT governance process and the related IT/Business alignment process.

In a concluding section, the construction and implementation of an IT BSC within a leading Canadian financial group is described and discussed. It is shown that such a project is a time consuming activity and certainly an evolutionary project that has to be matured over time and that is characterized by different stages as is illustrated by the IT BSC Maturity Model introduced in this chapter.

REFERENCES

Broadbent, M. (1998). Leading governance, business and IT processes: the organizational fabric of business and IT partnership. *Findings GartnerGroup*, 31 December, document # FIND-19981231-01.

Gold, C. (1992). Total quality management in information services – IS measures: a balancing act. *Research Note Ernst & Young Center for Information Technology and Strategy,* Boston.

Gold, C. (1994). US measures – a balancing act. *Ernst & Young Center for Business Innovation,* Boston.

Kaplan, R. and Norton, D. (1992). The balanced scorecard – measures that drive performance. *Harvard Business Review,* January-February, 71-79.

Kaplan, R. and Norton, D. (1993). Putting the balanced scorecard to work. *Harvard Business Review,* September-October, 134-142.

Kaplan, R. and Norton, D. (1996a). Using the balanced scorecard as a strategic management system. *Harvard Business Review,* January-February, 75-85.

Kaplan, R. and Norton, D. *(1996b). The balanced scorecard: translating vision into action.* Harvard Business School Press, Boston.

Ministry of International Trade and Industry (1999). Corporate approaches to IT governance. March, http//www.jipdec.or.jp/chosa/MITIBE/sld001.htm .

Pucciarelli, J., Claps, C., Morello T. and Magee, F. (1999). IT management scenario: navigating uncertainty. *Strategic Analysis Report GartnerGroup*, 22 June, document # R-08-6153.

Shleifer, A. and Vishny, R. (1997). A survey of corporate governance. *The Journal of Finance*, June, 737-783.

Paulk, M., Curtis, B., Chrissis, M.-B., and Weber, C. (1993). Capability maturity model for software, version 1.1. *Technical Report Software Engineering Institute,* CMU/SEI-93-TR-024, ESC-TR-93-177, February.

Saull, R. (2000). The IT Balanced Scorecard – A roadmap to effective governance of a shared services IT organization. *Information Systems Control Journal* (previously *IS Audit and Control Journal*), Volume 2, 31-38.

Van der Zee, J. (1999). Alignment is not enough: integrating business and IT management with the balanced scorecard. *Proceedings of the 1st Conference on the IT Balanced Scorecard,* Antwerp, March, 1-21.

Van Grembergen, W. and Van Bruggen, R. (1997). Measuring and improving corporate information technology through the balanced scorecard technique. *Proceedings of the Fourth European Conference on the Evaluation of Information technology,* Deflt, October, 163-171.

Van Grembergen, W. and Timmerman, D. (1988). Monitoring the IT process through the balanced score card. *Proceedings of the 9th Information Resources Management (IRMA) International Conference,* Boston, May, 105-116.

Van Grembergen, W. (2000). The balanced scorecard and IT governance. *Information Systems Control Journal* (previously *IS Audit & Control Journal),* Volume 2, 40-43.

Van Grembergen, W. and Saull, R. (2000). *Aligning business and information technology through the balanced scorecard at a major Canadian financial group: its status measured with an IT BSC Maturity Model.* April (paper submitted for publication).

Venkatraman, N. (1999). *Valuing the IS contribution to the business,* Computer Sciences Corporation.

Willcocks, L. *(1995). Information management. The evaluation of information systems investments,* Chapman & Hall, London.

Chapter XIV

Using a Balanced Scorecard Framework to Leverage the Value Delivered by IS

Bram Meyerson
QuantiMetrics, South Africa

INTRODUCTION

Measurement as a Catalyst for Change

The pace of business and technological change continues to accelerate and the gap between business requirements and the capability of Information Systems (IS) groups to deliver, is getting wider. Very often both business and IS executives fail to understand the value of their IS investments and the factors that underpin IS performance. What is required to address these issues is a broad range of metrics to gauge both IS value and performance.

The fundamental principle underlying the approach described in this chapter is that measurement should be used as a catalyst for change. "*You cannot fully understand a subject until it is measured,*" is a cliche from a famous physicist. The subject matter under review is that of the overall effectiveness and efficiency of an IS group in meeting the business needs. The emphasis is more on the information systems that provide business functionality than on technology processes and technical infrastructure issues.

IS management is particularly challenging as it usually lacks mature measurement. This is paradoxical as the IS industry promotes the use of information systems to more effectively manage businesses.

Scope and Objectives

The objective of this chapter is to discuss a broad range of IS and business metrics, within the Balanced Scorecard (BSC) framework. This chapter does not specifically describe the transformation actions often associated with the results of assessments. The assessment outputs and related metrics do however form an integral part of a strategic framework for action. This chapter uses the term IS which is regarded as a subset of broader IT activity.

Understanding IS Value

Many organizations are questioning the value that they receive from their IS investments. Most organizations nowadays are not only reliant on IS but have realized that IS can give a business a significant competitive advantage.

Some organizations have chosen to manage their own internal IS departments while others have opted for various outsourcing options. These options range from total outsourcing of all IS activities to outsourcing IS infrastructure, outsourcing systems development, outsourcing the desktop environment, for example, to selective project outsourcing. Whether IS is run in-house or outsourced it nevertheless forms a significant portion of an organizations annual expenditure. Some institutions spend up to 20 percent of their total expenditure budget on IS. Often, e-commerce ventures spend far more on IS. Another popular metric is IS spend as a portion of assets. This can be as much as half a percent.

Few organizations understand how to measure the value of IS. It is not unusual for IS stakeholders and IS executives to have different perceptions on the concept of value. The value that IS delivers to the business will therefore be based on perceptions of the role that IS plays in the organization.

There are many reasons why the value of IS is questioned. In some cases the issue is high IS cost. In many cases however the question is whether IS is, in fact, supporting the business in the best way possible. Where IS is the lifeblood of an organization and forms the core of the business' development strategy, executive management must ensure that not only does IS deliver value but IS is leveraged in the best possible way to maximize business advantage.

It has often been said that value is in the eyes of the beholder. In this context it is difficult to provide a definitive definition of value. Value is a negotiated convention and the business, and the IS group should enter into dialogue and define their own appropriate value definitions and value propositions. To some organizations this may mean operational excellence where IS is used to drive down business expenditure, to improve efficiencies of various processes and to ensure that information is provided to management and to customers in a timeous manner and that this information is error free. To other organizations the IS value proposition may be more strategic. Many examples of this abound in the arena of E-commerce where IS replaces the intermediary between the business and the customer base.

Taking a process View of IT

To position IS within a broad IT framework, we describe the work that IT does in terms of three generic process areas. These are:
- the day-to-day delivery processes
- the software development processes
- the innovation or discovery processes.

The delivery domain contains all the infrastructural IT processes such as the management of the data centre, networks and support activities. It is in this domain that operational excellence is usually the value proposition of choice where the focus is on driving up transaction rates, improving processing efficiencies, maximizing network availability, and ensuring that the information resource is kept current and is error-free. Some have described the delivery domain as those IT activities that take place in the basement of the building.

The development domain is perceived to deliver more value to the business than the delivery domain. Delivery encompasses all systems delivery activities that are defined as those activities that begin with an idea or a concept and end with the implementation of a business solution. Projects could be the design and development of a new business system, the enhancement of an existing application to meet the ever-changing business needs, or the implementation of a package.

New systems delivery projects are usually initiated by the business. These projects then follow a systems development lifecycle that begins with the definition of the business requirements. Definition is usually performed by the IS group in conjunction with the business and is often managed as a separate project. It is during the requirements definition phase that the potential strategic benefits of the system will be described by the business and documented by IS in a manner that will allow the subsequent construction phases to be successfully executed.

Requirements definition and analysis is typically followed by a design phase where the technical solution to address the business needs is defined and documented. This is followed by a construction phase where the requirements are developed by the programming and construction team. This phase includes quality assurance or testing activities. A systems delivery project ends on implementation where the delivered product is complete, installed and handed over to the business.

The value derived by the systems delivery process is multidimensional. Firstly, the success of a project is dependent on an effective and efficient delivery process. Secondly, the success of the project is dependent on the quality of the requirements definition. Very often it has been said that the business "gets the IS that it deserves." If the business hasn't been engaged in defining specifically what their requirements are, the results of the systems delivery process will be less than desirable. Thirdly, the value derived by the systems delivery process is dependent on an element of innovation and out-of-the-box thinking by the IS group. IS may suggest ways of re-engineering the business so as to derive greater economic benefits.

The development domain relies on operational excellence. The systems delivery process should be executed efficiently, which means quickly, cost effectively and with the desired levels of quality. The value propositions of product leadership and customer intimacy also come into play in that the systems delivery process often delivers solutions that enable the business to become product leaders in some cases and to become more customer intimate in others.

The discovery domain is about innovation - finding new ways for the business to leverage value from IS. Many examples of this abound, the most common being in the areas of e-commerce, electronic banking and enterprise resource management. The actual process activities that take place in the discovery domain are usually informal, opportunistic and not always well managed. Organizations that seek to maximize the value derived from IS should allocate explicit budgets and resources to innovation and indeed manage the innovation process in a more structured manner.

The ratio of spend on the delivery, development and discovery domains often reveals the degree to which IS of strategic significance to the organization. Many large IS groups spend about 80 percent of the IS budget on day-to-day operational delivery activities. About 18 percent is spent on development and no more than 2% on discovery.

Reducing the delivery spend and reallocating this to development and discovery will add to the value that IS delivers to the business. Very often there is a backlog of applications to be developed, and the more of these that are successfully completed, the greater the benefit to the business.

IS management must realize the importance of the prioritization process as this backlog of application requests may not indicate inefficiencies in IS. It may be found that many of these application requests are indeed ones that will deliver little value to the business. The prioritization process should be one where the return on investment is seriously analyzed and priorities allocated to requests based on the economic benefit to the business, rather than to those originating from those business groups that shout the loudest.

Organizations should seek to understand the ratio of expenditure and resource allocation across the delivery, development and discovery domains and in so doing strive to allocate resources optimally.

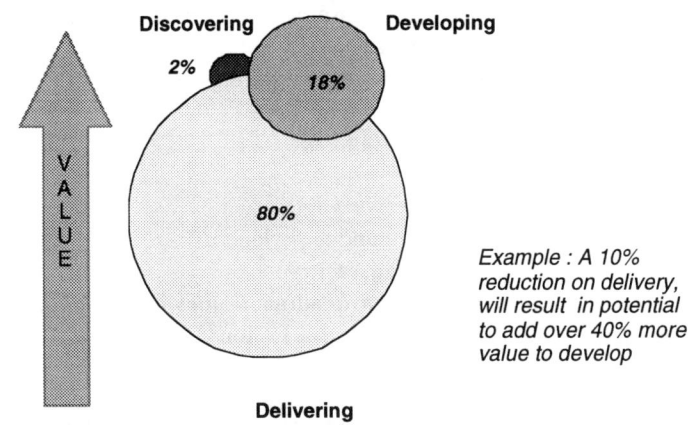

Figure 1

Example: A 10% reduction on delivery, will result in potential to add over 40% more value to develop

Figure 1 illustrates an example of an organization that is not realizing the potential value of IS because of a traditionally strong operational focus

IS Transformation

Balanced IS assessment goes beyond gauging and understanding the value that IS delivers to the business. Such assessments often form an integral part of an IS transformation process.

Many organizations have, from time to time, embarked on an IS transformation initiative. These projects seldom deliver the results anticipated by business executives. Rather, they often result in organizational restructuring, which alone cannot deliver breakthrough results. One of the fundamental weaknesses in typical transformation projects is that they lack the appropriate metrics. In the context of a major IS transformation program, metrics should be used to:

- Determine the truth about today, i.e., describe the current status quo.
- Benchmark the truth about today against best practice, where appropriate. Such benchmarks are more relevant for process measures and less so for perception based measures.
- Set future target performance.

A transformation or migration program should be based on the above metrics results and used together with a clear understanding the future business imperatives for IS. This will provide a case for action and the catalyst for change.

Balanced Scorecard IS assessment

This chapter describes a number of frameworks used in BSC IS assessments. These frameworks have been used to promote dialogue between IS executives and business management and have encouraged change. The techniques are drawn from the QuantiMetrics Value Improvement Program (QviP) range of services. Many of the assessment techniques described were originally researched and developed within Computer Sciences Corporation (CSC) Research Services division. This section focuses on some of the metrics deployed.

The scope of the QviP does not attempt to cover all aspects of the traditional BSC strategic framework for action (Kaplan and Norton, 1996). Many of these will follow from such an assessment. It does however assist in:

- Clarifying and translating Vision and Strategy, by articulating a vision and gaining consensus on vision and strategy;
- Strategic Feedback and Learning, by communicating a shared vision;
- Planning and Setting Targets, by setting targets;
- Communicating and Linking, by setting goals.

QviP addresses:
- IS expenditure and staffing,
- business alignment,
- process and service delivery,
- technology deployment and experience,
- partnership and user satisfaction.

Using the well-known four-quadrant framework (Kaplan and Norton, 1996), a BSC is developed for an IS group (typically in-house). Traditional BSC domains and the QviP domains are compared in Figure 2.

THE LANDSCAPE OF IS BSC MEASURES
The Business Domain : IS Cost Modelling and IS/business Alignment

This domain contains two major groups of measures. The first addresses hard financial aspects and the second, perceptions of alignment.

IS Cost Modeling

There is much debate around what an organization spends on IS and how this compares to others. What is probably more important is how the spend is allocated. The QviP IS Cost model addresses the following:
- What does the organization spend on IS?
- What does it spend it on?
- How does this compare with industry norms and best practice?

The input to this assessment is both business income and expenditure figures and analysis of IS cost allocations over an annual period. The IS expenditure is based on operational spend and as such excludes capital items, but does include a depreciation component where appropriate.

Figure 2. Domain Comparison

Business BSC *(Kaplan and Norton)*	IS BSC *(QMAA)*
Financial	**Business**
To succeed financially, how should we appear to our shareholders?	*How should IS deliver value to the business that it serves?*
Internal Business Process	**IS process**
To satisfy our shareholders and customers, what business processes must we excel at?	Which IS processes must IS excel at in order to deliver maximum value
Learning and Growth	**Learning and Growing**
To achieve our vision, how will we sustain our ability to change and improve?	What does IS need to do to continue to learn and improve its ability to change and help the business move forward?
Customer	**Users**
To achieve our vision, how should we appear to our customers?	What constitutes value in the eyes of IS users?

The key measures and indices used include:
- IS expenditure level related to business assets.
- IS expenditure level related to business gross income.
- IS expenditure level expressed as a proportion of business expense.
- IS costs per employee.
- IS staffing per employees.
- Ratios of expenditure across infrastructure, applications, research and management.
- Ratios across central IS spend and that in lines of business.
- Ratios of spend across customer, supplier, HR, performance, finance and strategy processes.
- Various IS staff distributions.

One of the most useful ratios is that of expenditure across research, strategic, differentiated, commodity and obsolescent initiatives.

These are described below:
- *Research* is speculative activity
- *Strategic* relates to developing or implementing products and services of strategic importance
- *Differentiated* is associated with competitively differentiating the business
- *Commodity* is associated with maintaining and supporting the business
- *Obsolescent* is associated with exiting from old business activities

In the absence of actual cost data, the use of full time equivalent (FTE) staff percentages can be used as a surrogate, to estimate relative spend in each category. Figure 3 below, shows an example where IS spend 'significance' leads business spend, yet strategic IT spend is on the low side versus best practice. This means that in this organization the IS group are forward thinking and preempting business needs. The best-practice comparisons are based on the experience of practitioners of the QviP in various companies.

The results of IS cost modelling can be used to monitor IS spending trends and to make decisions related to the allocation of resources. Examples include a simple analysis of the ratio of strategic versus operational expenditure as per the definitions described above. If IS is regarded, as it should be, of strategic significance and importance to the business, then one would expect that differentiated spend exceeds commodity spend. One would also expect to find at least 10% allocated to strategic and research projects. It is important to point out that strategic projects today will become differentiated initiatives tomorrow and that results

Figure 3

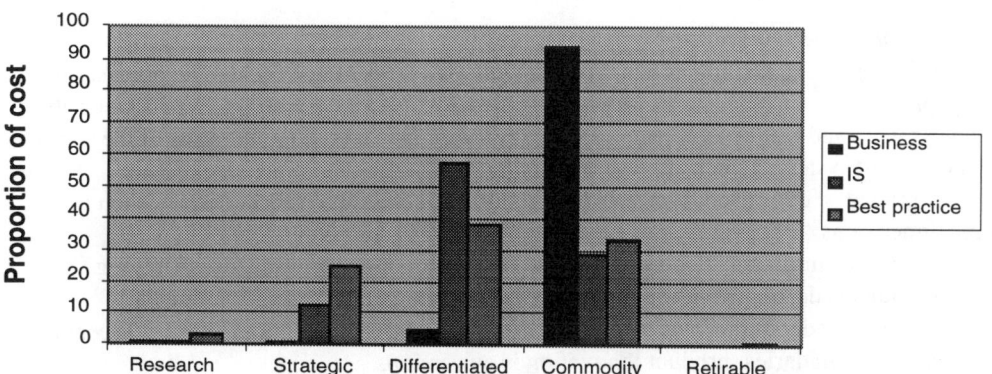

of successful research projects today will indeed become strategic projects tomorrow. If too little is spent on research and strategic initiatives then there will be a deficiency in strategic and differentiated initiatives in the future.

Management must strive to find ways of reducing commodity expenditure. This is accomplished typically through improved efficiency. Consideration should be given to outsourcing commodity activities and in so doing, focusing attention on differentiated strategic and research initiatives. Unfortunately, due to the legacy of old products and systems, little can be done to reduce obsolescence spent until these old products are eliminated or incorporated into newer ones.

Consideration should also be given to establishing benchmarking projects where the IS expenditure is benchmarked against other organizations in a similar manner. For example, insurance companies should seek to understand the ratio of their IS spend against the ratio of IS spend in other insurance organizations based on defined criteria such as the categories mentioned above. This is often challenging as this information is regarded by some as confidential. In such cases an organization should seek to benchmark with non-competing organizations, for example, in other countries or territories. Another idea is, for example, to benchmark against organizations in a different industry sector where IS processes are regarded as similar.

IS Business Alignment

QviP determines the extent to which IS reflects the needs and imperatives of the business it supports. Poor alignment results not only in wasted or incorrectly directed resources, it also leads to an inability to respond to new market demands with speed or agility.

In businesses where IS is of strategic importance, appropriate alignment between the IS service provider and the business leads to a capacity to differentiate the business within limited windows of opportunity. Poor alignment leads to follower status. In contrast, early movers are able to create new market opportunities and possibly "sweep the board" before a follower can respond.

This domain of assessment addresses current IS proficiency and determines how well aligned this is to both today's and tomorrow's business needs. The assessment employs the following alignment attributes:
- Value proposition.
- Goal alignment.
- Penetration.
- System Condition.

Value Proposition Alignment

This alignment attribute is based on the work of Tracey and Wiersma. It addresses the value proposition of business areas today and tomorrow and the alignment between the business view and that of IS. Current IS capability is also considered in terms of the focus of the current systems which support a business area.

A graded scale is used to measure perceptions of the value proposition in each of its three dimensions:
- Customer intimacy: Focuses on what specific customers want, on satisfying their unique needs, on providing the best solution and on cultivating relationships
- Product leadership: Focuses on offering the best products, those that push performance boundaries, product innovation is continuous.

Using a Balanced Scorecard Framework

- Operational excellence: Focuses on providing typically middle-of-market product but at the best price and least inconvenience—low price and hassle-free service.

Scoring rules are used in order to ensure one predominant point of view (a business cannot be all things to all people).

Various alignment profiles are analyzed, contrasting business and IS views both for today and tomorrow. Often problems of misalignment are highlighted when the focus shifts to that of future value proposition.

When evaluating the value proposition among a strategic business unit one often finds a difference in opinion of the management of that strategic business unit. This is the first indication of potential alignment problems as if the business unit cannot articulate a clear value proposition themselves, then how can IS hope to understand the value proposition of their business customer? A useful technique is to compare and contrast the perception of the business-defined value proposition and the value proposition as understood by IS. This is important as IS needs to intimately understand their business customers in order for them to be able to deliver desired systems and services to those customers. The value proposition framework has been found to be a very good way of articulating this understanding.

Another useful application for the value proposition framework is for business management to define the support of existing applications for the value propositions. For example, how does an existing insurance claim system support the value propositions? Such systems usually support the operational excellence proposition as the value derived by the business of such applications is enhanced if these applications promote cost reduction of these business transactions and improve cycle time to complete these transactions together with information accuracy.

Another example might be a centralized customer information system which ideally should major on the customer intimacy proposition where this system would allow cross-functional business domains access to up-to-date customer information and trends to enable them to get closer to the customers and what they desire, and in so doing, encourage better relationships with them.

Figure 4, shows the value propositions of a business unit within a large insurance company. The figure illustrates that the business intends to emphasize customer intimacy more than in the past.

Penetration Alignment

This attribute addresses the proportion of business activity, which is susceptible to IS. An attempt is then made to determine the breadth and depth (penetration) of business activities achieved by current information systems.

Figure 4. Value Propositions of a Business Unit.

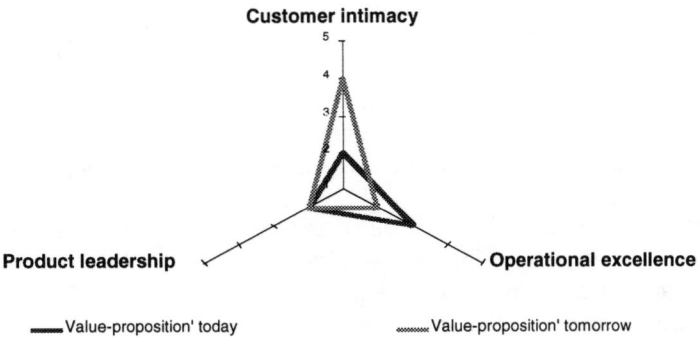

Both business and IS views are polled and compared. This technique essentially allows the business to express what investment they are prepared to make in IS.

Goal Alignment

This alignment metric requires the comparison of the business key performance indicators (KPIs) and the potential contribution of IS projects. Once again business and IS perceptions are contrasted.

System Condition

This alignment attribute determines the condition of business systems in two dimensions:
- Technology fitness (e.g., maintainability, quality of design, appropriateness of technology and systems performance).
- Business value. This dimension is typically analyzed four times. The first is based on broad application value perceptions of the business. Then perceptions of support for Operational Excellence, Customer Intimacy and Product leadership are analyzed.

The target domain is Renew or Refresh that means high value and high technology fitness. Problem areas are highlighted if systems fall into the other 3 domains of system condition: Retire (low value and low technology fitness), Redevelop (high value and low technology fitness) and Reassess (low value and high technology fitness).

It should be noted that some systems could be perceived as low value if they do not contribute to strategic imperatives. Examples would include infrastructural systems, budgeting systems, and human resource systems. Providing the technology fit is high, and the relative proportion of the systems portfolio is considered appropriate, such systems should be considered necessary building blocks or cornerstones of the applications portfolio.

Results are typically presented in graphical Boston Matrix form. These often highlight shifts in system condition perceptions as the business changes direction.

Understanding the systems condition of a portfolio of applications is one of the first steps to understanding the value that IS delivers to the business. This is because the legacy portfolio usually represents significant investment. Recently a major financial services institute appointed a new Chief Information Officer (CIO). This CIO inherited a backlog of systems delivery requests. He needed to define a systems provisioning strategy. Understanding the systems condition as described in the above refresh, redevelop, reassess and retire framework helped the CIO understand the extent of legacy applications that needed to be enhanced and, in some cases, replaced.

A project was undertaken to define the system condition of the legacy portfolio. This started off by documenting an inventory of systems and their major business functions. For each major business function key business users were identified, together with key application owners and IS technical support personnel. These three groups were surveyed using the QviP survey instruments to understand their perceptions of the systems and major business functions. The business personnel were asked to rate the extent to which the systems addressed the value propositions and were asked a number of short questions relating to their satisfaction with these systems. The IS personnel were asked to describe the technology fitness as described above, of these systems and the subsystems supporting the business functions.

The results were analyzed and presented using a series of Boston Matrices. The results painted the following picture. Firstly, as far as the business's level of satisfaction with existing systems was concerned, no major problems were identified. The major proportion of systems fell into the renew category with some falling into reassess and a few into redevelop

Using a Balanced Scorecard Framework 221

and none into the retire category. These results were pleasing to both business and IS management. Using this analysis alone, the CIO realized that the backlog of applications was perhaps unwarranted.

The analysis then continued to describe the application support for both current and future value propositions. Support for operational excellence was acceptable. Some of the previously categorized refresh, redevelop and reassess applications were downgraded and fell into the retire domain. Some previously categorized refresh applications were downgraded into the reassess domain. Nevertheless, the majority of applications remained in the refresh domain. This indicated that support for the operational excellence proposition was still satisfactory yet if this was to be the proposition of the future additional work, namely, enhancements to a number of applications would have to be done and indeed some applications would have to be replaced. This started to raise some alarm bells.

Our analysis then continued and focussed on the product leadership domain and what was found was that far more applications fell into the retire domain and far more also fell into the reassess" domain. This again indicated that if product leadership was the way forward, then far more work would have to be done in terms of future systems provision.

Finally, the analysis focussed on the customer intimacy proposition. This is where the assessment showed a major shortfall. Many applications were downgraded to reassess and additional ones now were categorized as retire. Clearly, much work has to be done in order for the business to be able to support the proposition of customer intimacy. This was a major concern as the business was beginning to embark on a major get closer to the customer programme. Having quantified the extent to which the portfolio of applications met current needs and would meet future needs, the CIO was equipped to begin defining the applications provisioning strategy.

Figures 5, 6 and 7 illustrate how the assessment of the portfolio of applications showed a worsening trend as the company moved towards a more focussed market value proposition.

The IS Process Domain : Process Measures and Benchmarks

There are numerous ways of defining IS processes. The following list, developed by CSC Research Services, suggests a number of key processes. The two highest value-adding processes are:

Figure 5: Most systems are rated as having both high value and high technical fitness–Average rating is weak-refresh—Some opportunities for redevelop and one for retire are evident.

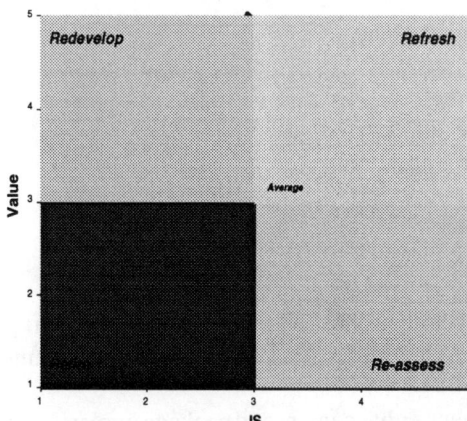

Figure 6: If business is to focus in on Operational Excellence, then only half the systems should be refreshed—More opportunities now exist to redevelop, reassess and even retire.

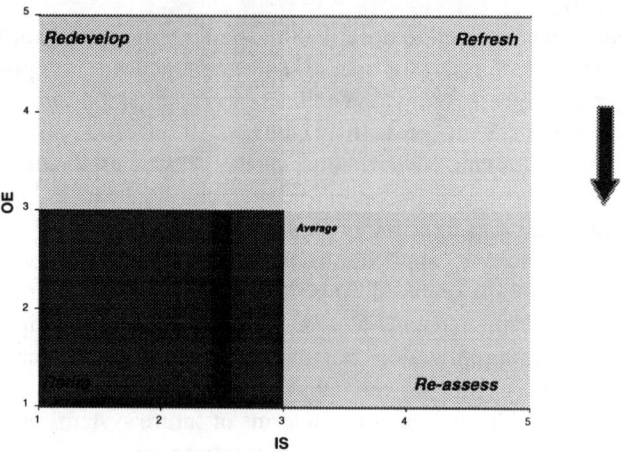

Figure 7: If business is to focus in on Product Leadership, suggestions are that most systems should be retired or reassessed

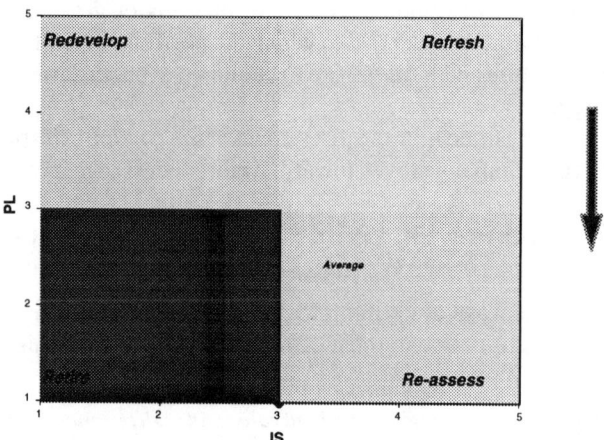

- Design and develop business solutions.
- Support users.

Other IS processes include:
- Manage new and emerging technology.
- Identify and prioritize business solutions.
- Manage business and technical models.
- Manage technical infrastructure.
- Manage human resources.
- Manage IS performance.

The high-value design and develop business solutions and user support processes are assessed and benchmarked via the QuantiMetrics Performance Enhancement Program (QPeP).

QPeP measures new development, enhancement and package implementation projects.

The assessment approach is independent of development method (project process), technology and platform. Application support baselining measures the efficiency of ongoing application support. This activity typically consumes the majority of IS resources, which could otherwise be deployed in higher value adding roles.

Hard measures and benchmarks allow for the establishment of objective performance baselines and for the setting of best practice targets. By assessing an IS organization's process efficiency, opportunities for improvement are objectively identified. These typically involve the project process itself, the development technology used and the people working on the project. When addressed, these can bring about significant process improvements, which include:
- Reduced cycle times.
- Increased productivity.
- Reduced project costs.
- Improved quality (freedom from error).
- Greater conformance to plans.
- Reduced support and maintenance costs.
- Lower residual defect density.

When improved, these result in higher perceptions of IS value by the business.

Assessing project and application support processes should include elements of objective benchmarking. As this domain focuses on hard measures, benchmarking is indeed meaningful. QPeP suggests three levels of benchmarking. These are:
- Internal project vs. project and application vs. application comparisons.
- Internal project and application comparisons over time.
- External project, application and company comparisons against representative records in broad industry databases.

Statistical analysis of the QPeP benchmarks reveals trade-offs between:
- Speed and quality.
- Productivity and size.
- Time pressure (business imposed) and productivity.
- Time pressure (business imposed) and quality.

QPeP assessments very often highlight the areas on which the IS group needs to focus so that they can boost their systems delivery performance. A recent assessment showed a 400 member systems development group that their speed of delivery was on par with that of their competitors but that their technical quality lagged behind. Technical quality is defined as the absence of errors or bugs in operation. The challenge for this group was to find ways of enhancing their technical quality without compromising their speed of delivery or productivity. Because of the nature of their competitive market place, speed of delivery is a major business focus. The assessment clearly indicated that speed of delivery was being pushed at the expense of quality. While this might be acceptable in the short-term, the costs of correcting the defects identified in live production are beginning to mount, and resources that are currently concentrating on delivering new systems are being deployed on older deliveries to rectify problems. This staff turnover or churn is not only disruptive but will begin to negatively impact the productivity of existing projects.

Another recent assessment highlighted the inconsistency within a particular organization systems delivery process. Using statistically proven normalization techniques the assessment showed over a five-fold productivity difference from the worst performing projects within this group to the best performing projects in this group. When the results were presented to the client, no good reasons could be understood as to why the poorer performing projects performed worse than the better performing projects. The only conclusion was that the process that was followed was ill-defined and dependant on the calibre of resources allocated to the projects. This inconsistent process was deemed to be unacceptable to the business as the business had no assurance at the beginning of a project whether that project would be a success or a failure. This organization clearly has to focus on tools, techniques and methods to bring consistency to the process.

An example similar to the one described above is one of an IS department within a large bank where the assessment exposed the fact that the level of project slippage was in excess of 50%. Project slippage is the deviation of the planned project schedule to the actual project time.

One of the critical success factors of projects delivery is that it is delivered on time and within budget. This organization then embarked on a project to improve their estimating capabilities. The approach that is now followed is that all project requirements are sized using the function point analysis (FPA) method. The function point count describes the amount of functionality requested by the business. It is a sizing technique that can be compared to techniques used by quantity surveyors in the construction business. Once these requirements have been objectively sized they can then be converted into resource requirements based on the statistics derived from previously completed projects. The benchmarks in the PEP database are also used as a means of converting the function point counts into estimates of project effort and duration. This organization has now been able to contain its slippage to within the 10% level for a period of a number of years since this initial problem was identified. This process also improved the relationship between the IS group and the business that it serves, because as the business was now more comfortable with the predictability of the projects that they requested. It also helped in defining requirements as these had to be clear in order for the function point counts to be performed. As in most engineering disciplines, the organization enforced the principle of revised estimating and quotations if the requirements in any way were modified during the course of the project.

Successful systems delivery is reliant on three elements, namely, people, process and technology. The organization has to ensure that it has adequately skilled and motivated resources. The capability assessment as described below will address this further. The processes followed by a project has been found to be one of the major influences as to the success of the project. A recent assessment showed that a major financial services group was one of the top performing groups in the entire PEP benchmark repository. The technology deployed on their projects is in no ways superior or "more leading edge" than their competitors. The people employed by this organization are skilled but are not regarded as industry leaders. It is the processes that this organization follows that proved to be the influencing factor resulting in their high performance. This organization uses strict project management disciplines and a formal release management approach. It appears as though these techniques have helped in optimizing their systems delivery performance.

Statistical analysis of the international PEP database reveals that PEP clients reduce their project effort by about 12%, reduce their duration by about 10% and reduce their error rates year on year.

Figure 8, describes a typical PEP performance profile. The broader the overall profile the better the overall performance. Where the profile exceeds the inner (PEP median), performance is good. Where the profile dips inside the inner circle, performance is poor. Each spoke of the wheel or radius describes a different important PEP measure. These include efficiency, speed, quality, costs and conformance to plan. The illustration shows a clear trade off between

Figure 8. Overall, higher than baseline efficiency, due mainly to short duration—Expensive staffing style, dampens producitivty—Testing ineffective or underdone–Deadlines met at the expense of budget overrun—unit costs are lower than baseline.

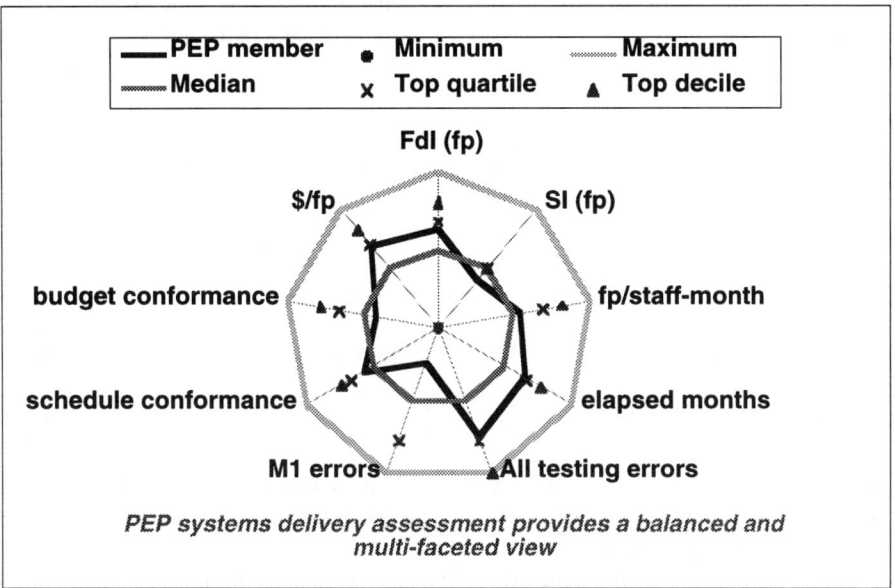

Source : QuantiMetrics Performance Enhancement Programme

efficiency and quality. Quality has been compromised by inadequate testing. Another trade-off is evident, where staffing style is expensive and resulted in budget overruns.

The Learning and Growing Domain : IS Capabilities

The QviP assessment in this domain is modeled on the CSC PEP Research Report, entitled Building the Capable Enterprise. It contrasts business capability today and that required tomorrow, with IS capability today. Business capability is assessed in terms of six domains. These are:
- Agency – organizational empowerment and learning ability, achieved through self-directed teams, economic value delivery teams etc.
- Agility – the ability and ease of product, process and structural change, achieved through selective outsourcing, resource pooling, process modeling/simulation etc.
- Speed – the speed of operation of business processes (including time to market) achieved through reusable components, assembly, fast cycle methods etc.
- Completeness – the ability to provide any product and process, any time, anywhere through distributed systems and data, cross boundary systems, workflow etc.
- Connectivity – the ability to directly connect with anyone (customers/suppliers/remote staff) through the deployment of technologies such as voice mail, electronic conferencing, mobile and wireless connection.
- Intelligence – the ability to master information and extract timely insight and intelligence through technologies such as data warehouse/marts, information agents. Rating rules are used to ensure that a clear capability focus emerges.

IS capability is assessed from an analysis of numerous tools, techniques and methods ranging from selective outsourcing, resourcing policy, centre of excellence, RAD, time boxing, CASE tool, client/server development etc. The level of capability for each is determined as research, pilot, fully implemented and no action.

This may be assessed perceptually or alternatively more objectively through the analysis of in-house skill databases. The percentage of the total resource pool as a full time equivalent (FTE) with the commensurate skill is estimated and then compared with baselines.

Several aspects of the SEI Capability Maturity Model (CMM) could be also deployed in this domain of the BSC.

The CMM for software was produced by the Software Engineering Institute (SEI) and is used worldwide to improve the way that software is built and maintained. The SEI is funded by the US Federal Government and administered by the Carnegie Mellon University in Pennsylvania. The CMM describes 5 levels of maturity of a software organization.

The first level is described as initial. This level is characterized by unpredictable and poorly controlled processes. Processes do at times produce good software but usually by overcoming process deficiencies with near-heroic efforts.

Level 2 is defined as repeatable. This means that projects can repeat previously mastered tasks. At level 2 the process is under basic management control and there is a management discipline so that successful projects in terms of costs, schedule and requirements are the norm.

Level 3 is called defined. Here processes are characterized and fairly well understood. At this level what works and what does not work is understood and the organizations processes are standard and consistent.

Level 4 is defined as managed where the processes are measured and controlled.

Level 5 is called optimized where the focus is on process improvement. At level 5 the organization's software processes operates well as a matter of routine. This frees up people to focus on continuous improvement.

The CMM is regarded as a qualitative means of understanding where an organization is at in terms of process maturity. The CMM also provides guidelines on what needs to be done to move from one level of maturity to the next. Recent research showed that over 70% of organizations are at initial level 1, less than 20% at level 2 and approximately 10% collectively at levels 2, 3 and 4.

Qualitative measurements, such as CMM and quantitative measurements such as PEP assessments, go hand in hand as PEP identifies in an objective manner, opportunities for improvement. CMM recommends steps that need to be taken in order to improve maturity. PEP quantifies improvements in process performance. Organizations that are serious about process improvement use combined qualitative and quantitative assessment and improvement techniques.

Domain User : Customer Satisfaction and Partnership

QviP addresses various aspects of user satisfaction with IS services and the partnership between IS and the business. Through the use of survey instruments, this domain assesses the following:
- Satisfaction with existing applications.
- Satisfaction with recently completed projects.
- Satisfaction with IS services.

These are difficult to evaluate because they are often intangible. An IS service is particularly difficult to evaluate, because in addition to the intangible nature of the service, the customer is intimately involved in the service itself and therefore affects the outcome.

Creating and adding value through effective use of information technology requires effective internal and external partnerships. For the CIO and the development executive, partnerships have become an essential ingredient in the information technology success formula. As IS becomes a primary engine for business change – and a core business competency – a partnership between technical experts and business experts is emerging as a pivotal success factor.

The CSC PEP report, entitled, "Creating and Maintaining Partnerships," describes such partnerships in terms of maturity levels. In summary, they define four levels of maturity. These are:
- Order Taking
- Basic
- Anticipatory
- Cross Boundary

The assessment rates the level of partnership against the model.

At level 0 the supplier awaits a command or order from the customer. Only once the order has been received is the emphasis on cheap and quick execution of the order. This is not really a partnership as no future requirements are anticipated.

At level 1 or basic, there is usually a regular demand from the customer over time. This enables the supplier for the IS group to meet the customer's basic needs which they understand. This level is below the threshold of a true partnership although the IS group has proven itself in terms of being a reliable source of products and services.

At level 2 needs are anticipated. This is where the true definition of partnership comes into play, as the supplier is sufficiently aware of the customer's business that it can anticipate requests and make suggestions to the customer. At level 3 it becomes difficult to identify who is the supplier and who is the customer. This is ultimately a true partnership.

To move from order-taking to basic, IS needs to achieve operational reliability and consistency. This was described in the process domain of the balanced scorecard. To move from order taking to basic, IS also needs to build credible skills, both technical and business, to enable IS staff to understand business issues. To move from basic to anticipatory requires the development of deep business and industry knowledge and a very good understanding of technical issues. IS and the business need to jointly understand the concept of value and what this means in terms of the products and services that they supply. IS also needs to demonstrate success in anticipating business needs and understanding the most important technology options that will help their client.

One often finds that existing organizational structures do not promote partnerships. Networks and in particular e-mail and the Internet are promoting partnerships. These virtual communities are becoming the focal point of economic value creation for the business.

Figure 9, shows some examples of the issues that the QviP partnership assessment explores. Both the business and IS are surveyed.

THE TYPICAL ASSESSMENT PROCESS

An organization that is serious about undertaking an alignment assessment needs to carefully plan this assessment as it would any important IS project. There are generally three main phases to the assessment process.

The first phase involves knowledge transfer and data collection. The assessment planning should address the communication of the assessment objectives to all concerned. Communication of why the assessment is being undertaken, what it is going to address and what it has hoped to achieve is critical to mitigate uncertainty and to promote the benefits of the assessment to the organization. Communication from the executive sponsoring the

Figure 9: The business and IS typically share views on partnership issues

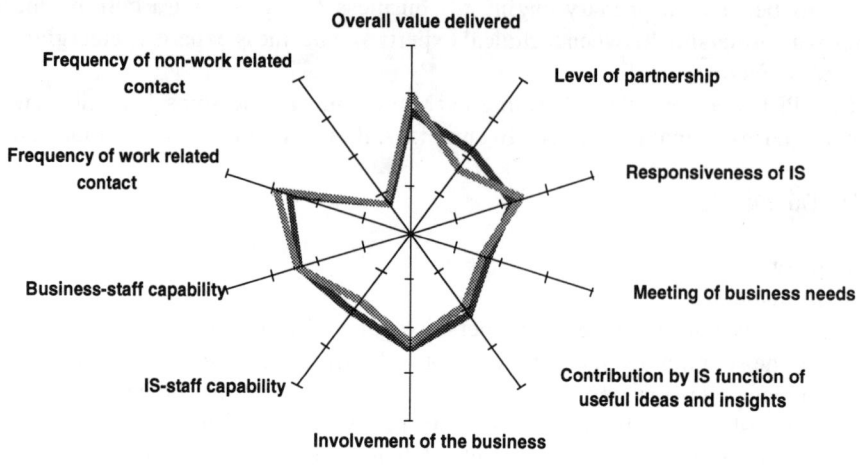

assessment should stress that the assessment is not being undertaken as a witch-hunt but is being undertaken to identify opportunities to improve IS and to improve the relationship between IS and the business. It is also being undertaken to identify ways of aligning IS and the business more closely, to ultimately find ways for IS to deliver greater value to the business. Communication should preferably be done by way of presentations rather than written memorandums. This provides the sponsors of the assessment exercise an opportunity to address any concerns that may be raised from various quarters. Interviewers then need to be trained as to the content of the assessment data collection instruments. These are well-defined forms, which are used during a prearranged interview session. The client assessment coordinator should arrange a series of interviews with the relevant parties and should document these on a project plan. The interviewers will conduct these interviews based on the forms and submit the responses to the coordinator.

The second phase is the analysis phase. This is undertaken by the assessment consultants who will, typically off-site, analyze the results of the assessment. The key to this activity is to highlight the major findings and to document these with reference to a more detailed finding. The consultants will also identify the issues arising from the assessment.

The third phase is the feedback phase where the consultants will present the results of the assessment. This is typically performed at least twice. The first such engagement is where the draft results are presented and debated with the client. The client is encouraged to provide feedback on the findings so as to help prioritize the findings and gain consensus for the key issues. The assessment reports and deliverables may be updated after this first informal debriefing. The main activity in the feedback phase is a management workshop. This workshop is convened with the key decision makers within the organization. These are those people who have the ability to bring about change as a result of the assessment findings. The management workshop typically takes about three hours. The consultant presents the results of the assessment, describes all the concepts and frameworks used and highlights the key issues. The workshop, as its name implies, is interactive, and the client is encouraged to debate the findings and issues identified. The main deliverable of the workshop is an action plan for improvement. This should be the responsibility of the client and accountability and timing should be allocated at the workshop. Following the workshop it is useful to provide feedback to all who provided information into the assessment and presentation should be organized to present these findings together with a summary of the actions identified at the workshop.

Figure 10 describes the typical QuantiMetrics assessment process.

Figure 10: The Assessment Process

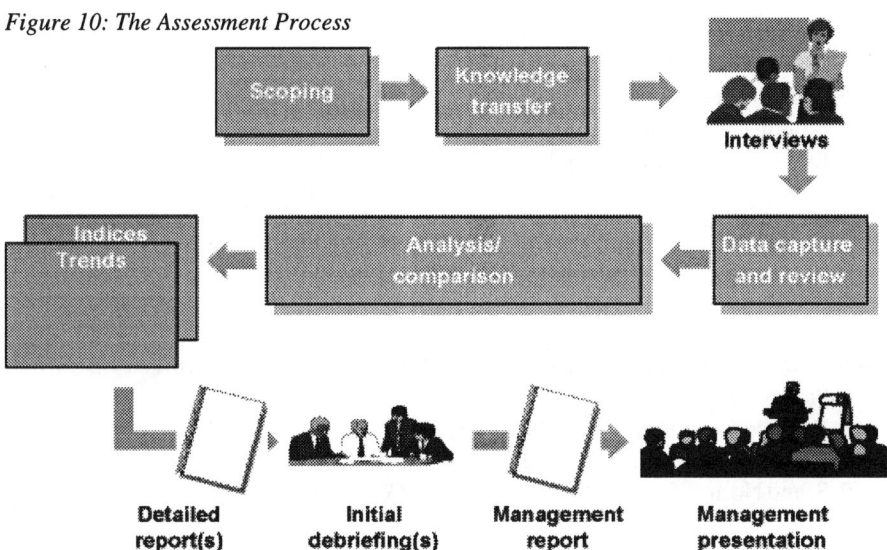

LESSONS LEARNED

Balanced IS assessments using the QviP have been performed for a number of organizations.

These have served to:
- assist in understanding the *value* that IS delivers to the businesses it serves,
- define the truth about today,
- define the future vision,
- highlight numerous opportunities for strategy development, particularly around finding ways to transform the way in which IS delivers value to the business.

BSC assessment alone cannot leverage value delivered. BSC assessment projects should be coordinated within a transformation context. This should also include an understanding of the role and value of IS to define a *case for action*, which should be used as input to the transformation program.

The widening gap between business capability and that of IS, often is not the fault of the IS alone. Although IS may point to their year-on-year improvements in efficiency, they now need to demonstrate the closer alignment between business and IS. IS groups need to focus on the distribution of their expenditure and move away from an operational center of gravity and become more strategic in their thinking. They also need to look at ways to more effectively partner with their business customers.

The business often gets the IS it deserves. Business executives also need to improve their understanding of the issues that impact IS delivery and capability and modify their behavior accordingly.

As Einstein once said, "not everything that can be counted counts." This chapter described a number of useful frameworks that can indeed lead to counting. Used appropriately and in a balanced fashion, they can provide the information that will act as a catalyst for change. Without such measures, the case for action may not be convincing nor will objective targets be defined. Gilb once said, "projects without clear goals will not achieve their goals clearly."

Transforming IS value means achieving sustainable improvement in the performance and capability of IS in support and in anticipation of the business' direction. To make this real, IS must be measured in business terms. The BSC approach has provided a holistic set of measures as a precursor and input to IS transformation.

QviP has been successfully deployed in assessing the alignment of IS organizations and the businesses that they serve, determining how appropriately IS resources have been spent, assessing the maturity of partnership between IS and its customers and in so doing determining the overall value delivered by IS to the business.

REFERENCES

Boggis, P.M. Personal dialogue, client reports and presentations.

Computer Sciences Corporation (CSC)(a). Creating and Maintaining Partnerships Research Services. *CSC PEP Research report.*

Computer Science Corporation (CSC)(b). Building the Capable Enterprise. *CSC PEP Research Report.*

Computer Science Corporation (CSC)(c). Valuing the IS Contribution to the Business. *CSC Foundation Operational Excellence Report.*

Kaplan, R.S. and Norton, D.P. (1996). *The Balanced Scorecard.* Harvard Business School Press.

QuantiMetrics. QuantiMetrics Alignment Assessment service, frameworks, data collection forms, analyses and clients results and presentations.

Prosser, G. Personal dialogue, client reports and presentations.

Tracey and Wiersma. The Discipline of Market Leaders.

Chapter XV

Management of Large Balanced Scorecard Implementations: The Case of a Major Insurance Company

Peter Verleun, Egon Berghout and Maarten Looijen
Delft University of Technology, The Netherlands

Roel van Rijnbach
Nationale-Nederlanden, The Netherlands

In this chapter, established information resource management theory is applied to improve the development and maintenance of large balanced scorecard implementations. The balanced scorecard has proved to be an effective tool for measuring business performance. Maintaining a business-wide balanced scorecard measurement system over a longer period implies, however, many risks. An example of such a risk is the excessive growth of scorecards as well as scorecard metrics, resulting in massive data warehouses and difficulties with the interpretation of data. This is particularly the case in large organisations.
This chapter proposes balanced scorecard management framework that is illustrated with the experience gathered from the company-wide balanced scorecard implementation in the insurance company Nationale-Nederlanden in the Netherlands.

INTRODUCTION

An increasing number of organisations use the balanced scorecard to measure and control their business performance. The dynamics of their business environment requires instant responses and traditional, financially orientated, budgeting and performance measurement systems are too slow in producing adequate management information. It simply takes too much time before business changes actually appear in financial figures.

The balanced scorecard is an obvious option for these organisations. However, management systems based on the balanced scorecard can become quite complex and less effective, particularly when used over a longer period. In the last chapter of their book,

Copyright © 2001, Idea Group Publishing.

Kaplan and Norton (1996) already emphasize that a strategic management system, which is supported by the balanced scorecard, should be adequately managed in order to capture all benefits. However, they do not present a method on how to maintain their strategic management system in the operational stage.

In this chapter the framework for information resource management of Looijen is applied to manage a large balanced scorecard implementation (Looijen, 1998). A comprehensive list of task areas that need to be addressed is given on basis of the distinction between: functional management, application management and technical management. Furthermore, an associated organisational model for large balanced scorecard models is described.

Both tasks and organisational model are illustrated with the experiences gained from the balanced scorecard implementation at insurance company Nationale-Nederlanden (NN). NN has implemented a balanced scorecard based reporting system as their company-wide management information system. From this case study, it is concluded that the resource management based approach appears to be an adequate method to manage large balanced scorecard implementations.

Descriptions of both the balanced scorecard methodology and the applied information systems resource management framework are kept to a minimum in this chapter. Readers can obtain extensive information in the two references. In this chapter we focus on the new combination of the two areas and the case study at NN.

This chapter has the following outline. First, a brief introduction to the balanced scorecard is given. Second, Looijen's framework for information resource management is presented. Third, this framework is illustrated given the situation of NN. This Chapter ends with conclusions and recommendations.

THE BALANCED SCORECARD

This section provides a brief introduction to the balanced scorecard. Given the fact that this technique is already well known we will particularly focus on characteristics of the scorecard that increases the complexity of scorecard implementations.

Balanced Scorecard and Performance Measurement

Lynch and Cross give many examples of organisations that under-perform due to an inadequate performance measurement system (Lynch and Cross, 1991). Through measuring on four distinct perspectives, the balanced scorecard already proved to be an excellent measurement system. These four areas are (Kaplan and Norton, 1996):

- Customer perspective;
- Internal organisation perspective;
- Financial perspective;
- Innovation and growth perspective.

The four perspectives are illustrated in Figure 1.

Once businesses have built their initial scorecard, the scorecard should be embedded in the organisations ongoing management system. For fully using the scorecard potential the scorecard should be tied to a number of management processes, such as budgeting, alignment of strategic initiatives and set-

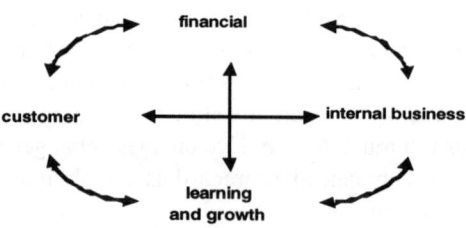

Figure 1: The four perspectives (Kaplan and Norton, 1996)

ting of personal targets. The following processes are identified and should be completely integrated with the balanced scorecard (Kaplan and Norton, 1996):
- Planning and target setting;
- Communicating and linking;
- Clarifying and translating strategy and vision;
- Strategic feedback and learning.

The processes and their relationship with the balanced scorecard are illustrated the Figure 2.

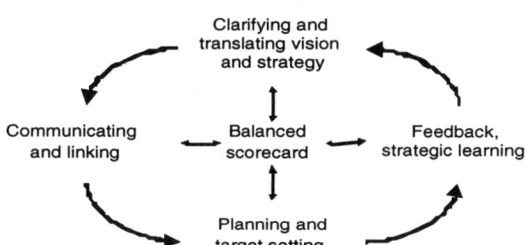

Figure 2: Management System (Kaplan and Norton, 1996)

Complexity of large balanced scorecard implementations

Establishing and controlling budgets is already a complex activity in any organisation of some magnitude. Theoretically, the balanced scorecard increases the size of traditional budgeting by four. Although this rule of thumb will seldom be exactly true, balanced scorecard implementations will indeed normally require more data processing than standard budget procedures.

A second characteristic that contributes to this complexity is the fact that a balanced scorecard reflects the strategic vision of the organisation and all members of the organisation tend to thrive for a prominent position in this strategy. "I am measured, so I exist", seems to be the slogan. The balanced scorecard has no explicit mechanism to reduce the number of measurements.

A third factor that adds to complexity is time. As time evolves, new measurements will be required, however cleaning-up old measurements appears to be a difficult task. A task that needs to be managed carefully, as there will often not be consensus about which measurements can be excluded. Again, removal of old measurements is not an explicit task.

Another complexity adding factor is the fact that scorecards are used at multiple levels and locations within the organisation. The way the different scorecards are used and integrated in day-to-day business results in additional challenges for the maintenance of the company-wide scorecard. Collecting all the necessary data and making this data ready for company-wide use is, as with other datawarehousing projects, a major challenge.

After designing and building a balanced scorecard, a maintenance organisation is required to keep the implementation operational and handle the above issues. In this chapter a framework is presented to do so.

A FRAMEWORK FOR IT MANAGEMENT

In this section Looijen's management, control and maintenance (MCM) framework for information resource management is described (Looijen, 1998). First, a brief introduction to the underlying concepts is given. Second, these concepts are explained in the context of the balanced scorecard.

Framework Concepts

The framework is first based on the life cycle of information systems. Starting at the information policy and ending at the moment the information system is taken out of order.

The following life cycle states are distinguished:
- State IPP (information policy and information planning). In this state information policy and information planning are determined, which leads to the information system being developed.
- State D (development). In this state the information system is designed and built.
- State AI (acceptance and implementation). In this state the information system is either accepted or not. If not accepted, it goes back to the previous state. If accepted, implementation takes place, whereupon it will be put into use as well as exploited.
- State U (utilisation). In this state the functions of the information system are being used.
- State E (exploitation). In this state the information system is kept operational and exploited for utilisation.
- State M (maintenance). In this state the information system, or part of it, is being maintained, initiated from either state U or E.

The life cycle states and their relationships are illustrated in Figure 3.

For each of the states U, M and E a list can be made of corresponding management tasks. Comprehensive lists of management tasks are given by Looijen (1998). Besides tasks, three management units are discerned to perform the various tasks. These management units are:

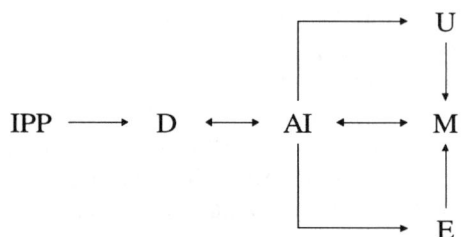

Figure 3: State Model (Looijen, 1998)

- Functional management (FM);
- Application management (AM);
- Technical management (TM).

FM is responsible for the maintenance and control of the functionality of the system. This is the main issue in the scope of utilisation of the system. AM is responsible for the maintenance and control of the application software (non-basic software) and the database. TM is responsible for the maintenance and control of the operationalisation of the information system.

Furthermore, for each unit, three levels of management at which the unit operates are identified. These levels are strategic level (policy-determining), the tactical level (translating policy and managing the resources) and the operational level. Each unit can have its own staff departments, for example personnel management. The management tasks are illustrated in Figure 4.

This approach for management of information systems has been used in practice for many years and has proven to be successful.

Figure 4: Triple Model of MCM (Looijen, 1998)

MCM framework and balanced scorecard

The framework for managing information systems described in the previous section, can also be applied to manage balanced scorecard implementations. In this section such an implementation is described in more detail.

Application Management

Application Management (AM) is, as already stated, responsible for the maintenance and control of the actual balanced scorecard application. In most cases the scorecard information system will be an application that includes or uses some form of data warehouse.

One of the challenges for AM is to deliver usable data. The data originating from all the organisational disciplines should comply with the business data definitions as specified by FM. This has proven to be a complicated task.

Furthermore, AM is responsible for adding new metrics and, particularly also, for removing metrics.

Maintenance of the application and database management are important tasks for AM. Maintenance of the application consists of the following tasks:
- Programming and testing new software. Changing a scorecard might result in changing queries on the data. AM should implement and test the new queries.
- Maintenance of the technical design. New and changing strategies and scorecards can result in new demands on the application.
- Performance and integration testing. Periodically, the system and the data should be tested for compliance, quality etc.

Database management consists of the following tasks:
- Maintenance of the database structures. Changing demand in the business domain will often result in changes in the structure of the database.
- Conversion of data. When changing database structures and demands.

The AM unit should be organised including a strategic level, which is responsible for the AM policy and the relation with other organisational units, a tactical level, which is responsible for the translation of the policy and the human, technical and financial resources at operational level and an operational level, which is responsible for all the tasks mentioned above.

AM at a strategic level normally requires a top-management steering committee. Issues considering the contents of the balanced scorecard can only be resolved at this level.

An IT-project manager can be made responsible for the tactical level. This person will be responsible for translating the directions and policy formulated by the AM steering committee into actions. IT-teams will be made responsible for the different tasks.

Functional Management

Functional Management (FM) is concerned with utilisation management and functional maintenance. Regarding utilisation management a number of task fields are discerned (only the most important ones for the balanced scorecard are described here):
- User support. Advising on the use of the balanced scorecard and training users. The fact that managers should be trained and advised on how to use their scorecard in day-to-day business is an aspect that often does not get the attention it deserves.
- Management of the business data. Management of data sets, managing authorisation and information supply on ad hoc basis are important tasks for data warehouse systems.

The other task area within functional management is functional maintenance. Functional maintenance relates to the maintenance of procedures, specifications and the definitions for which the user themselves are responsible:
- Maintenance of procedures. Many procedures are required to gather, store and distribute the balanced scorecard metrics. FM is responsible for maintaining these procedures.
- Data definition control. This concerns the maintenance of the definitions of business data. This should be done in consultation with the relevant owners. Control of the consistent use of business data definitions for the various scorecards and the data delivery information systems.

The FM unit itself should be organised by creating a strategic level, which is responsible for the FM policy and the relation with other organisational units, a tactical level, which is responsible for the translation of the policy and the human, technical and financial resources at operational level and an operational level, which is responsible for all the tasks mentioned above.

As Kaplan and Norton (1996) state, it will normally be difficult to make a single manager responsible for managing the scorecard process. A steering committee consisting of a number of senior managers will, therefore, be required. This way, commitment of all necessary disciplines within the organisation can be guaranteed.

At the tactical level a project manager should be installed. This person will be responsible for translating the steering committee's policy into operational tasks and directions. The operational level will consist of a number of multidisciplinary project teams, each responsible for particular tasks.

Technical Management

Technical Management (TM) is responsible for keeping the system running and for the actual operationalisation of the database, the application and the queries. Although TM is important, the associated tasks can normally be handled by the existing technical IT organisation.

Summary of MCM framework and balanced scorecard

The MCM framework for managing information systems provides an adequate starting point for managing a company-wide scorecard. The FM, AM and TM tasks can easily be translated to a balanced scorecard implementation. The number of tasks that should be managed regarding a balanced scorecard implementation is significant. Experiences from other information systems learn that less than a third of the effort is spent, when the information system is built and implemented. Most effort is in maintenance.

Creating strategic, tactical and operational levels also positions the tasks of senior management in the management process.

CASE STUDY AT NN

In this section the concepts introduced in the foregoing section will be described in the context of insurance company NN. First, the NN organisation is introduced. Second the organisational framework of the balanced scorecard will be described. Third, a number of special issues that require additional attention are given.

Description of the case

NN is the largest insurance company in the Netherlands with premium revenues of $3.9

billion. During 1994-1997, NN, part of ING Group, reorganised itself from a product/functional-driven organisation to a market/process-driven organisation based on a new strategy of integrated financial services. In 1997 the balanced scorecard was introduced for internal reporting. Gradually, periodical management reporting will be based on the balanced scorecard. The company, business units and departments should all have their own hierarchically layered and interlinked scorecards. This situation is scheduled to be complete in the summer of 2000.

The phased introduction of the BSC started with two pilots: The BSC as a performance measurement system, the first pilot, produced a cohesive set of strategic performance indicators. The BSC as a management system, the second pilot, produced not only a cohesive set of strategic performance, but also a strategic action plan for the next three years.

After the successful implementation of the two pilots, NN decided at the end of 1997 to implement the BSC for the whole organisation. During the first six months of 1998 the NN-scorecard was designed and translated to each level of management. After that, NN decided to implement the scorecard at all levels of management.

Organisational framework

During the implementation phase of the scorecard the following organisation existed. A steering committee, including senior management, was responsible for the entire implementation. Project management and project teams handled the development, realisation and management of the scorecards. This approach was experienced as being successful, since most scorecards were designed and embedded within the particular departments and problems could be adequately dealt with because of the senior management commitment to the project.

In the beginning most emphasis was on design. This gradually shifted to realisation. A realisation team was responsible for developing and realising a controlled data flow to fill the scorecards. The first generation of (45) scorecards was realised through the implementation of a user interface (dashboard) using spreadsheets. The second generation of scorecards implied that the already existing user interface had to be transformed and connected to a scorecard-data-mart, in order to guarantee data delivery from the data warehouse, that matches the scorecard business definitions.

A BSC Management Team was responsible for the ongoing management/ scorecard process. Uniform use, consistent high quality data delivery, scorecard maintenance, automating the scorecard and even more important the evaluation process.

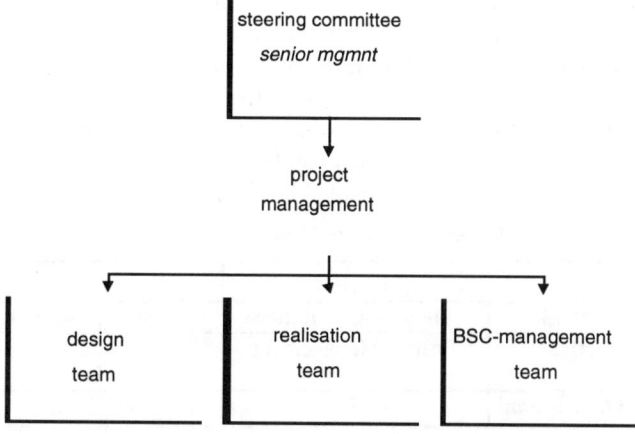

Figure 5: Organisation of scorecard development

Organisational problems

Actually delivering the required data has proved to be one of the major challenges of the scorecard implementation team. Data originating from different departments and different levels of the organisation had to be merged. This often required amending the data in order to comply to the business data definition. However, this also raised questions, such as, what is the valid business definition and who is responsible for these definitions?

Another problem is the completeness of the data. Given the fact that the data is derived from a significant number of operational systems, how can the scorecard-team guarantee that the data is complete?

Currently, the second generation scorecards are being implemented. The next step will be to further integrate the scorecard into the management processes. Scorecards should be linked to the four management processes and should be used at the different locations and different levels in a similar way. This indicates that continuous training and evaluation should be introduced. Because of the growing importance of the scorecard/management aspect, the organisational structure, as mentioned above, should evolve into an organisation in which managing the scorecard process is the main goal.

NN and the MCM model

According to the MCM model, management of information systems should be divided into three areas, being, functional management, application management and technical management. All of these three areas can be assigned to different organisational units, each of which consisting of three different levels of management; strategic, tactical and operational.

In an earlier section a structure is proposed, where a steering committee, consisting of senior management, is responsible for functional management at the strategic level. For the other two disciplines the steering committee can consist of IT managers/specialists. The steering committee should be aware that all three disciplines are equally important.

At the tactical level a project manager can be appointed. This person is in fact the scorecard manager. The areas application management and technical management could well be integrated in the existing IT organisation. Particular IT departments or teams consisting of IT-specialists can be made responsible for AM and TM at the operational level (for example by using SLAs). This organisation is illustrated in Figure 6. A possible organisation structure is given in Figure 7.

Remaining challenges

By organising scorecard management as proposed before, a number of challenges remain. A first challenge is how to manage the data definitions. Balanced scorecards frequently require data originating from various information systems that apply dissimilar data definitions. A solution might be that all the different domains in the organisation are held responsible for their own definitions. For example, the marketing department is responsible for the definition of customer satisfaction, the finance department is responsible for the definition of revenue, etc. In this case, FM remains responsible for the fact that the definitions are indeed managed by the decentralised departments.

Figure 6: Implementation of MCM framework

	Functional	Application	Technical
Strategic	Steering Committee	IT/Steering Committee	IT/Steering Committee
Tactical	Project Management	Project Management/ IT Specialist	Project Management/ IT Specialist
Operational	FM Team	AM Team	TM Team

The same approach can be taken regarding the data delivery process. In this case operational data should be transformed to data that complies with AM definitions. AM remains responsible that all information system owners deliver the data in compliance to the data definitions.

The maintenance of the scorecard itself is one of the most important tasks in the scorecard management process is. Managers need to be trained in how to use the scorecard and how to integrate the scorecard into day-to-day management. FM can, for example, detect metrics that are no longer used.

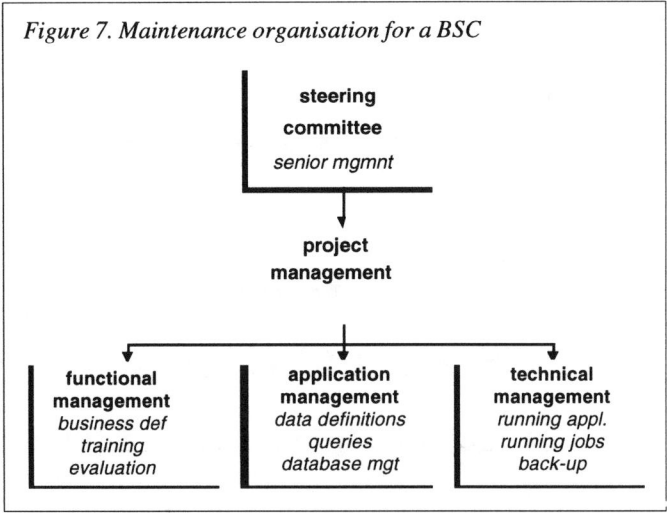

Figure 7. Maintenance organisation for a BSC

CONCLUSIONS

Establishing a company-wide balanced scorecard reporting system is obviously not an easy task to implement. However, maintaining such a system seems to be even more complicated. In practice, the number of measurements in a balanced scorecard implementation tends to increase rapidly. What really needs to be measured needs to be planned carefully.

Kaplan and Norton state "Most organisations today have a leadership void for this system (balanced scorecard). No executive in a traditional organisation has the responsibility or prospective to manage a strategic management process, and it's unclear who should assume this responsibility." (Kaplan and Norton, 1996). In this Chapter the ongoing scorecard management process is compared with the MCM model of Looijen (1998). The MCM model is a proven method for dealing with the challenge of managing and maintaining large information systems. By using this method, the management and maintenance of the multiple scorecards is divided into three management areas, functional management, application management and technical management. Each area has its own tasks and particularities.

In the case of insurance company NN, the MCM model was experienced as an adequate framework to organise the management of a balanced scorecard implementation. Inadequate organisation of the maintenance function will probably create an unworkable situation within two years after the enthusiastic launch.

REFERENCES

Kaplan, R.S., and Norton, D.P. (1996). *The balanced scorecard: translating strategy into action*, Harvard Business School Press.

Looijen, M. (1998). Management, control and maintenance of information systems. *Kluwer Business Sciences (international edition).*

Lynch, R.L. and Cross, K.F. (1991). *Measure up: yardstick for continuous improvement.* Basil Blackwell.

Chapter XVI

Integrating the Balanced Scorecard and Software Measurement Frameworks

Nancy Eickelmann
Motorola Labs, USA

INTRODUCTION

This chapter describes the integration of the Capability Maturity Model (CMM) and ISO-9126 software measurement frameworks with the National Aeronautics and Space Administration Independent Verification and Validation Facility (NASA IV&V). Balanced Scorecard IV&V is a unique aspect of software development practice as it provides a service of independent and objective lifecycle evaluation of the software product and processes used for development. To accomplish this rigorous task a sophisticated measurement program is desirable. This chapter describes the application and integration of strategic measurement (BSC) with organizational measurement (CMM) and product measurement (ISO-9126). The CMM is a measurement model of ordinal ranking of an organization's software process variability and repeatability. As an organization's process becomes more mature it may traverse the scale from a level one to a level five organization. The CMM provides a basis for collecting accurate and timely measures of process performance. The international standard ISO/IEC-9126 focuses on information technology and software product evaluation through measurement of software quality characteristics. The development of a core set of metrics for implementing the Balanced Scorecard is the most difficult aspect of the approach. Developing metrics that create the necessary linkages of the operational directives with the strategic mission prove to be fundamentally difficult as it is typical to view organizational performance in terms of outcomes or results. The metrics must address performance drivers or the measures that provide feedback concerning day to day progress.

The Balanced Scorecard Institute has identified 12 criteria to guide the development of metrics (1999):
1. Leading indicators: forecast future trends inside and outside the agency
2. Objective and unbiased
3. Normalized - so they can be benchmarked against other agencies
4. Statistically reliable - small margin of error
5. Unobtrusive - not disruptive of work or trust

6. Inexpensive to collect - small sample sizes adequate
7. Balanced - qualitative/quantitative, multiple perspectives
8. Appropriate - measurements of the right things
9. Quantifiable - for ease of aggregation, calculation and comparison
10. Efficient - can draw many conclusions out of data set
11. Comprehensive - show all the significant features of agency's status
12. Discriminating - small changes are meaningful.

This list represents the ideal of a set of metrics, in practice it is very difficult to develop metrics that achieve all twelve criteria. For that reason the metrics that are chosen should be evaluated as to their degree of conformance to the ideal and evaluated for inclusion or exclusion to the organizational metrics program.

This chapter follows this organization. Background is provided that describes independent verification and validation. A high level overview of the tasks and activities conducted by IV&V are described. The BSC architecture is described as if being constructed in conjunction with BSC leading and lagging measures. The customer themes are described and their corresponding measures stated. The integration of the ISO-9126 process measures and the CMM infrastructure or learning and growth measures are discussed. The chapter concludes with a summary of lessons learned.

BACKGROUND

IV&V is a set of technical and managerial activities performed by someone other than the developer of a system to improve the safety, quality and reliability of that system and to assure that the delivered product satisfies the user's operational needs. These activities can provide the benefits of defect avoidance and defect detection at early stages of the lifecycle. A primary contribution of IV&V is the early discovery of defects, thus preventing their propagation into the later stages of the development process where they are far more costly to correct. Independent Verification and Validation is characterized by the notion that the activities associated with IV&V are performed by a group other than the developers of the system. The degree to which the IV&V group is independent from the developers is assessed across three areas technical independence, managerial independence and financial independence. In this section we describe the meaning of independence and the key activities of V&V.

Technical Independence

Technical independence is characterized by the notion that the IV&V group has a perspective of the system that is unbiased by the developers of the system [9]. Thus, it follows that the IV&V personnel should be people not involved in the development process. This promotes the need for the IV&V group to formulation its own understanding of the problem itself and how the system is solving that problem. With respect to software tools, ideally the IV&V group should develop its own set of testing tools separately from the developers testing tools. In cases where development of a separate version of tools would be too expensive, the IV&V group may share some of the developer's tools. If the IV&V group is to share any developer's the IV&V group must perform qualification tests of said tools. These tests are to ensure that there are no errors in the tools themselves that may mask errors in the software.

Managerial Independence

To achieve managerial independence, the responsibility for the IV&V effort must rest with an organization that is distinctly separate from both the development management as well as the program management. The management of the IV&V activities must have autonomy to select the portions of the system to test and analyze and the technique(s) that are to be used to perform the analysis and/or test. The IV&V management will set its own schedule for IV&V activities to analyze the specific issues it has selected to act upon. To maintain managerial independence the IV&V group must report directly to the development and program management with out developer interference. An effective IV&V effort must be allowed present results directly to management in a timely fashion with being restricted by developer approval.

Financial Independence

The financial independence of the IV&V group is achieved by placing the budgetary controls for these activities with an organization that maintains independence from the development organization. This safeguards the IV&V process from failing to meet its objectives of timely feedback to the project management for reason of funds being diverted due to pressures and/or influences in other areas of the project.

Verification and Validation

Verification is a process by which a system and the products associated with that system are shown to meet the requirements for all activities during each life cycle process [9]. Compliance with requirements includes checks for correctness with regard to the existing requirements, completeness of the requirements set and consistency among the requirements and the resulting system. A second concern of verification is to assess the degree to which standards, practices, and conventions are satisfied during the life cycle processes. Typical life cycle processes include acquisition, supply, development, operation and maintenance. Finally, verification seeks to establish a basis for assessing the completion of a given life cycle activity and the initiation of other life cycle activities.

Validation is a process that seeks to determine and provide evidence that the system solves the correct problem. With regard to software, validation assesses the extent to which the requirements that are allocated to the software portion of the system are met by the resulting software that is developed. While verification focuses more on the system working correctly, validation focuses on the systems ability to solve the problem that it was intended to solve.

The high level tasks that are performed in V&V include phase dependent tasks and phase independent tasks, see Figure 1. The phase dependent tasks include software requirements analysis,

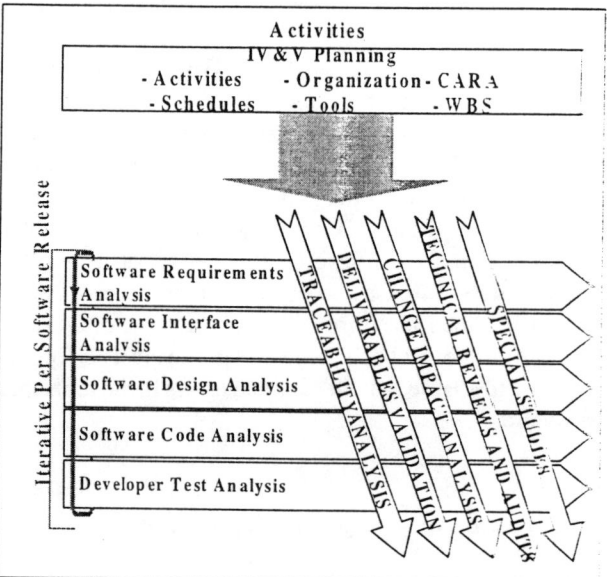

Figure 1. Independent Verification and Validation Phase (in)dependent activities.

software interface analysis, software design analysis, software code analysis and developer test analysis. These activities are iterative with every software release. Phase independent activities include traceability analysis, deliverables validation, change impact analysis, technical reviews and audits and special studies. The inputs to these activities are the program milestone products and schedules, requirements repositories, developer documentation, software test plans and procedures, software development folders, source code and problem reports. The outputs of the process include a master software IV&V plan, a listing of high risk and critical functions, technical reports that describe details of the activities, software problem reports, the IV&V traceability matrix for lifecycle artifacts, findings and recommendations, IV&V metrics and monthly progress and status reports. The planning activity must take into account not only the activities and resources required by IV&V but it must also integrate the organizational aspects of the development context. The IV&V schedule must be developed based on when the artifacts that are inputs to the IV&V process are to be made available from the development organization as well as their format. Artifacts may be delivered in a format that requires manipulation prior to analysis and would therefore contribute to additional planning contingencies. In addition, the scoping of the IV&V is required to detail how much of the software will be placed under IV&V review. Those components in the system that are most critical and require more focused IV&V are identified and placed under rigorous review. The NASA IV&V organization is integrating the BSC, CMM, and ISO-9126 measurement frameworks to guide day-to-day operations towards achievement of the organization's strategic goals. In the next section we discuss the strategic goals for NASA IV&V and describe the construction of the BSC core metrics.

BSC Objectives

The BSC architecture provides a framework that facilitates translating the strategic plan into concrete operational terms that can be communicated throughout the organization and measured to evaluate the day-to-day progress towards success. The BSC draws its focus from the organizational strategy and uses the organizational strategic plan for direction. The strategic plan contains the vision, goals, mission and values for the organization. The Government Performance and Results Act, GPRA requires all federal agencies to establish strategic plans and measure their performance in achieving their missions. The vision and goals for the NASA IV&V Center are stated below.
- Vision: To be world-class creators and facilitators of innovative, intelligent, high performance, reliable informational technologies that enable NASA missions.
- Goals: To become an international leading force in the field of software engineering for improving safety, quality, reliability, and cost performance of software systems; and to become a national Center of Excellence (COE) in systems and software independent verification and validation.

The BSC is not the organizational strategy but rather a measurement paradigm to provide operational and tactical feedback. The organizational strategic vision and goals are the foundation upon which the measurement framework is constructed. There are two categories of measures used in the BSC the leading indicators or performance drivers and the lagging indicators or outcome measures. The performance drivers enable the organization to achieve short-term operational improvements while the outcome measures provide objective evidence of whether *strategic objectives* are achieved. The two measures must be used in conjunction with one another to link measurement throughout the organization thus giving visibility into the organization's progress in achieving strategic goals through information resource management and process improvement initiatives. The three prin-

244 Eickelmann

ciples of building a balanced scorecard that is linked through a measurement framework to the organizational strategy include:
1) Defining the cause and effect relationships,
2) Defining the outcomes and performance drivers,
3) Linking the scorecard to the financial outcome measures.

The initial steps of cause effect graphing for the BSC engage in the construction of a set of hypotheses concerning relationships among objectives for all four perspectives of the balanced score card. The measurement system makes these relationships explicit therefore, they can be used to assess and evaluate the validity of the BSC hypotheses.

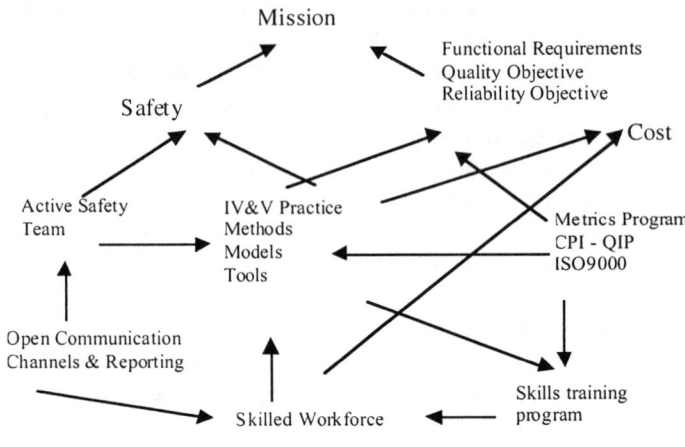

Figure 2. Influence diagram of IV&V BSC objectives.

We formulate hypotheses concerning the value of information technologies required by IV&V in the context of the Space Shuttle IV&V activities. The hypotheses are based on inferred or known relationships documented in prior studies reviewed under the first phase of our ROI project (see Figure 2).

Hypothesis 1: Information management and information analysis as a result of IV&V are essential IT to achieve mission safety requirements.

Hypothesis 2: Information management and information analysis as a result of IV&V results in measurable reduction of risk of mission failure.

Hypothesis 3: The benefits of defect detection and defect prevention are enhanced by IV&V through support of continuous process improvement.

Hypothesis 4: This is fundamentally a unique system that is developed using so-

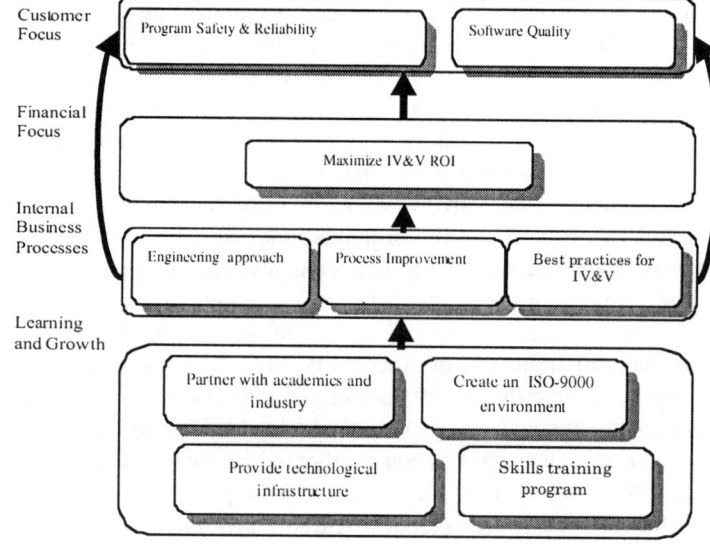

Figure 3. BSC objectives hierarchy.

phisticated reuse. This requires us to view the system as generating shuttle "builds" from an investment of core assets. The benefits are primarily derived in the reusability and rapid extensibility of the shuttle code.

Hypothesis 5: Adherence to an architecture enables system safety, reliability and quality standards to be imposed and verified for the core assets of the shuttle. Acceptable degrees of variability to extend functionality are approved by a team of architects and systems engineers that includes the IV&V team.

We map our hypothesis to a set of objectives concerning the value of IV&V and the necessary and sufficient factors to creating value for the organization in terms of the strategic vision and goals. The BSC is segmented into four categories of objectives customer, financial, internal business processes and learning and growth segments see Figure 3. The objectives for the four segments are the following.

- Customer segment objectives correspond with the high level goals of mission success through high quality, reliability and safety.
- Financial segment objectives focus on cost reduction, efficient asset utilization and high ROI values of IT investments.
- Internal process objectives relate to specific software and systems engineering approaches such as product-line development paradigms, CPI and QIP efforts, and test technologies and best practices as defined for IV&V.
- Learning and growth objectives include technological infrastructure for distributed development, workforce training programs, skills assessment program, and ISO-9000 process structure.

Customer Measures of Mission Success

The customer focused objectives of improved mission safety, improved mission systems and software reliability, improved systems and software quality each have specific measures and targets that are used to evaluate whether or not strategic goals are achieved and to what degree. This requires identifying viable measurement strategies for software IV&V in the NASA context.

- Safety is defined as freedom from accidents or losses. This is an absolute statement; safety is more practically viewed as a continuum from no accidents or losses to acceptable levels of risk of loss.
- Reliability is defined in terms of the probabilistic or statistical behavior, that is the probability that the software will operate as expected over a specified period of time.
- Quality is defined in terms of correctness and number of defects. Correctness is an absolute quality. It is also a mathematical property that establishes the equivalence between the software and its specification.

The measurement of the contribution of IV&V in improving safety, reliability and quality while reducing cost is discussed in the following sections. The contribution of IV&V to shuttle safety is difficult to measure directly. It is therefore necessary to make assumptions concerning those factors that would impact safety and to what degree. It is assumed that a reduction in the probability of failure is a contribution to increased safety. A reduction of the number of In Flight Anomalies (IFAs) of a severe nature due to IV&V identification and removal is a contribution. An independent evaluation of potential failure modes that results in identifying previously unidentified hazards is a contribution. The elimination or mitigation of hazardous states or their potential is quantified relative to probabilistic measures of hazard occurrence and the likely severity.

Figure 4 Balance Scorecard core metrics set

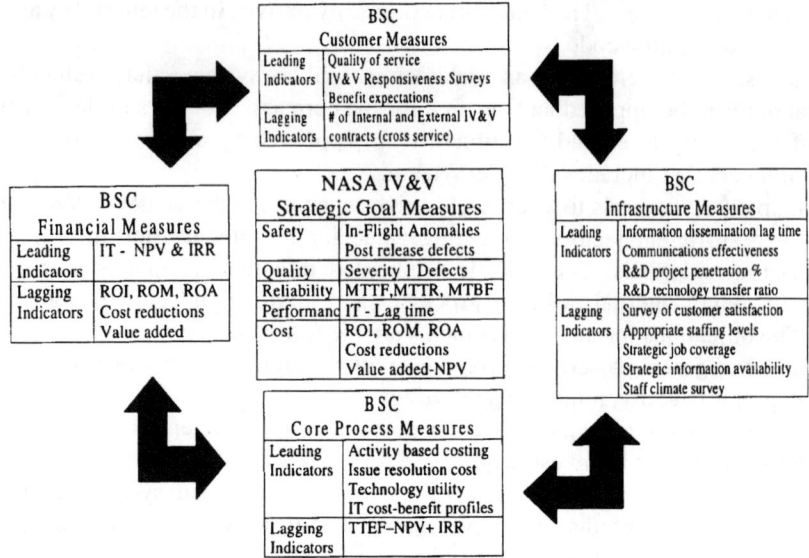

The contribution of IV&V to shuttle reliability is more directly attributed to the specific verification activities that are applied during the shuttle software development process towards defect management. Research investigating the ramifications of testing strategies for reliability provides quantification of benefits relative to specific IV&V activities. A minimization of estimated residual faults is provided according to the sequence of testing strategies and the duration of those test executions, for example the number of defects detected by applying functional, decision, data flow and mutation test methods in sequence. The CPU execution time or the number of test cases can measure test effort. As the test effort increases defects detected can be optimized through applying more optimistic or pessimistic test strategies. The resulting increase in reliability is measured by increased MTTF or improved failure intensity profiles and is quantified as a reduction in the distance from the reliability targets of subsystems undergoing IV&V.

The contribution of IV&V to shuttle quality is measured as a reduction of defect density trends through process improvement paradigms such as traversing the CMM stages from levels 2, 3, 4 to level 5. The intuition behind this model is that the measurable impact of process improvement is in the reduction of the cost of rework. In addition, the rework cost avoidance of detecting defects of severity 1, severity 2, and severity 3 can be quantified relative to phase of detection and level of severity. The reduction of defect density is measured as a reduction of distance from the overall quality objective measured in defect density according to severity.

The shuttle software safety, reliability and quality are measured according to the BSC core metrics of IFAs, post release defects, severity 1 & 2 defects, mean time to failure (MTTF), mean time to recovery (MTTR), and mean time between failures (MTBF). A key software engineering practice that is in part responsible for the software's high degree of assurance is that of reuse. The shuttle program has a sophisticated approach to the sustaining of the core functionality of the systems and software while providing controlled evolution to support new missions. The measurement of reusability of key architectural, design, code and test artifacts is currently under study as part of the BSC research.

Table 1 Customer focus metrics definition.

Customers (Internal External)	Objectives	Measures	Targets	Initiatives
	No Losses	# Severity 1 &2	Remove < FRR	Formal Methods
	Reduce Risk	# IFA's	No Severity 1	Risk Management
	Manage Risk	Fault tolerance	Performance	Risk Mitigation

The objectives are used in the selection of a minimum set of required metrics to measure day-to-day performance and outcome or results metrics. This aspect of the framework focuses on development of leading and lagging indicators. An example of a customer focused objective would be the improvement in overall safety due to IV&V activities. A leading indicator for this objective could be the number of identified potential hazardous states resulting from a safety impact analysis or a tracking of the hazard rate during development. A result measure or lagging indicator could be the number of IFAs that are documented. The leading and lagging indicators must be assigned desired or normative values. These values become targets or target ranges for the metrics collected. Finally, the initiatives that have been sponsored to achieve the objective are identified and reevaluated with respect to the quantitative and qualitative evidence of success relative to the target values (see Table 1).

Process Measures

The NASA IV&V BSC further uses the ISO-9126 metrics to map the strategic goals to operational activities based on required product qualities. ISO-9126 provides guidelines for evaluating software quality. The process is iterative and begins with a quality requirement definition. Relevant to the quality definition metrics are selected and value ranges assigned. The development products are measured and compared to the ranges of acceptable values and assigned a rating. The products' rated level is assessed as to its degree of conformance. This translates measured values to an ordinal ranking that is assessed by management as either acceptable or unacceptable. The ISO-9126 is a measurement framework that mirrors the BSC in defining an objective (ISO-9126 defines a quality requirement) that maps to a metric that is assigned a range of possible values. A target range of values is assigned to each metric value (ISO-9126 defines an ordinal ranking of the target values). An interpretive guideline is provided to management as decision criteria for accept or reject decisions of the measured product or process.

The ISO-9126 standard documents six high-level software qualities including functionality, reliability, usability, efficiency, maintainability and portability. These high-level qualities are mapped to 24 sub-characteristics. Metrics are proposed to measure the high-level software qualities relative to the sub-characteristics. This ISO standard could provide the necessary metrics to measure operational processes under the process aspect of the BSC, relative to the application of product line reuse, and map them to the high-level goals. Of particular interest in this standard is the definition of reusability as the combination of maintainability and portability. It will be of interest to analyze the appropriateness of the standard in measuring reuse for the shuttle, specifically, reuse across a vertical product line that incorporates domain engineering, architecture-based reuse, and reusable test technologies.

The BSC requires that measurement is linked from operational activities to high level strategic goals. The CMM and ISO-9126 initiatives both have key process areas that address technologies and there impact (see Table 2). The CMM has specific KPAs that are directed

Table 2 ISO/EIC- 9126: Six qualities included in the standard for quality requirements definition.

	ISO/IEC - 9126
Functionality	Suitability, Accuracy, Interoperability, Compliance, Security
Reliability	Maturity, Fault tolerance, Recoverability
Usability	Understandability, Learnability, Operability
Efficiency	Time behavior, Resource behavior
Maintainability	Analyzability, Stability, Testability
Portability	Adaptability, Installability, Conformance, Replaceability

to project level processes and test practices. This section provides the results of integrating CMM process measures and the ISO-9126 measures into the BSC framework.

The CMM is structured in a five tier hierarchy that addresses many project level KPAs in the first three tiers (see Figure 5). Specifically, level one is a chaotic process. Level two has several KPAs directed at the project level including project planning, software configuration management and software quality assurance. Independent verification and validation activities are a part of the level two KPAs. Level three KPAs directed at the project level may include software inspections, software testing, software configuration management and software process definition. The organizational focused KPAs for level three include software standards and a software engineering process group SEPG. The organizational focused KPAs of level three and above do not fit well in the core process section of the BSC. The Capability Maturity Model decomposes the essential control elements required to conduct software engineering in an efficient and effective manner. These elements include documented requirements, communications with the customer(s) and user(s), agreed-to commitments, planning, documented process and work breakdown structure WBS. These elements are embedded in the ordinal hierarchy of the CMM. The CMM goals for each KPA are abridged and listed by maturity level below. An organization that is maturing in the key process areas of the CMM should be simultaneously reducing process variability and increasing quality and productivity.

Level Two KPAs: Repeatable

Level Two is characterized by project management, product assurance, and change control.

Requirements Management
Goal 1- Baseline system requirements allocated to software
Goal 2- Consistency management of plans, products, tasks

Software Project Planning
Goal 1- Use estimates in planning and tracking
Goal 2- Document plans and activities
Goal 3- Agreed-to commitments among groups

Software Project Tracking and Oversight
Goal 1- Track results to actuals

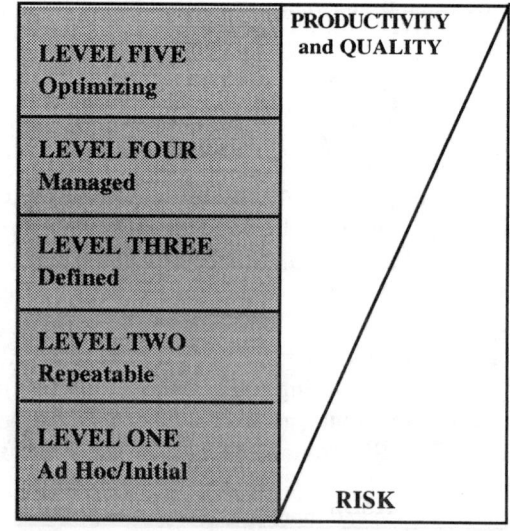

Figure 5. CMM five tier hierarchy

Goal 2- Act to correct deviations from plans
Goal 3- Changes to commitments are by mutual agreement
Software Subcontract Management
Goal 1- Use qualified subcontractors
Goal 2- Agreed-to commitments among all contractors
Goal 3- Effective communication channels
Goal 4- Prime contractor tracks the subcontractors progress against commitments
Software Quality Assurance
Goal 1- Software quality assurance activities are planned
Goal 2- Product adherence to standards
Goal 3- SQA activities are known
Goal 4- Senior management intervenes with conflict resolution issues as necessary
Software Configuration Management
Goal 1- SCM is planned
Goal 2- Selected software products are made persistent and retrievable
Goal 3- Product change is controlled
Goal 4- Software baselines are appropriately disseminated

Level Three KPAs: Defined

Level Three is characterized by a focus shift to an organizational level from a project level. The establishment of an software engineering process group (SEPG), establish a process architecture, and a software engineering development methods and technologies.
Organization Process Focus
Goal1- CMI activities are integrated across the organization
Goal 2- Actual processes are evaluated to a standard
Goal 3- Organization level process improvement activities are planned
Organization Process Definition
Goal 1- A standard development process is maintained
Goal 2- Project's collect relevant information for review to standard conformance
Training Program
Goal 1- Planned training activities
Goal 2- Skill and knowledge development focused training program
Goal 3- SEPG training
Integrated Software Management
Goal 1- Project processes are based on organizational standards
Goal 2- Project planning and management are consistent with the process
Software Product Engineering
Goal 1- Software engineering tasks are consistent
Goal 2- Software work products are consistent
Intergroup Coordination
Goal 1- Agreed to customer requirements
Goal 2- Agreed to commitments among engineering groups
Goal 3- Resolution of intergroup issues
Peer Reviews
Goal 1- Peer review activities are planned
Goal 2- Defects are identified and removed

Level Four KPAs: Managed

Level Four, once the process is defined it can be managed, measured, and improved. This requires that a core set of metrics be defined, and data collection and analysis infrastructure be in place.

Quantitative Process Management
- Goal 1- Quantitative process management is planned
- Goal 2- Process is controlled quantitatively
- Goal 3- Process capability is known and quantified

Software Quality Management
- Goal 1- Software quality management is planned
- Goal 2- Measurable goals are defined
- Goal 3- Progress towards quality goals is tracked

Level Five KPAs: Optimizing

Level Five, this level represents a well-defined, process with task variability minimized and monitored. A fine-grained level of control is provided to management in measuring the efficiency and effectiveness of the processes used, the resources allocated, and the quality of the resulting product.

Defect Prevention
- Goal 1- Defect prevention activities are planned
- Goal 2- Common causes of defects are identified
- Goal 3- Common causes of defects are prioritized and eliminated

Technology Change Management
- Goal 1- Incorporation of technology changes are planned
- Goal 2- New technologies are evaluated to determine their effect on quality and productivity
- Goal 3- Appropriate new technologies are transferred into practice

Process Change Management
- Goal 1- Continuous process improvement is planned
- Goal 2- Organization wide process is improved
- Goal 3- Standard processes are improved continuously

The NASA IV&V facility uses a subset of the CMM KPAs that apply to the overall organizational core processes that are relevant for IV&V activities. The integration of the BSC core metrics drawn from the CMM and ISO-9126 is applied across both business process and learning and growth tiers of the BSC hierarchy.

Infrastructure (Learning and Growth)

The learning and growth objectives are the drivers for achieving desired results in the other three areas of the scorecard. This is the BSC category that supports the creation of the necessary infrastructure to achieve the strategic goals of the organization. Three key factors in this perspective of the strategy are identified as: employee capabilities, information systems capabilities and employee motivation, empowerment, and alignment.

Employee capabilities require that the strategic skill base necessary to meet organizational objectives be well documented and understood. Employee's personal goals are aligned with organizational goals to allocate retraining, skill enhancement, job enrichment, and promotion opportunities. Information systems capabilities provide the necessary infrastructure to allow employees to see their personal linkages to organizational goals. Information is a key resource to create an employee base that can make informed decisions concerning operational efficiencies. The CMM provides infrastructure to support key processes in the creation or management for the software engineering process improvement through the SEPG, software metrics groups and software quality assurance groups. These structures focus support efforts on key functional areas for continuous process improvement. The measures used by NASA IV&V is to enumerate staffing levels, strategic job coverage, strategic information availability and dependability. The core measures for learning and growth focus on the employee. There are three measures employee satisfaction, employee retention, and employee productivity. The relationships among these measures are hierarchical with employee retention and productivity dependent on degree of employee satisfaction. The enablers to employee satisfaction include employee skill base of core competencies, technological infrastructure, and the general work climate.

SUMMARY AND LESSONS LEARNED

The Balanced Scorecard (BSC) Framework provides the necessary structure to evaluate quantitative and qualitative information with respect to the organization's strategic vision and goals. There are two categories of measures used in the BSC the leading indicators or performance drivers and the lagging indicators or outcome measures. The performance drivers or leading indicators enable the organization to quantitatively track whether or not the organization is achieving short-term operational improvements. The outcome measures or lagging indicators provide objective evidence of whether strategic objectives are achieved and to what degree. The two measures must be used in conjunction with one another to link measurement throughout the organization thus giving visibility into the organizations progress in achieving strategic goals through process improvement.

The development of a core set of metrics for implementing the Balanced Scorecard is the most difficult aspect of the approach. Developing metrics that create the necessary linkages of the operational directives with the strategic mission prove to be fundamentally difficult as it is typical to view organizational performance in terms of outcomes or results rather than focus on metrics that address performance drivers that provide feedback concerning day-to-day organizational progress.

The results from prior case studies are applied that identify key factors that differentiate the use of the BSC for nonprofit government organizations (NPGO) versus industry or for-profit organizations. The steps to build the scorecard do not change significantly but do differ with respect to linking the four views. This is due to the required change in the BSC geography of making the customer objectives the top tier of the hierarchy and the financial objectives the second tier. The BSC geography is mutated to accommodate the thematic focus of the customer. This enables quantifying the financial value of IT investments relative to their contribution towards the organization's progress towards achieving their strategic vision and goals through effective processes and efficient use if IT.

A measurement framework is necessary to bridge the gap between strategic measures of improved reliability, safety, and quality at reduced cost and operational measures of optimization of resource allocations applicable to daily activities to achieve these goals. The BSC provides a means of measuring the efficiency of resource allocations for the operational processes of software and systems verification and validation activities that must then be linked to the high level goals of mission success at reduced cost. In applying the BSC we have learned many lessons of value concerning our strategic planning as it relates to the activities conducted to accomplish daily operational goals. First, we have found that a customer focus of the strategic themes provides the necessary linkages in the BSC to measure our leading and lagging indicators successfully. We have also learned that the CMM and ISO-9000 initiatives are split across the core process tier and the infrastructure tier of the BSC hierarchy. These two findings are essential in applying the BSC to a government or not-for-profit organization such as the NASA IV&V Facility.

REFERENCES

Attewell, Paul (1992). Information Technology and Productivity Paradox. *Report IST 8644358, version 3.1*, Department of Sociology, Graduate Center of the City University of New York.

Boehm, B. (1981). *Software Engineering Economics*, Englewood Cliffs, Prentice-Hall.

Crosby, P. B. (1979). *Quality is Free*. McGraw-Hill.

Crosby, P. B. (1985). *Quality without Tears*. McGraw-Hill.

Eickelmann, Nancy S. (1999). A Comparative Analysis of BSC as Applied in Government and Industry Organizations. *Information Technology Balanced Scorecard Symposium*, Antwerpen, Belgium, March 15-16.

Eickelmann, Nancy S. (1999). Evaluating Investments in Emerging Test Technologies. The Proceedings of the *Sixteenth International Conference on Testing Computer Software: Future Trends in Testing*. Bethesda, MD, June 16-18.

Hetzel, B. (1993). *Making Software Measurement Work*. John Wiley and Sons.

Humphrey, Watts, S. (1990). *Managing the Software Process*. SEI Series in Software Engineering, Addison-Wesley Publishing Company, Pittsburgh, PA.

Humphrey, W. (1989). *Managing the Software Process*. Addison-Wesley.

Humphrey, W., Snyder, T., and Willis, R. (1991). Software Process Improvement at Hughes Aircraft. *IEEE Software*, July.

Jenner, M. (1995). *Software Quality Management and ISO 9000*. John Wiley and Sons.

Jones, C. (1991). *Applied Software Measurement*. McGraw-Hill.

Kaplan, Robert, and Norton, David, (1996). *The Balanced Scorecard: Translating Strategy Into Action*. Harvard Business School Press, Boston, MA.

Keiso, Donald and Weygandt, Jerry (1986). *Intermediate Accounting*. John Wiley and Sons, USA.

McGrath, R. and MacMillan, I. (1995). Discovery-Driven Planning. *Harvard Business Review*, July-August.

Strassman, Paul (1990). *The Business Value of Computers: An Executive's Guide*. The Information Economics Press, New Canaan, Connecticut.

The Balanced Scorecard Institute, (1999). http://www.balancedscorecard.org/default.html .

Chapter XVII

A Comparative Analysis of the Balanced Scorecard as Applied in Government and Industry Organizations

Nancy Eickelmann
Motorola Labs, USA

INTRODUCTION

Organizations have become increasingly dependent on information technologies to conduct daily operations, achieve competitive advantage and to create and penetrate new markets. This dependence has come at a high price, in 1990 U.S. companies spent over $154 billion on information technologies. However, organizations have found it difficult to measure the value added from these investments. Survey results found four significant barriers to measuring financial performance related to information technologies including:
- Difficulty of measuring economic benefits
- Inability to determine returns
- Lack of good metrics
- Incomplete records/accounting of investments

The Balanced Scorecard framework provides part of the structure required to overcome these barriers. How organizations can overcome these barriers and successfully measure performance with respect to achieving strategic plans is the focus of this chapter.

This chapter provides a comparison of results of two case studies regarding the use of the Balanced Scorecard measurement framework. The application of the Balanced Scorecard (BSC) is evaluated for a Fortune 500 information technology organization and a government organization. Both organizations have a business focus of software development. The BSC framework is applied and reviewed in both contexts to provide insight into unique organizational characteristics for government and contract software environments. A specific focus is to inform the use of financial measures such as Return On Investment (ROI) in the government context.

The BSC framework provides the necessary structure to evaluate quantitative and qualitative information and identify the critical linkages between financial measures of past performance and key measures of future performance.

Copyright © 2001, Idea Group Publishing.

Specific objectives of this chapter include:
- A better understanding of applying strategic measurement to measure economic value;
- A characterization of how to choose a core metrics set required for a BSC;
- A characterization of BSC framework requirements in industry settings versus government settings;
- An identification of open measurement issues and further research.

The chapter organization follows a description of two case studies that examine how BSC is applied in an industry versus a government context. First, a brief overview of each organization's strategic vision and goals is given. An analysis of key differences among financial perspectives, customer perspectives, internal business-process perspectives, and learning and growth perspectives for industry versus government organizations is discussed. A unifying thread of the study is to evaluate the use of measurement for the operational, managerial, and strategic purposes of an organization. These case studies provide additional insight into applying BSC in a not for profit government environment. The results also provide a basis for understanding and applying historical cost accounting measures such as ROI in conjunction with measures of long-range competitive capabilities.

BACKGROUND

The BSC architecture provides a framework for translating the strategic plan into concrete operational terms that can be communicated throughout the organization and measured to evaluate its day-to-day viability. The three principles of building a Balanced Scorecard that is linked through a measurement framework to the organizational strategy include:

1) Defining the cause and effect relationships,
2) Defining the outcomes and performance drivers,
3) Linking the scorecard to the financial outcome measures (Kaplan and Norton, 1990).

The initial steps of BSC engage in the construction of a set of hypotheses concerning cause and effect relationships among the stated objectives for all four perspectives of the Balanced Scorecard; A financial, customer, internal business processes, and learning and growth infrastructure perspectives. The measurement system makes these relationships explicit. Therefore, they can be used to assess and evaluate the validity of the BSC hypotheses. The questions asked in each category of the four perspectives are based on the objectives and provide a segue into the cause effect diagramming activity. The measures chosen based on the cause effect relationships provide quantification of the differences between industry and government organizations. There are two categories of measures used in the BSC, the leading indicators or performance drivers and the lagging indicators or outcome measures. The performance drivers enable the organization to achieve short-term operational improvements while the outcome measures provide objective evidence of whether strategic objectives are achieved. The two measures must be used in conjunction with one another to link measurement throughout the organization thus giving visibility into the organizations progress in achieving strategic goals.

The strategic goals of not-for-profit government organizations vary from industry in significant ways. Due to differences the Balanced Scorecard, as applied in industry and government, is approached from two very disparate viewpoints. Industry is very aware of the importance of financial performance measures in managing an organization. Publicly held companies must be responsive to market and shareholder demands. Market share, share price, dividend growth, and other significant results oriented financial measures have been used historically to evaluate an organization. Government organizations must respond to regulatory and legislative acts. One such legislative act is the Government Performance and Results Act (GPRA) passed by Congress and signed by the President in 1993. This act provides a new tool to improve the efficiency of all Federal agencies.

The goals of GPRA are to:
- Improve federal program management, effectiveness, and public accountability;
- Improve congressional decision making on where to commit the nation's financial and human resources;
- Improve citizen confidence in government performance.

A significant difference in the implementation approaches of BSC is due to regulatory influences. Industry is adopting the new BSC methodology by retraining management in implementing a new approach. Government (specifically the DoD 1988 Performance Plan) suggests cost reduction as a strategic goal. A means of accomplishing this is to apply the new BSC approach in terms of what is currently applied i.e., Business Process Improvement (BPI). This saves the cost of retraining, and if effective, creates cost reduction as a strategic goal. A specific difference between government and industry is explicit in the government's focus on cost reduction as compared to industry's focus on revenue generation and profitability. The Balanced Scorecard provides a framework to refocus the organization in evaluating the overall performance of the organization through a linked hierarchy of specific performance drivers and outcome measures.

The case studies are contrasted in the following sections. A discussion of the four perspectives of the BSC provides a foundation to discuss the essential differences and issues in applying the BSC in industry versus government organizations. The organizational vision and goals are described in the next section. This provides a foundation to discuss the measures associated with both organizations according to the BSC.

ORGANIZATIONAL VISION AND GOALS

The application of the Balanced Scorecard (BSC) is evaluated for a Fortune 500, information technology organization and a government organization. Both organizations have a business focus of software development. The strategic vision and goals are described for each organization and are noted as either industry or government. The strategic vision and goals of the organizations are taken from public domain documents.

Industry

Industry has long recognized that strategic planning to define the strategic vision, goals and desired outcomes is an essential best practice for survival in a competitive environment. The vision statement and goals for the organization have been recast in language that preserves the meaning but provides no disclosure of identity.

Vision: To be a world class leader in the development, delivery, and support of advanced technological business solutions for the global telecommunications marketplace.

Goals: To become the number one provider of computer platform solutions based on fundamentals of reliability, data integrity, scalability, reduced cost, and open interfaces.

Government

The strategic plan contains the vision, goals, mission and values for the organization. The Government Performance and Results Act (GPRA), requires all federal agencies to

Table 1. Balanced Scorecard Government and Industry Mappings

	BSC Vision and Strategy Mapping to Operational Focus	
	Government	Industry
Financial	How can we reduce costs and not compromise our mission?	To succeed in financial terms, what results do our shareholders expect?
Customer	To achieve our vision, what customer themes do we need our customers to rely on us?	To achieve our vision, how should we appear to each category of our customers?
Internal Business Process	To satisfy our customers, what business processes do we need to excel at to differentiate us and create a COE?	To satisfy our shareholders and customers, what business process are critical for us?
Learning & Growth	What infrastructure do we need to sustain our ability to change and improve?	What do we need to promote the necessary change and improvement required to meet our vision?

establish strategic plans and measure their performance in achieving their missions. The NASA IV&V Facility vision and goals are stated below.

Vision: To be world-class creators and facilitators of innovative, intelligent, high performance, reliable informational technologies that enable NASA missions.

Goals: To become an international leading force in the field of software engineering for improving quality, safety, reliability, cost, and performance of software systems; and to become a national Center of Excellence (COE) in systems and software independent verification and validation.

The government and industry strategic vision and goals provide the initial high level questions as stated in Table 1. The BSC for industry focuses on linking strategic measures through the hierarchy from the learning and growth tier to the business process tier then to the customer tier and finally to the financial tier. This naturally draws the focus to the financial objectives of the organization and the outcome measures of operational results. The BSC for a not-for-profit government organization focuses on linking strategic measures through the hierarchy from the learning and growth tier to the business process tier then to the financial tier and finally to the customer tier. This changes the focus to what Kaplan calls customer themes. The NPGO has financial objectives of eliminating redundant costs and optimizing resource allocation and utilization to achieve financial constraints. The financial measures and process measures are both linked to the customer themes. The next sections identify the performance drivers and outcome measures for each of the four segments of the BSC: financial, customer, process, learning and growth.

FINANCIAL MEASURES

Government agencies and other nonprofit organizations exist to accomplish a mission. In contrast, private-sector companies are driven by profit as a performance metric that is easy to measure and closely related to the strategic success of the company. Nonetheless government organizations have a long-term interest in financial performance that cannot be ignored. Government agencies that are not responsive to budgetary constraints, and ever shrinking resources are subject to public scrutiny and critical review. This fact makes internal efficiency or cost effectiveness one of the key strategic goals of a government agency. Therefore the financial measures for government must be viewed as a balance between mission success and effective cost containment or reduction. The core financial measures for the Balanced Scorecard include return-on-investment, economic value-added, profitability, revenue growth/mix and cost reduction productivity. Financial performance is only one indicator of success. Government organizations must contain or reduce costs while maintaining the ability to complete their mission to the satisfaction of their customers. The financial objectives must be linked with the overall scorecard through cause and effect mappings documented by objective measurements.

A conceptual framework for financial reporting is a useful tool in evaluating financial measures. The 1976 FASB publication of a Discussion Memorandum entitled *Conceptual Framework for Financial Accounting and Reporting: Elements of Financial Statement and Their Measurement*, provides a measurement framework for financial reporting practice. There have been four subsequent documents issued as *Statements of Financial Accounting Concepts*, taken together these documents provide a foundation for defining financial performance drivers and outcome measures (Keiso and Weygandt, 1986). There are three levels to the framework, creating a hierarchical structure. Level one identifies the objectives for collecting financial information. Level two defines the qualitative characteristics of the information collected and the elements from which they are derived. Level three defines the measurement and recognition concepts for financial reporting.

Level One: Objectives

Level one objectives provide financial information that is useful in
1) making investment decisions,
2) assessing future cash flows, and
3) identifying enterprise resources, claims to resources, and changes in the status of resources.

Level Two: Qualitative Characteristics and Elements

Level two provides a two-tiered enumeration of qualitative characteristics required for objective financial measures and the elements that are measured.
- Primary qualitative characteristics include:
  1) relevance based on predictive value, feedback value, and timeliness,
  2) reliability based on verifiability, representational faithfulness, and neutrality.
- Secondary qualitative characteristics include:
  1) comparability and
  2) consistency
- The elements that are measured include assets, liabilities, equity, investment by owners, distribution to owners, comprehensive income, revenues, expenses, gains and losses.

Level Three: Recognition and Measurement Concepts

Level three of the framework identifies the essential recognition and measurement concepts including:
1) Assumptions of economic entity: monetary unit, going concern, and periodicity.
2) Principles of historical cost: revenue recognition, matching and disclosure.
3) Constraints: cost-benefit, materiality, conservatism and industry practice.

This framework represents the key factors required for computing financial performance measures. The complexity of developing the measures for the Balanced Scorecard is an artifact of the complexity of developing measures that adhere to all the required properties and provide the desired information. The leading and lagging indicators or performance drivers and outcome measures for each organization are given in Table 2.

Industry

The industry organization in this study was a producer of information technology artifacts but did not manage information technology as a key component to attaining a Balanced Scorecard. The information management and analysis systems were primarily focused on manual data collection and team-based dissemination through face-to-face meetings. The financial officers of the company used information technology very effectively and tracked financial measures closely. The chief financial officer (CFO) of the organization recognized the need for visibility and leading indicators of nonconformance in the development of the BSC. This may be a direct result of the CFO having the most timely information in the organization through a highly efficient use of information technology in monitoring day to day operations from a financial perspective.

Government

The government outcome measures (see Table 2) provide insight into the duality of the financial goals of the government organization. The monetary measures of cost reduction are achieved through efficient resource allocation and utilization and avoidance of redundant efforts or rework, all of these efficiencies must not be achieved at the expense of the identified customer themes. This requires a high degree of visibility into the customer's strategic mission and goals, as well as their internal business processes. Cost reductions must be evaluated relative to the ability to successfully achieve strategic goals. It is true that information technology and its effective use must play a key role in achieving the strategic mission. The increased visibility into business processes provides a means to identify and measure core competencies required in the daily organizational activities. Cost reduction must result from the appropriate allocation of resources towards the most efficient use of capital, personnel and technology.

CUSTOMER MEASURES

Government and industry-based organizations may have a diverse set of customers, yet the core measurement areas for customer focus are the same. The core measures include market share, customer retention, customer acquisition, customer satisfaction and customer profitability. These five core measures are used in conjunction with one another to evaluate and profile the status of the customer base of an organization. Many of the core customer

Table 2. Balanced Scorecard Government and Industry Mappings

	BSC Financial Performance Measures	
	Government	Industry
Financial	"How can we reduce costs and not compromise our mission?"	To succeed in financial terms what results do our shareholders expect?"
Performance Drivers	Activity based process-centered costing Technology utility	
Outcome Measures	Return on Investment/Payback Economic value added Cost reduction	Revenue growth Operating margin Incurred expenses

measures are not applied in the same context. The differences in the customer focus between the industry and government customer base is evidenced by the segmentation and differentiation of customers. In addition, the time context for the strategic measures of an industry organization is considerably shorter than that of the government organization. Industry monitors customer behavior and responds to marketing opportunities much more readily than the government bureaucracy. This is particularly evident when the primitive measures that support the outcome measures are identified. The performance drivers and outcome measures for both organizations are given in Table 3.

Industry

The industry organization segments its customers by three major market segments and then by product groups as assigned to primary customers. The primitive product-related measures include volume usage, end use, benefit expectations, brand loyalty, and price sensitivity. The product for the industry-based organization is commercially marketed to a specialized market segment in telecommunications software and products. The primary contact with the key customers is a knowledgeable and highly trained specialized sales force. The sales force is not used as strategic source of information in marketing and management. Information technology is not used to connect the widely distributed sales force to the corporate knowledge base and operational information. The focus is on selling the technology, not using the technology to facilitate sales.

Government

The government organization segments its customers according to whether the customer is internal or external to the government. Internal customers must belong to one or more of eleven NASA mission centers. Inclusion in an external federal or state government agency, academia, or industry further segments external customers. The product is a service and is differentiated as independent verification and validation, independent assessments, applied and basic software research and special services. The NASA FY1999 Final Performance Plan outlines selected measurements to evaluate progress the Agency intends to make in FY99 toward the achievement of its goals. The revisions in

Table 3. Balanced Scorecard Government and Industry Mappings

	BSC Customer Measures	
	Government	Industry
Customer	"To achieve our vision how do we want our customers to perceive us?"	To achieve our vision how should we appear to each category of our customers?"
Performance Drivers	Quality of service as defined by the customer Responsiveness as defined by the customer Customer surveys Benefit expectations relative to customer themes	On-Time Delivery (OTD) Number of outages Duration of outages Number of high severity defects Volume usage, benefit expectations, brand loyalty, and price sensitivity, end use
Outcome Measures	Achieving customer themes Acquisition/Retention of customer base by segment	Acquisition/Retention of customer base versus plan Acquisition/Retention of customer base by segment

the FY1999 Final Performance Plan address inputs from our customers requesting that we:
1) provide better linkage to our budget,
2) incorporate additional information establishing the reasonableness of our performance targets;
3) better identify the means we will use to verify our performance.

This focus provides a customer driven linkage from the customer to the financial and mission measures. The mission themes identified are safety, reliability and quality.

INTERNAL BUSINESS PROCESS MEASURES

The internal business process measures for government and industry have focused on similar key factors and have introduced the concepts in much the same way. Process improvement frameworks have included Total Quality Management (TQM), the Software Engineering Institute's Capability Maturity Model CMM, and more recently ISO-9000 Certification. All of these efforts share a customer focus of measurable business process improvements that result in cost reductions and cycle time improvements. This foundation is used by both organizations as a starting point for developing the additional framework of strategic measures from multiple perspectives. A key resource in deploying effective business processes is information and information technology.

Industry

The industry organization viewed information technology primarily as an end use of their product line (see table 4). There were two exceptions in their application of IT in business processes: they deployed metric dissemination using web-based technology and as previously mentioned, the financial data collection and reporting information resources were highly developed. A key factor in business process improvement is the presence of an effective communication structure that provides the necessary flow of information throughout the organization. Effective process improvement based on employee participation requires the communication channels of the organization to support bottom-up as well as top-down information flows. The ease of collecting and disseminating information vertically and horizontally in the organization is key to quick response and adjustment of operational focus to align with strategic initiatives. The industry organization had a predominantly top-down, unidirectional communication structure.

The use of information and information technology as a strategic tool to improve corporate competitiveness was not explicit. The central use of process improvement efforts concentrated on measures to estimate the effort required to develop a product and counting the defects in the product as it progresses to delivery status. Specific software measures that were identified as requiring more information from the business processes were software estimation measures and models. A key requirement to make the measure representational would be to base it on product complexity or functionality versus product size. This would require the organization to characterize their product in terms of complexity relative to functionality. This focus on product was necessary but insufficient as it ignored the critical strategic component of using information technologies in measuring core customer measures of profitability, satisfaction and retention.

Government

The government organization views information and information technology as a key resource and a high cost center in achieving their strategic missions. Using information and information technology to measurably improve business processes must be managed carefully. Strassman (1992) and Attewell (1990) have both noted the productivity paradox of IT and caution organizations to invest and use IT judiciously. NASA seeks continuous process improvement of information resources management by evaluating the progress toward achieving the objectives of the NASA *Information Resources Strategic Plan.* The mechanism for integrating the measurable improvement into day to day activities is through the *Information Resources Self-Review Process Guide.* The guide is a team developed annual review method and documentation to provide feedback to the organization concerning the effectiveness of the their use of IT. The focus of the assessment is on identifying the implementation opportunities to progressively maintain best practices at all NASA Centers. Information technology is viewed as a critical resource that is measured relevant to its contribution to strategic goals. To measure the use of IT for the BSC the guide identifies best practices, accomplishments, areas for improvement, a risk rating for IT, assessment of the guide itself and the performance measures. The guide identifies best practices relative to success in achieving continual improvements in service to customers, improved efficiency and cost effectiveness, and standardization of practice. Accomplishments are identified for the prior fiscal year that apply to information management, information resource oversight, information technology security oversight, and MIS management. Accomplishments are then measured according to cost savings, cost avoidance, operational efficiencies, labor savings, cycle time decreases, mission performance, and standardization improvements.

The independent self-assessment results are integrated into the development of the Annual Integrated Program Assessment (AIPA). This report is used to assess the agency-wide areas of excellence, areas of risk and identification of best practices. The self-review process engages in two-way communication of information through the post-review teleconference. This teleconference is the communication channel to facilitate the dissemination of best practices, areas needing improvement, and performance measures across the agency. This practice creates a critical internal business process structural link for the BSC. The government organization uses information technologies in its operations and recognizes the need to measure the effectiveness of IT as deployed across the organization to achieve strategic objectives as well as a product to be used in a mission. This is a sophisticated level of understanding concerning the use of IT in an organization and its strategic role in success

Table 4. Balanced Scorecard Government and Industry Mappings

BSC Internal Business Process Measures		
	Government	Industry
Internal Business Process	"To satisfy our customers what business processes do we need to excel at to differentiate us and create a COE?"	To satisfy our shareholders and customers what business process are critical for us?"
Performance Drivers	Activity based process-centered costing Activity-based issue resolution count Technology utility measures Technology cost-benefit profiles Diminishing effort/issue ratio	Quality metrics Productivity Schedule slippages
Outcome Measures	Documented technological effectiveness Quality of services delivered and technologies applied	Frequency of post release faults Software Reliability Engineering, SRE reliability goal

or failure of the overall mission. Specific measures are used to track and evaluate the progress towards consolidation, modernization, interoperability, standardization, increased use of commercial off the shelf COTS products, high customer satisfaction, training goals and high return on investment ROI for the agency (see Table 4). These measures represent linkages from one component of the BSC to another. It is this mixture that provides the integration of the cause and effect graphing essential for constructing the BSC.

LEARNING AND GROWTH MEASURES

The learning and growth objectives are the drivers for achieving desired results in the other three areas of the scorecard. This is the BSC category that supports the creation of the necessary infrastructure to achieve the strategic goals of the organization. Three key factors in this perspective of the strategy are identified as: employee capabilities; information systems capabilities; employees' motivation, empowerment, and alignment.

Employee capabilities require that the strategic skill base necessary to meet organizational objectives be well documented and understood. Employees' personal goals are aligned with organizational goals to allocate retraining, skill enhancement, job enrichment, and promotion opportunities. Information systems capabilities provide the necessary infrastructure to allow employees to see their personal linkages to organizational goals. Information is a key resource to create an employee base that can make informed decisions concerning operational efficiencies. Edward Lawler discusses the ramifications of a learning and growth infrastructure for participative management practices and identifies the appropriate investment in information technology as a key component to support the necessary information flows for distributed and pushdown decision making approaches. Employee empowerment is the authority to make decisions, motivation is linking responsibility to accountability through the reward structure. Recognition of contributions must recognize that not all employees will consider all incentives to be a true reward.

The core measures for learning and growth focus on the employee. There are three measures: employee satisfaction; employee retention; and employee productivity. The relationships among these measures are hierarchical with employee retention and productivity dependent on degree of employee satisfaction. The enablers to employee satisfaction include employee skill base of core competencies, technological infrastructure, and the general work climate. This perspective of the BSC is often where inexperienced managers cut costs and undermine the organizational strategic plan. Cost cutting in this area typically results in short term improved financial results. The detrimental effects of low employee morale and diminished productivity and creativity are not evidenced immediately, but surface gradually over time.

Industry

The industry organization provided three focused programs to address this perspective of the BSC (see Table 5). The use of a single day scheduled annually to allow employee input to the company direction. A training program administered by human resources required 40 hours per year of training relative to assigned job skills. Employees were also encouraged to contribute to measurable process improvement through suggestions. Suggestions were tracked and evaluated. These three efforts were identified as insufficient to attain the high level goals for the organization. In addition, it was noted that they had ignored the need for information technology to support and enable significant employee contributions to be communicated. Employee retraining and educational opportunities were not evaluated with respect to the individual goals of the employee. This aspect of the BSC may not show

Table 5. Balanced Scorecard Government and Industry Mappings

BSC Learning and Growth Measures		
	Government	Industry
Learning & Growth	"What infrastructure do we need to sustain our ability to change and improve?"	"What do we need to promote the necessary change and improvement required to meet our vision?"
Performance Drivers	Communications effectiveness measure R&D project penetration % R&D technology transfer ratio Information review	Business mix Loss ratio
Outcome Measures	Positive scores on staff climate survey Survey of customer satisfaction with staff and support personnel Appropriate staffing levels in all functional areas Alignment of individual work plan activities and results Strategic job coverage Strategic information availability Staff climate survey	Customer satisfaction based on Quarterly review measure Web-based communication results and metrics data Well defined strategic plan and translation process Employee survey

detrimental effects in its absence but there is surely an opportunity cost. Aligning the organizational efforts with personal goals of employees creates a unique work climate that can be used as a competitive advantage. Harvey Mackay describes the use of this strategy in his book "Beware of the Naked Man Who Offers You His Shirt". The employee satisfaction survey was used to measure the overall effectiveness of learning and growth relative to the general work climate. The employee survey results documented potential areas for improvement as well as areas that demonstrated accomplishment.

Government

The government focus addresses employee capabilities, information systems capabilities and employee motivation, empowerment, and alignment (see Table 5). Employee capabilities are addressed through: an aggressive hiring of new personnel in the research and project management areas including public information, business development, metrics, education, and technology transfer; balancing of the work load among all staff; providing resources needed to fulfill staff roles and responsibilities. Information systems capabilities focus on supporting the distributed work context of NASA facilities. An area of targeted improvement is accounting and reporting systems, both between Ames and the IV&V facility and within the facility. This application of information and information technology recognizes the need of the distributed and co-located personnel to have timely information concerning their available resources and appropriateness of resource allocations. To support employee motivation, empowerment, and alignment the following initiatives are implemented. A review and implementation of effective policies and procedures will support staff needs (e.g., travel, training), improve efficiency of the contracting processes, create new contract vehicles that provide greater flexibility, ensure that required skill sets are in place and kept current for both NASA staff and contractors, and provide staff with cross-training opportunities. Specific activities that are designed to implement these initiatives include:
1) Common effective accounting and reporting systems across programs
2) Responsive and flexible administrative procedures

3) Hiring of additional contract support specialists
4) Local contracting process mechanisms in place
5) Facility-specific employee handbook
6) Skills assessment completed
7) Skill set profiles for hiring staff documented
8) Formal training plan created
9) Training handbook updated

Measures of success that are leading indicators and important barometers of strategic direction are positive scores on staff climate surveys, surveys of customer feedback related to satisfaction with staff and support personnel, fully functioning performance evaluation processes (internal and external critical successes).

BSC CHARACTERIZATION FOR GOVERNMENT VERSUS INDUSTRY

The comparative analysis of the Balanced Scorecard framework for a Fortune 500, information technology organization and a government organization provides insight into unique organizational characteristics for government and contract software environments. The BSC framework provided the necessary structure to evaluate quantitative and qualitative information contained in the critical linkages between financial measures of past performance and key indicators (measures) of future performance.

A foundation for the BSC framework requirements in industry settings versus government settings is provided through a brief overview of each organization's strategic vision and goals. The following section provides an analysis of key differences among financial perspectives, customer perspectives, internal business-process perspectives, and learning and growth perspectives for industry versus government organizations (see Table 6).

Financial Perspective

A key difference in the financial perspective appears to be the result of a duality in the financial goals for government, specifically, the goal of mission success at lowest cost or reduced cost. This is significantly different from the industry perspective that focuses specifically on singular financial goals. The government organization had financial performance drivers related to the cost reduction strategy. The cause and effect mapping of the BSC has an added linkage from the financial perspective to the internal business process perspective. The key aspect is the efficient allocation of advanced IV&V technologies that provide a least cost solution to the customer. This is in sharp contrast to the industry scorecard that tended to focus on financial outcome measures.

Customer Perspective

The customer perspective is contrasted based on what, at times, were subtle differences in the areas of operational and managerial focus. The customer perspective was distinct in the strategic focus for government by targeting customer themes. We found that altering the "geography" of the BSC by placing the customer focus at the top of the hierarchy and moving the financial focus to the second tier enables the strategic focus to emphasize measuring customer thematic. This is appropriate as the primary concern for NASA is mission success not financial success. This focus provides a customer driven linkage from the customer to the financial measures. Thus providing the necessary foundation to graph the cause and effect relationships from one component of the BSC to another, integrating the focus of

Table 6. Government and Industry BSC Strategic Measurement Focus

BSC Perspective	NPGO/Industry	BSC Measurement Focus		
		Operational Focus	Managerial Focus	Strategic Focus
Financial	Government	Cost	Efficient resource utilization	Mission Success
	Industry	Profit	High investment returns	Financial Success
Customer	Government	Monitor public sentiment	Achieving customer goals	Customer themes
	Industry	Target profitable customer base	Delivering on customer needs	Customer satisfaction
Business Process	Government	Efficient resource allocation	Business process improvement	Process efficiency
	Industry	Low cost producer	Continuous process improvement	Process optimization
Learning and Growth	Government	Workforce skills and training IT architecture IT infrastructure	IT productivity enhancements IT Utility	Knowledge management
	Industry	Workforce skills and training IT architecture IT infrastructure	IT performance IT utility	Enterprise resource planning

mission success. The top-tier, customer perspective, captured aspects of the NASA IV&V strategic focus areas of safety, reliability, quality, and cost performance. The customer objectives represented the key IT services, information management and information analysis that the NASA IV&V facility delivers for its customers. The financial objectives became the enablers for helping achieve the overall customer objectives, which included services at reduced cost.

Internal Business Processes

The internal business processes have a foundation in the core business process improvement (BPI) strategies popularized in the late 70s and 80s The application of quality initiatives and defect reduction productivity strategies are both central to core values expressed in the BSCs of the two organizations. The government organization is differentiated from the industry organization in the following regard. The management of the government organization is trained in business process improvement approaches. The cost reduction strategy of the government extends to applying the BSC, using the tacit understanding of the management by introducing guides and materials that apply BSC but use the familiar terminology and BPI. This approach to BSC eliminates the need for retraining in new methods as the BSC structure is incorporated into the process materials introduced.

Learning and Growth Perspective

The learning and growth perspective of the BSC is the foundation for long term success. There are not significant differences in the application of this perspective between government versus industry organizations. The primary differences are a reflection of regional and strategic differences that are not attributable to either type of organization. Key differences in strategy were the focus on hiring of specialized personnel for the government and a belief in industry that a 40-hour per annum training program would fulfill their needs. This is a difficult perspective for the BSC, especially in unstable economic times where employee morale may be low due to external factors that are negatively impacting an organization. Examples of this might include congressional deadlock on budget resolution causing a government shutdown or a reduction in product profitability due to external price competition. Measurements of the learning and growth perspective must filter for external impacts that are not reflective of the internal strategic alignment of the BSC with operational effectiveness and managerial competence.

CONCLUSIONS AND FUTURE RESEARCH

This case study is designed to contrast BSC implementation issues for government versus industry organizations. The literature has not made this distinction explicit previously and has often combined BSC implementation issues for industry and government measures and results. This may be due to an unclear perspective as to which differences are significant and how they impact the implementation and success of a BSC. The results of our case studies are explained in the context of the key practices to build a BSC as described in the GAO report "IT Performance Measurement Guide". This report was prepared under the direction of Dave McClure and is an example of a series of case studies of BSC implementations from both government and industry organizations. The report is organized to provide generic BSC implementation guidelines and caveats. The fundamental practices are described in a proscribed sequence of five key practices:

1) Create and follow an information technology "results chain"
2) Follow a balanced approach focusing on the four perspectives and narrowing the measures to vital objectives in each tier
3) Create target measures, results, and accountability at decision-making tiers
4) Build a comprehensive measurement data collection, and analysis capability
5) Strengthen IT processes to improve mission performance

These generic guidelines can be used across diverse organizations, however, the success of the BSC implementation may be dependent upon recognition of the BSC differences and their potential impact for each type of organization.

Specific findings of this study demonstrate significant differences in applying the key practice areas identified by the GAO. The first key practice, to create and follow an information technology "results chain," requires that the strategy be defined according to what the organization will do and will not do. Results outside the scope of the strategy should be noted and the chain evaluated for strategic relevance.

The second key practice area, is to follow a "balanced" approach focusing on the four perspectives. This requires narrowing the measures to vital objectives in each tier. The objectives that are vital are a function of the mapping of the four tiers of the scorecard. It is shown that the BSC for government requires adjustments to the scorecard geography (e.g., moving the customer tier to the top level, followed by the financial tier, process tier and finally the innovation and learning tier). The reverse mapping of the financial tier to the customer tier is critical for government organizations and creates a unique relationship with internal business processes, which map resource utilization efficiency directly to the financial tier and success of customer themes to the top tier. This is clearly shown when the financial measures are moved to the second tier position of the scorecard allowing the customer measures to be the first tier of the framework. This alteration of the positioning or geography of the BSC provides the means to link the financial measures into the customer themes as expressed in the strategic vision and goals. Customers are not stratified for the government organization, and the strategic goals are stated in terms of thematic success relative to the customer base.

The third key practice, creating target measures, results, and accountability at decision-making tiers, provides an insight into the impact of BSC differences. Decision-making criteria polarize around financial issues for both types of organization. Therefore, the differences for the two types of organizations in applying the BSC are predominantly related to the duality of the financial goal structure and are evidenced in many aspects of the BSC implementation. Financial measures for industry focus on outcome measures. Financial measures for government focus on both leading and lagging indicators. This difference becomes obvious when the decision structure of the two types of organizations are evaluated. A government organization is constrained by issues of cost and cost containment, while industry is focused on profitability.

The fourth key practice, building a comprehensive measurement, data collection, and analysis capability is impacted the least by whether an organization is governmental or industrial. This key practice is more a factor of the organizational structure, degree of automation in core business processes, and degree of decentralization of operations.

The fifth key practice, strengthening IT processes to improve mission performance, requires that both types of organizations apply appropriate technologies in achieving their missions. For industry this aspect is critical as it often separates the industry leaders from their competitors.

There are more subtle differences between industry and government organizations that will require additional investigation. One area of interest is the viability of reducing BSC implementation costs through documentation vehicles and leveraging of historical organizational knowledge. Another research focus that might prove extremely fruitful is development of a corollary framework to the measurement framework identified for financial measures. The financial measurement framework provides a three level hierarchy of objectives, qualitative characteristics and elements, and recognition and assumptions. There currently is no such measurement framework for the customer perspective, the internal business process perspective or the learning and growth perspective. This is identified as an open measurement issue requiring further research.

REFERENCES

Attewell, Paul, (1992). Information Technology and Productivity Paradox, Department of Sociology, Graduate Center of the City University of New York, *Report IST-8644358, version 3.1.*

Eickelmann, Nancy S., (1999). Software Measurement Frameworks to Assess the Value of Independent Verification and Validation. In the *Proceedings of the 24th Annual Software Engineering Workshop,* NASA GSFC Software Engineering Laboratory.

Eickelmann, Nancy S., (1999b). A Comparative Analysis of BSC as Applied in Government and Industry Organizations. *Information Technology Balanced Scorecard Symposium*, Antwerpen, Belgium.

GAO/AIMD-97-163 (1997). *Executive Guide: Measuring Performance and Demonstrating Results of Information Technology Investments.* General Accounting Office Accounting and Information Management Division. September.

Humphrey, Watts, S., (1990) *Managing the Software Process.* Addison-Wesley Publishing Company, SEI Series in Software Engineering, Pittsburgh, PA.

Kaplan, Robert, and Norton, David, (1996) *The Balanced Scorecard: Translating Strategy Into Action.* Harvard Business School Press, Boston, MA.

Kaplan, Robert, and Norton, David, (1992). The Balanced Scorecard—Measures that Drive Performance. *Harvard Business Review*, Vol. 74, No. 1., January-February.

Kaplan, Robert, and Norton, David, (1993). Putting the Balanced Scorecard to Work. *Harvard Business Review*, Vol. 71, No. 5, September-October.

Keiso, Donald and Weygandt, Jerry, (1986). *Intermediate Accounting*. John Wiley and Sons, USA.

Strassman, Paul, (1990). The Business Value of Computers: An Executive's Guide. *The Information Economics Press,* New Canaan, Connecticut.

About the Authors

EDITOR

Wim Van Grembergen is a professor at the Business Faculty of UFSIA (University of Antwerp) and is a guest professor at the University of Leuven (KUL). He teaches information systems at undergraduate and executive level, and researches in business transformations through information technology, audit of information systems, and IT evaluation. Dr. Van Grembergen presented at the European Conference on Information Systems (ECIS) in 1997 and 1998 and at the Information Resources Management Association (IRMA) Conferences in 1998, 1999 and 2000. He is Track Chair "IT Evaluation Methods and Management" for the 2000 and 2001 IRMA conferences. He published articles in journals such as *Journal of Strategic Information Systems, Journal of Corporate Transformation, Journal of Information on Technology Cases and Applications, IS Audit & Control Journal* and *EDP Auditing (Auerbach)*. He also has several publications in leading Belgian and Dutch journals and published in 1997 a book on business process re-engineering in Belgian organizations and in 1998 a book on the IT Balanced Scorecard. He is engaged in the development of CobiT 3rd Edition. Until recently he was Academic Director of the MBA Program of UFSIA and presently he is coordinator of a master program on IT-audit. Professor Van Grembergen has consulted with a number of organizations and is member of the Board of Directors of an IT company servicing a Belgian financial group. His e-mail address is: wim.vangrembergen@ufsia.ac.be.

CONTRIBUTING AUTHORS

Egon Berghout is an associate professor of Information Management at Delft University of Technology and associate of the M&I/PARTNERS group of IT strategy consultants. He is specialised in improving the efficiency and effectiveness of the information function (valuation). He published more than 40 articles, four books and is a frequently invited conference speaker. He is chairman of the Dutch Information Economics Working Group and member of the executive committee of the European Conference on the Evaluation of Information. Currently, he is a visiting researcher at the London School of Economics. He holds a PhD in Informatics (Delft University of Technology), MSc in Information Management (Tilburg University).

Rodney Carr is a lecturer at Deakin University, in the School of Management Information Systems. His specialty is statistics and data analysis. As a statistician, Rodney has been involved in numerous projects and has wide experience with analysis of data for many different types of applications. His involvement also covers methods of data collection and questionnaire and survey design. Rodney is the author of a number of texts and software packages, mainly for mathematics and data analysis. A number of his products have gained widespread recognition and are used by hundreds of users in many organisations nationally and internationally.

Nancy Eickelmann is currently a research scientist for Motorola Labs and is leading the Motorola software and system test process measurement and evaluation research initiative. Prior to joining Motorola she was program manager at the NASA/WVU Software Research Laboratory. Her research focused on integrating the Balanced Scorecard into the NASA context to provide a measurement framework for software test technology improvements. Before joining NASA she was a member of the Advanced Programs Research Group at MCC where she developed a measurement framework for guiding the decision-making process in product line development. Dr. Eickelmann began her research career as a member of the technical staff at Hughes Research Laboratory (HRL) in Malibu, California while completing her doctorate at the University of California, Irvine. She was named a Hughes Doctoral Fellow while working with the Formal Methods group at HRL and with Dr. Debra Richardson's Formal Methods and Software Testing Group at UCI. Dr. Eickelmann has collaborated internationally on research projects for defense systems, space station applications, space shuttle and global software development.

Bill Hewett is a senior academic in the School of Management Information Systems at Deakin University on the Warrnambool campus. He has been actively involved as a consultant and as an executive in the IT industry for more than 35 years. Since joining Deakin in 1991 Bill has maintained his involvement with commercial applications of information technology. He has acted as a consultant specialising in the evaluation of IT services and more recently in the evaluation of ecommerce initiatives. Bill's research expertise in the areas of managing the information systems function and the strategic implications of ecommerce has given him a valuable perspective on ecommerce in rural and regional areas.

Soo Kyoung Lim, PhD, is a principal for Consulting SSU of LG-EDS Systems Inc. in Korea. Dr. Lim is responsible for driving the e-solutions across private and public sector.

Key areas of focus are e-business services, e-solutions trends, and e-business models. Before joining LG-EDS, Dr. Lim worked at NCA (National Computerization Agency), where she was a director of IT evaluation and auditing division. She had played an important role as an auditor to a number of IT projects sponsored by government. She also participated in a project subject to evaluate IT levels of city governments in Korea. Dr. Lim holds a bachelor's degree in Industrial Engineering from Korea University and PhD from KAIST (Korea Advanced Institute of Science and Technology). She also has studied in University of Wisconsin at Madison as a post-doctor.

Chad Lin is completing a PhD in Information Systems at Curtin University of Technology, and has published papers at conferences such as ACIS, IRMA and WAWISR.

Professor Dr. Maarten Looijen is professor of Management of Information Systems at Delft University of Technology. He published more than 100 articles and 10 books, among others, the best-selling book *Information Systems, Management, Control and Maintenance (Kluwer)*. He participates in many committees, such as, the Dutch Scientific Technical Council on ICT and is involved in many projects in developing countries (Kenya, Tanzania, Lesotho, Mozambique, Rwanda, Botswana, Swaziland, Namibia, The Philippines, India, China, Bulgaria and the Ukraine). He is chairman of the editorial board of *ICT Management Select*. He holds a PhD from Eindhoven University of Technology and a MSc from Delft University of Technology

Peter Marshall is an associate professor in the School of Management Information Systems at Edith Cowan University. He lectures at master's and doctoral levels in electronic commerce, strategies for e-business, business integration and organisational transformation using IT to postgraduate students both in Australia and throughout the SE Asian region. He is an active researcher in the field of electronic commerce and new organisational forms, and his research activities are currently being supported by a number of research grants to look into developments in virtual organisations and electronic commerce initiatives in Australia. Peter is a committed action researcher, and guides action research initiatives in government and industry whenever suitable opportunities arise.

Judy McKay is a senior lecturer in the School of Management Information Systems at Edith Cowan University. She teaches a number of master's and MBA-level units in the school, specializing in information management, e-business, the realisation of benefits from IT investments and information requirements determination. She is an active researcher in the fields of electronic commerce, business integration, strategic business network information systems planning and new organisational forms, with a particular interest in the issues, challenges and benefits associated with operationalising electronically supported networked organisations. In addition, she is very interested in the practice of action research, and has written and published extensively on this particular subject. Her research interests frequently find her undertaking action research projects in government and industry.

Bram Meyerson is the founder of QuantiMetrics, a company dedicated to improving IT performance (efficiency and effectiveness) of its clients and maximising the value of their IT investments. QuantiMetrics also conducts research into emerging trends in the software development industry and manages an international benchmarking network. For the last seven years Bram has performed system-delivery performance assessments and benchmarks for some of the largest IT groups. His clients currently include major banking, insurance, logistics, telecommunication and utility organisations. Bram has also run a number of projects which gauge the effectiveness of IT divisions by assessing the alignment between IT and the business(es) they serve. Bram participates in international conferences and has presented a number of papers at such events. Bram has degrees in Applied Mathematics and Computer Science and is also a Certified Function Point Specialist (CFPS).

Kenneth E. Murphy (Ken) holds a PhD from Carnegie Mellon University in Operations Research and has been employed by Florida International University since 1994. Dr. Murphy's research interests are varied, spanning the quantitative methods and technology arenas. Specifically, he has worked and published in the areas of machine and personnel scheduling as well as in the organizational learning literature. More recently, his focus has shifted to the value of implementing large-scale packaged software in global organizations and the implications of this trend for education. Dr. Murphy has published in *Operations research, Naval Research Logistics* and *Communications of the ACM journals.* He is an active member of The Institute for Operations Research and Management Science (INFORMS) and the Decision Sciences Institute (DSI).

Dr. Nandish V. Patel lectures in Information Systems in the Department of Information Systems and Computing at Brunel University, England, where he is a member of the Centre for Strategic Information Systems. He is advisor to companies in software quality management and telecommunications market research. He is known for originating the concept of tailorable information systems and was invited to prepare a position paper on the subject for the international ACM Special Interest Group on Computer Supported Co-Operative Work. Dr. Patel has written numerous nationally and internationally referred journal and conference papers and book chapters, and has spoken at conferences on information systems in the USA and Europe. His work on evaluating evolutionary information systems has received a citation of excellence for practical implications.

Graham Pervan is Professor of Information Management in the School of Information Systems at Curtin University of Technology in Perth, Western Australia. He is currently the Asia-Pacific Editor for the *Journal of Information Technology* and IT Management editor for the *Australian Journal of Management.* He has more than 100 publications in journals such as *DSS, JIT, AJIS, JCIS* and others, and conferences such as ECIS, ACIS, WITS, IRMA and many others.

Mahesh S. Raisinghani, is the founder and CEO of Raisinghani and Associates, a diversified global firm with interests in software consulting and technology options trading. As a faculty member at the Graduate School of Management, University of Dallas, he teaches MBA courses in information systems and e-commerce, and serves as the Director of Research for the Center for Applied Information Technology. As a global thought leader on e-business and global information systems, he has been invited to serve as the local chair of the World Conference on Global Information Technology Management and the track chair for E-Commerce Technologies at the Information Resources Management Association. He has published in numerous leading scholarly and practitioner journals, presented at leading world-level scholarly conferences and has recently published his book *E-Commerce: Opportunities and Challenges.* He has been invited to serve as the editor of the special issue of the *Journal of Electronic Commerce Research on Intelligent Agents in E-Commerce*. Dr. Raisinghani was also honored to be selected by the National Science Foundation after a nationwide search to serve as one the panelists on the Information Technology Research Panel for awarding $500,000 grants to appropriate proposals. He serves on the editorial review board for leading information systems publications and is included in the millennium edition of Who's Who in the World and Who's Who in Information Technology. He can be contacted at mike@uta.edu.

Theo J.W. Renkema is working as an Executive IT Consultant in the Rabobank Group and is employed as a Senior Research Fellow in the Department of Technology Management of Eindhoven University of Technology, both in the Netherlands. He previously worked with a large financial services firm and was a senior management consultant in an electronics conglomerate. As well as gaining a cum laude master's degree in Business Economics, Theo Renkema has a PhD in Industrial Engineering & Management Science. Dr. Renkema is a leading authority in information economics and as such has authored several books, many articles, and regularly lectures in the areas of IT investment appraisal, business value assessments, and benefits management. His most recent title is *The IT Value Quest* (2000), published by John Wiley & Sons.

Michael Rosemann received his master's in Business Administration (1992) and his PhD in Information Systems (1995) from the University of Munster, Germany. From 1992-1999 he worked for the Department of Information Systems at the University of Munster. Since 1999 he has been working as a Senior Lecturer for the School of Information Systems at the Queensland University of Technology (QUT). He is also lecturing at the Northern Institute of Technology, Hamburg. As an Associate Director of QUT, he is involved in research in the areas of enterprise systems, process management, knowledge management, customer relationship management and IT evaluation. Michael Rosemann published various books, book chapters and journal papers in these areas. Furthermore, he has comprehensive consulting expertise.

Ron Saull, MBA, CSP is the Senior Vice-President and Chief Information Officer of the Information Services Division of Great-West Life, London Life and Investors Group headquartered in Winnipeg, Canada. He has more than 25 years' experience as an information systems professional and manager in both the public and private sectors. Mr. Saull is a Vice-President of the International Board of Trustees of ISACA (Information Systems Audit and Control Association) consisting of over 20,000 information systems professionals worldwide. He is the past chairman of ISACA's Research Board and is currently active as a member of the Board of Directors, and a member of the IT Governance Board. He was recently appointed to the CobiT (Control Objectives for Information and related Technology) Steering Committee. This Committee is currently engaged in the development of CobiT 3rd Edition, which will include management guidelines for the evaluation and control of the IT function. He has published an article on the IT Balanced Scorecard in the *Information Systems Control Journal* and presented on this topic at ISACA conferences. His e-mail address is: ron.saull@investorsgroup.com.

Vassilis Serafeimidis is an ebusiness strategy specialist with extensive experience in the management and governance of information systems. He is presently a consultant with KPMG Consulting in London. He specialises in electronic business and estrategies and the management/governance of information technology in a wide variety of industry sectors and European countries. In addition, Dr. Serafeimidis is an Associate Lecturer and a Visiting Fellow in Information Systems at the Department of Computing of the University of Surrey (UK). Dr. Serafeimidis is an established researcher. He has published his research in many academic journals and he has been a speaker in many international conferences and executive seminars around the world. Dr. Serafeimidis is a member of the editorial board of three academic journals. His research interests include information systems strategies, information technology in developing countries, socioeconomic aspects of information technology and organisational transformation. Dr. Serafeimidis's academic background comprises a PhD in the area of information technology investment evaluation, an MSc in Analysis, Design and Management of Information Systems, both from the London School of Economics. He also holds a first class honours degree in Applied Informatics from the Athens University of Economics and Business.

Steven John Simon is a professor in the Business School of Florida International University. He received his PhD from the University of South Carolina, specializing in MIS and International Business. Before entering the doctoral program he spent 18 years in the private sector in management/computer operations and was owner/operator of seven McDonald's franchises. His current research interests include information determinants of international business structures, enterprise information systems, IS training and learning issues, electronic commerce in the international environment and organization change and learning. He has extensive ERP experience having worked with companies such as IBM on SAP implementation projects. Dr. Simon is also an officer in the United States Naval Reserve assigned to the directorate of logistics for Unites States Atlantic Command. His past Navy assignments included serving as Information Resource Management Officer to the Commander of the Second Naval Construction Brigade. He has consulted and lectured extensively in Korea, Hong Kong, Malaysia, Singapore, and the People's Republic of China. He has previously published in *Information Systems Research, Journal of Applied Psychology, Communication of the CAM, Database, The Journal of Global Information Technology Management, The Journal of Global Information Management* and *The Information Resources Management Journal.*

John Thorp is President of the Thorp Network Inc., and a Consulting Fellow with DMR Consulting, where he leads DMR's Centre for Strategic Leadership. He is the author, with DMR, of The Information Paradox, published by McGraw Hill in 1999. John can be reached at john_thorp@thorpnet.com. John Thorp is an internationally sought-after management consultant with more than 35 years' experience in the information management field, including business management, systems delivery and operations. He is the author of *The Information Paradox*, and a frequent speaker on various aspects of strategic planning, information as a strategic resource, and the effective management of information technology (IT). John addresses and advises leaders of the world's largest organizations in the United States, Canada, Europe and Asia-Pacific, including Fortune 100 companies. His opinion is regularly sought and published by major magazines and newspapers throughout North America. John's focus is on realizing the benefits of IT. He consults with organizations in the area of identifying the opportunities IT provides to transform themselves, selecting the opportunities with the greatest potential, and managing their investments in these opportunities to realize maximum benefit as they strive to remain competitive in today's rapidly changing global market place.

Roel van Rijnbach received a master's degree in Business Economics and has post graduate degrees in Accountancy and EDP-auditing from the Free University of Amsterdam, the Netherlands. He has held several positions in the field of Financial, EDP and Operational Auditing. His current position is Vice President of Management Information Systems and Data Warehousing for Nationale-Nederlanden insurance company, which is part of the ING group.

Peter Verleun received his MSc in Informatics from Delft University of Technology, where he conducted research into organisational issues regarding the maintenance and control of company-wide balanced scorecard. One of the projects concerned the case described in this book. Peter Verleun is now a management trainee at ABN AMRO bank.

Carla Wilkin is an academic and post-graduate in the School of Management Information Systems at Deakin University on the Warrnambool campus. Her post-graduate studies have been involved with investigating the evaluation of quality of delivered IT systems. Her work has been presented at three international conferences as well as Australian conferences.

Index

A

accountability 27
action 37
active benefit realisation 19, 46
activity-based costing 186
analytic network process 185, 188
annual integrated program assessment 261
application area 83
application management 232
architecture 146
average accounting rate 85

B

balanced IS assessment 215
balanced scorecard 172, 187, 199, 231, 240, 253
Bedell's method 88
benchmarking 186
benefits 80
benefits from use 114
benefits management process 18
benefits realisation principles 17
benefits realization 27
benefits realization approach 27
budgeting 6
business contribution 200
business performance 231
business process improvement 255
business processes 132
business value 3, 220

C

clarifying and translating strategy and vision 233
communicating and linking 233
control evaluation 106
cost benefit analysis 157
cost/benefit, payback period 6
cost of IT 58
costs 6, 80
customer 175, 254
customer acquisition 258
customer intimacy 219
customer perspective 187, 232
customer profitability 258
customer retention 258
customer satisfaction 258

D

decision-making process 83
delivery deadlines 6
delivery domain 213
DeLone and McLean's success model 112
development domain 213
discounted cash flow/internal rate of return 6
discovery domain 214
divestment 80
DMR results chain 29

E

electronic taxation 146
enterprise application systems 171

enterprise resource planning 155, 171
enterprise systems 171
enterprise-wide systems 171
evaluation 59
evaluation and benefits management life cycle 46
evaluation criteria 83
evaluation of information technology 78
evaluation of investments 46
evaluation process 146
evaluation schema 102
evaluation stakeholder analysis 101
evaluation time 7
evolutionary information systems 130
expectation 119
expenditures 80
exploratory evaluation 107

F

financial 254
financial approach 82
financial impacts 79
financial independence 242
financial perspective 187, 232
financial/cost 175
four "ares" 33
full cycle governance 27, 34
function point analysis 224
functional maintenance 235
functional management 232
future orientation 200
future use 114

G

Government Performance and Results Act 255

H

hardware and software 80
high potential systems 12

I

identification 163
industry 258
information economics 85, 157
information economics approach 10
information policy and information planning 234
information quality 113
information strategy plan 144
information system 58, 80, 130, 146, 185
information systems evaluation 99
information systems planning 46
innovation 175
innovation and growth perspective 232
integrated evaluation lifecycle approach 8
integrated life cycle of activities 50
integrated standard software packages 171
integrated vendor software 171
internal business process measures 260
internal business processes 254
internal business processes perspective 187
internal organisation perspective 232
internal processes 175
internal rate of return 85
international accounting standard (IAS) 38, 159
interval scale 84
investment mapping 12, 89
investment portfolio 88
iInvestments 79
IS/IT 2
IS/IT investments evaluation 2
IS/IT planning 7, 46
IT assessment method 87
IT evaluation 111
IT investment decisions 44
IT investments 27
IT outcomes 58
IT planning 146
IT utilisation 44

K

key operational systems 11
Kobler unit framework 10

L

learning 175
learning and growth infrastructure 254
learning and innovation perspective 187
level one objectives 257

M

maintenance investments 80
managerial independence 241
manufacturing resource planning 155
market share 258
matching objectives, projects and techniques 9
measure 130
measurement 3, 27
measurement concepts 257
measurement of technology investments 157
measurement scale 84, 91
measures 200
MEDIC 37
methodologies 61
mission 199
multi-criteria approach 82
multi-objective multi-criteria technique 10, 139

N

net present value 6, 85
new economy 25
new technology 6
nominal scale 84
nonfinancial impacts 79

O

objectives 200
operational excellence 200, 219
options analysis 186
options theory 11
ordinal scale 84
organisational change 132
organisational context 105
organisational performance 45
organizational analysis 64
organizational resources 58

P

payback period 84
people 80
performance-based budgeting 186
planning and target setting 233
portfolio approach 83
portfolio management 27
present worth 6
procedures 80
procurement 146
product leadership 219
profitability 80
profitability index 6
program management 27
project perspective 175

Q

qualitative characteristics 257
quality 115
QuantiMetrics alignment assessment 215
QuantiMetrics performance enhancement programme 223

R

ratio approach 82
ratio scale 84
return 80
return on investment 6, 194
return on management (ROM) 9
return on management method 87
revenue activities 146

S

sacrifices 80
satisfaction 115
sense-making evaluation 107
service to public 146
SESAME 9
setting goals 187
SIESTA 87
social learning evaluation 107
strategic climate 7
strategic feedback and learning 233
strategic matrix 11
strategic objectives 243
strategic planning 50
strategic systems 11
strategic vision 187
support systems 11
system quality 113
system use 157

T

technical independence 241
technical management 232
technical/functional evaluation 62
technology fitness 220
timing 6
total-cost-of-ownership 176
training 146
transaction processing systems 156

U

user orientation 200
user satisfaction 157
utilisation management 235

V

validation 242
value 80
value analysis 11
value of IS 213
value of IT projects 156
verification 242
virtual communication space 188
virtual distribution space 188
virtual information space 188
virtual transaction space 188

Y

yieldings 80

NEW!

Community Informatics: Enabling Communities with Information and Communications Technologies

Michael Gurstein
Technical University of Brtish Columbia

ISBN 1-878289-69-1
(hardcover)
US$139.95 * 450 pages
Copyright © 2000

Community informatics is developing as an approach for linking economic and social development efforts at the community level to the opportunities that information and communication's technologies present. Areas such as SMEs and electronic commerce, community and civic networks, electronic democracy and online participation are among a few of the areas affected.

Community Informatics: Enabling Communities with Information and Communications Technologies is an introduction to the discipline of community informatics. Issues such as trends, controversies, challenges and opportunities facing the community application of information and communications technologies into the millennium are studied.

An excellent addition to your library

For your convenience, the Idea Group Publishing Web site now features "easy-ordering" for this and other IGP publications at http://www.idea-group.com

Idea Group Publishing

1331 E. Chocolate Avenue • Hershey, PA 17033 USA
Tel: (717)533-8845 • Fax: (717)533-8661 • cust@idea-group.com
http://www.idea-group.com